PRACTICAL TECHNIQUES FOR SAVING ENERGY

IN THE CHEMICAL, PETROLEUM

AND METALS INDUSTRIES

Practical Techniques for Saving Energy in the Chemical, Petroleum and Metals Industries

Marshall Sittig

NOYES DATA CORPORATION

Park Ridge, New Jersey, U.S.A.

1977

Published in the United States of America by
Noyes Data Corporation
Noyes Building, Park Ridge, New Jersey 07656

FOREWORD

In the very near future all industrial operations that have not yet reacted to the energy crisis, must be organized to institute a systematic approach toward conserving energy in all forms, through more efficient utilization of existing processes and careful and studied reduction of losses and waste.

As outlined in detail in the introduction to this volume, the chemical, petroleum, and metals industries are the most energy-intensive manufacturing industries in the U.S. They constitute an interacting group; they use many common raw materials and are simultaneously feeding products and by-products to one another.

Practical thinking about industrial energy conservation requires interceptive calculations of such material transfers to produce positive energy savings. The three major industries are subdivided into 39 individual processing industries. Energy-saving discussions and proposals are presented for all of them.

Under the Energy and Conservation Act the U.S. Federal Energy Administration is required to set energy conservation targets and goals to be attained by January 1980. Most industries discussed in this book fall within the categories considered. As can be seen from the large and impressive table of contents, entries have been arranged in an encyclopedic manner whenever possible.

A complete and detailed list of reports and references are found at the end of the book. In this list, titles are complete with their publishers and other sources of procurement and are never abbreviated. Major contributions to the manuscript are acknowledged from the energy target documents, particularly from those on the chemical, petroleum, and metals industries.

Advanced composition and production methods developed by Noyes Data are employed to bring our new durably bound books to you in a minimum of time. Special techniques are used to close the gap between "manuscript" and "completed book." Industrial technology is progressing so rapidly that time-honored, conventional typesetting, binding and shipping methods are no longer suitable. We have bypassed the delays in the conventional book publishing cycle and provide the user with an effective and convenient means of reviewing up-to-date information in depth.

CONTENTS AND SUBJECT INDEX

Contents and Subject Index

INTRODUCTION

This book is designed to present the latest in practical thinking about industrial energy conservation. The metals, chemicals and petroleum industries were picked for coverage here because they are the three largest consumers of energy in the United States economy and they constitute an interacting group using many common raw materials and feeding products and by-products to one another as shown in Figure 1.

FIGURE 1: MATERIAL INTERACTIONS BETWEEN METALS, CHEMICAL AND PETROLEUM INDUSTRIES

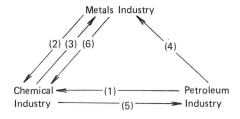

(1) Hydrocarbon raw materials
(2) Sulfur dioxide from ore roasting
(3) Oxygen, lime and other chemicals
(4) Fuel for furnaces
(5) Catalysts, reagents for treating
(6) Aromatics and other by-products from coke ovens

Under the Energy and Conservation Act, the U.S. Federal Energy Administration was required to set energy conservation targets for the ten most energy intensive manufacturing industries. Goals to be attained by January 1, 1980 were to be established by December 22, 1976 under the Act. Major contributions to the manuscript are acknowledged from the energy target documents on the metals (4), chemicals (5) and petroleum (6) industries.

TABLE 1: PROJECTED ENERGY OUTPUT COEFFICIENTS FOR THE SIX ENERGY INTENSIVE INDUSTRIES

	1971 Energy Requirements per $VA (MBtu/$VA)	1980 Energy Requirements per $VA (MBtu/$VA)	1971-1980 Percent Decrease (per year)	1990 Energy Requirements per $VA (MBtu/$VA)	1971-1990 Percent Decrease (per year)
Food and kindred products	31.98	26.29	17.8 (2.15)	26.47	17.2 (1.00)
Paper and allied products	115.80	94.31	18.6 (2.26)	81.15	29.9 (1.85)
Chemicals and allied products	95.70	84.02	12.2 (1.46)	72.76	24.0 (1.43)
Petroleum and coal products	451.00	373.55	17.2 (2.07)	370.57	17.8 (1.03)
Stone, clay and glass products	146.60	119.55	18.5 (2.24)	105.90	27.4 (1.70)
Primary metals industries	212.66	177.90	16.3 (1.96)	154.50	27.3 (1.67)
Six energy intensive industries	128.04	107.49	16.0 (1.93)	93.32	27.1 (1.65)

TABLE 2: HISTORICAL AND PROJECTED MANUFACTURING ENERGY REQUIREMENT 1954-1990 (10^{12} Btu)

	1954	1958	1962	1967	1971	1975	1977	1980	1985	1990
All manufacturing	9,826	10,698	12,501	14,765	16,085	19,029	20,614	22,224	25,777	30,361
Six high consuming industries	8,351	9,210	10,672	12,385	13,371	15,491	16,568	17,567	20,175	22,919
Food and kindred products	805	766	803	767	920	912	939	1,001	1,154	1,345
Paper and allied products	648	806	929	1,156	1,315	1,378	1,448	1,567	1,742	1,947
Chemicals and allied products	1,359	1,634	1,969	2,598	2,783	3,757	4,428	4,822	6,175	7,481
Petroleum and coal products	1,761	2,020	2,316	2,556	2,956	3,227	3,332	3,449	3,865	4,322
Stone, clay and glass products	905	945	1,056	1,229	1,367	1,407	1,457	1,566	1,702	1,900
Primary metals industries	2,873	3,049	3,599	4,080	4,030	4,810	4,964	5,162	5,519	5,924
All other manufacturing	1,475	1,488	1,829	2,380	2,714	3,538	4,046	4,657	5,602	7,442

Source: Reference (10)

Annual fuel utilization in the six largest fuel consuming industrial sectors in the early 1970s was characterized (see note below) as follows:

Chemical industry usage*	$1,160 \pm 120 \times 10^{12}$ kcal
Primary metals industry usage	$1,310 \pm 130 \times 10^{12}$ kcal
Petroleum industry usage	$766 \pm 80 \times 10^{12}$ kcal
Paper industry usage	$645 \pm 65 \times 10^{12}$ kcal
Stone-clay-glass-concrete in-dustry usage	$365 \pm 40 \times 10^{12}$ kcal
Food industry usage	$323 \pm 30 \times 10^{12}$ kcal
Total for the six sectors	$4,569 \pm 500 \times 10^{12}$ kcal*

Note: Purchased electricity is valued at 2,500 kcal/kwh.
*Process fuel utilization, 670×10^{12} kcal; feedstock fuel utilization, 490×10^{12} kcal.
**This amounts to 25 to 30% of the total energy consumption in the United States.

Some estimates of unit energy consumption in 1971 for these six most energy intensive industries with projections to 1980 and 1990 are shown in Table 1.

The total energy requirements for the same six industries are shown for past years back to 1954, again with projected future estimates, this time for 1980, 1985 and 1990, in Table 2.

It is interesting to compare the 1971 values for the various industries from the Project Independence study (10) as shown in Table 2 and as shown earlier from the Dow Chemical Company report (1). The comparative data in consistent units are as follows.

Industry	....Reference (1).... 10^{12} kcal	10^{12} Btu	..Reference (10).. 10^{12} kcal	10^{12} Btu
Primary metals	1,310	5,198	1,016	4,030
Chemicals	1,160	4,603	701	2,783
Petroleum	766	3,040	745	2,956
Paper	645	2,559	331	1,315
Ceramic	365	1,445	344	1,367
Food	323	1,282	231	920

Some close agreement (in petroleum and ceramic, for example) and some wide disagreement (in paper in particular and to a lesser but sizeable degree in chemicals) are evident in the comparative data.

Annual fuel utilization by unit operation in the six industrial sectors was characterized (1) as follows:

Direct heating of process streams	$1,780 \pm 400 \times 10^{12}$ kcal
Compression	$340 \pm 100 \times 10^{12}$ kcal
Distillation	$300 \pm 100 \times 10^{12}$ kcal

(continued)

Electrolysis	$340 \pm 50 \times 10^{12}$ kcal
Evaporation	$165 \pm 30 \times 10^{12}$ kcal
Drying	$270 \pm 50 \times 10^{12}$ kcal
Cooking, sterilizing, and digestion	$185 \pm 30 \times 10^{12}$ kcal
Feedstock	$490 \pm 50 \times 10^{12}$ kcal
Other or unaccounted for	$699 \qquad \times 10^{12}$ kcal
Total	$4,569 \pm 500 \times 10^{12}$ kcal

Annual fuel utilization by type of fuel was characterized (1) as follows:

Purchased electricity	$813 \pm 80 \times 10^{12}$ kcal
Coal	$853 \pm 85 \times 10^{12}$ kcal
Petroleum	$701 \pm 70 \times 10^{12}$ kcal
Natural gas	$1,602 \pm 160 \times 10^{12}$ kcal
Other	$600 \pm 60 \times 10^{12}$ kcal
Total	$4,569 \pm 500 \times 10^{12}$ kcal

The level of annual heat rejection from all processes was characterized (1) as follows:

Radiation, convection, conduction, other	$410 \pm 150 \times 10^{12}$ kcal
Below 100°C	$1,420 \pm 300 \times 10^{12}$ kcal
From 100° to 250°C	$728 \pm 200 \times 10^{12}$ kcal
From 250° to 800°C	$557 \pm 150 \times 10^{12}$ kcal
From 800° to 1800°C	$254 \pm 100 \times 10^{12}$ kcal

It was predicted by Dow Chemical Company in September 1975 (1) that energy conservation efforts should be capable of decreasing annual energy usage in the short run (less than 5 years) as follows:

Chemical industry	187×10^{12} kcal
Primary metals industry	208×10^{12} kcal
Petroleum industry	136×10^{12} kcal
Paper industry	170×10^{12} kcal
Stone-clay-glass-concrete industry	37×10^{12} kcal
Food industry	36×10^{12} kcal
Total	774×10^{12} kcal

Table 3 shows the estimated short term (less than 5 years) effect of applying recommended conservation approaches to the six big fuel consuming industries. Research and development on new processes, on increasing product yields, or on other areas might yield even more beneficial effects on fuel utilization. This conservation approach was not included in this analysis because the effects of research and development efforts are very difficult to estimate. The estimated effect of applying conservation approaches other than research and development is to decrease annual fuel usage by 774×10^{12} kcal. The order of effectiveness of conservation approaches is design modification, maintenance and insulation, process integration, waste utilization, process modification, operation modification, and market modification.

TABLE 3: ENERGY CONSERVATION IN THE SIX BIGGEST FUEL CONSUMING INDUSTRIES

Industry	Conservation Approach	Estimated Fuel Savings (10^{12} kcal/yr)	Industry Energy Usage (10^{12} kcal/yr)
Chemical	Waste utilization	14	
	Maintenance and insulation	50	
	Market modification	1	
	Operation modification	24	
	Design modification	78	
	Process integration	20	
	Total	187	670*
Primary metals	Waste utilization	35	
	Process integration	13	
	Process modification	36	
	Design modification	66	
	Maintenance and insulation	30	
	Operation modification	28	
	Total	208	1,310
Petroleum	Process integration	32	
	Design modification	40	
	Maintenance and insulation	40	
	Waste utilization	8	
	Operation modification	16	
	Total	136	766
Paper	Process integration	68	
	Marketing modification	3	
	Process modification	15	
	Design modification	5	
	Waste utilization	29	
	Maintenance and insulation	50	
	Total	170	645
Stone-clay-glass-concrete	Process modification	21	
	Maintenance and insulation	10	
	Design modification	6	
	Total	37	365
Food	Maintenance and insulation	10	
	Process integration	6	
	Design modification	20	
	Total	36	323
Grand Total		774	4,569

*Process energy only. This does not include feedstock energy usage.

Source: Reference (1)

The Dow Chemical Company study (1) predicted that energy conservation approaches should be capable of decreasing annual energy usage in the short run (less than 5 years) as follows:

Waste utilization	86×10^{12} kcal
Maintenance and insulation	190×10^{12} kcal
Operation modification	68×10^{12} kcal
Design modification	215×10^{12} kcal
Process integration	139×10^{12} kcal
Process modification	72×10^{12} kcal
Market modification	4×10^{12} kcal
Total	774×10^{12} kcal

Tables 4, 5 and 6 summarize briefly the energy conservation steps which can be taken to reduce energy consumption in the metals, chemicals and petroleum industries. These were taken from the Dow study for EPA (1); a much more detailed (and in some cases somewhat different) set of conservation measures is outlined in the later sections of this book quoting from Federal Energy Administration (FEA) reports dealing specifically with the metals (4), chemicals (5) and petroleum (6) industries. Table 7 gives some of the units of measure and conversion factors used by FEA for reference.

TABLE 4: ENERGY CONSERVATION IN THE PRIMARY METALS INDUSTRY

Conservation Approach	Estimated Fuel Savings (10^{12} kcal/yr)
(1) Waste utilization	
(a) Use 25% of the presently unaccounted for blast furnace gas as fuel.	15
(b) Increase domestic recycle of scrap steel. Decrease exports by 3×10^6 tons per year (~40% of exports in 1972).	10
(c) Increase old scrap recycle of aluminum from approximately 5% of aluminum production to 10%.	10
(2) Process integration	
Coproduce electricity and steam. If 50% of the process steam generated in manufacturing steel were coproduced with electricity, approximately 7×10^{12} kcal of electricity could be generated using an extra 8×10^{12} kcal of fuel. This amount of electricity typically requires 21×10^{12} kcal of fuel for its generation.	13
(3) Process modification	
(a) Replace the open hearth process for producing steel with the basic oxygen process. Assume that one-half of the open hearth portion of steel production (26% in 1972) is replaced with the basic oxygen process.	9

(continued)

TABLE 4: (continued)

Conservation Approach	Estimated Fuel Savings (10^{12} kcal/yr)
(b) Increase the use of continuous casting in the steel industry from 6% of raw steel cast in 1972 to 50% of raw steel cast.	15
(c) Increase the ratio of iron-ore pellets to sinter in the blast furnace charge. Reduce sinter charge to 20% of the total charge.	7
(d) Use Alcoa's newly developed aluminum process to produce 10% of the U.S. aluminum production.	5
(4) Design modification	
(a) Increase waste heat recovery by charging hot sinter, pellets and coke into the blast furnace. Assume that 30% of the heat from these materials can be salvaged.	9
(b) Preheat combustion air supplied to sinter and pellet furnaces. Assume that 25% of the heat from hot stack gases can be recovered.	5
(c) Increase the air blast temperature to $1100°C$ and the top gas absolute pressure to 210 kN/m^2 in the blast furnace on 50% of the furnaces. Coke savings of 20% on the charged furnace can be achieved.	35
(d) Assume that the off-gases from 50% of the basic oxygen furnaces are used for their fuel and sensible heat.	5
(e) Improve heat recuperators in open hearth furnaces, soaking pits, reheat furnaces and heat treating furnaces.	8
(f) Reduce electrolyte resistance in aluminum electrolysis cells by closer electrode spacing or modifying bath composition.	4
(5) Operation modification	
(a) Operate aluminum electrolysis cells at 20% lower current density.	20
(b) Closely control depth of aluminum pad, the distance between anode and cathode, and bath composition.	8
(6) Maintenance and insulation	
Improve maintenance and insulation of steam systems in all primary metals processes. This should result in savings of 10 to 20% in steam usage.	30
Total	208

Total primary metals energy usage $\sim$1,310$\pm$130 x 10^{12} kcal/yr.

Source: Reference (1)

TABLE 5: ENERGY CONSERVATION IN THE CHEMICAL INDUSTRY

Conservation Technique	Estimated Fuel Savings $(10^{12}$ kcal/yr)
(1) Waste utilization	
(a) Recover the fuel value of wasted by-product in chlorine process. Assume that 50% is now being wasted.	4
(b) Increase burning of other wasted by-products.	10
(2) Insulation and maintenance	
Improve maintenance and insulation of steam systems. This should reduce steam usage by 15 to 20%.	50
(3) Operation modification	
(a) Operate electrolysis cells at lower current densities.	9
(b) Closely control excess air to furnaces.	10
(c) Closely control the reflux on distillation columns.	5
(4) Design modification	
(a) Increase waste heat recovery from hot streams such as furnace stack gases or hot process streams.	50
(b) Design distillation columns to operate at a lower reflux.	10
(c) Convert the chlorine cells using graphite anodes (approximately 50%) to metal anodes.	8
(d) Replace inefficient compressors and motors with more efficient equipment.	10
(5) Process integration	
Increase efforts to coproduce steam and electricity.	20
(6) Market modification	
Substitute 50% NaOH in half of the applications now using 100% NaOH.	1
Total	187

Source: Reference (1)

TABLE 6: ENERGY CONSERVATION IN THE PETROLEUM INDUSTRY

Conservation Approach	Estimated Fuel Savings $(10^{12}$ kcal/yr)
(1) Process integration	
Coproduce electricity and process steam. At present only 10 to 15% of process steam production is combined with electric generation. Assume that this can be increased to 50%. Then an additional	

(continued)

TABLE 6: (continued)

Conservation Approach	Estimated Fuel Savings (10^{12} kcal/yr)
17×10^{12} kcal of electricity could be generated using 19×10^{12} kcal of fuel. Utilities typically require 51×10^{12} kcal of fuel to generate this quantity of electricity.	32
(2) Design modification	
(a) Increase heat recuperation from furnaces. Assume that air preheaters which will decrease fuel consumption 15% are installed on an additional 25% of industry furnaces.	16
(b) Increase heat interchange between process streams.	8
(c) Increase use of turbines to recover mechanical energy from high pressure process streams.	8
(d) Design distillation columns to require lower reflux.	8
(3) Maintenance and insulation	
Improve maintenance and insulation on steam systems. This should reduce steam consumption by 15 to 20%.	40
(4) Waste utilization	
Increase the use of flue gas from catalytic crackers as fuel.	8
(5) Operation modification	
Closely control steam stripping operations, use of H_2 in desulfurization operations, use of excess air in furnaces and reflux in fractionation operations.	16
Total	136

Total petroleum industry fuel usage $\sim 766 \pm 80 \times 10^{12}$ kcal.

Source: Reference (1)

TABLE 7: UNITS OF MEASURE AND CONVERSION FACTORS

Type of Material or Energy	Unit of Measure	. . . . . .Equivalent to–	
		Thousand Btu	Kilowatt Hours
Propane, butane and mixtures	Barrel	4,011	1,175
Middle distillates	Barrel	5,825	1,707
Residual fuel oil	Barrel	6,287	1,842
Chemical feedstock	Barrel	4,011	1,175
Other petroleum products			
Gasoline	Barrel	5,253	1,539
Kerosine	Barrel	5,670	1,661

(continued)

TABLE 7: (continued)

| Type of Material or Energy | Unit of Measure | . . . . . .Equivalent to– . . . . . . | |
		Thousand Btu	Kilowatt Hours
Lubricants	Barrel	6,065	1,777
Wax	Barrel	5,537	1,622
Asphalt	Barrel	6,636	1,944
Residual fuels, petroleum, coke, acid sludge	Barrel	6,006	1,760
Miscellaneous	Barrel	5,796	1,698
Coal	Short ton	26,200	7,677
Anthracite	Short ton	25,400	7,442
Bituminous	Short ton	26,200	7,677
Lignite	Short ton	14,770	4,328
Natural gas	Thousand ft^3	1,032	302.3
Fuels, NEC			
Coke oven gas	Thousand ft^3	550	161.2
Blast furnace gas	Thousand ft^3	92	27.0
Still gas	Thousand ft^3	1,501	439.8
Coke	Short ton	26,000	7,618
Coke screening and breeze	Short ton	20,488	6,003
Electrical energy	kwh	3.412	1.0
Steam, low pressure	Pounds	1.150*	336.9
Steam, high pressure	Pounds	1.500*	439.5

*Average net heat content value. For Btu to generate steam multiply by 1.25.

Source: Reference (5)

METALS INDUSTRY

SIC 33 is defined by the 1972 Standard Industrial Classification Manual, Office of Management and Budget, to include establishments engaged in the smelting and refining of ferrous and nonferrous metals from ore, pig or scrap; in the rolling, drawing, and alloying of ferrous and nonferrous metals; in the manufacture of castings and other basic products of ferrous and nonferrous metals; and in the manufacture of nails, spikes, and insulated wire and cable. It also includes the production of coke.

Except that all components deal with the smelting, refining, casting, or some other treatment of metals, the technologies in the various components differ greatly. For example, the technology of steel is much different from the technology for aluminum. Also, steel is coal-intensive while aluminum is electricity-intensive.

The different problems faced in their manufacture (and in the manufacture of the numerous other metals and products defined to be within SIC 33) are weighted into a set of conclusions in which SIC 33 which has been defined as an "industry." The definition of SIC 33 implies some degree of homogeneity. The element of homogeneity is that all components deal with metals, mainly the energy-intensive aspects of the production of metals.

Approximately 85% of the energy consumption within the primary metals category occurs in the manufacturing processes for steel and aluminum.

The SIC 33 target computations are dominated by the conservation activities in the steel plant component, which accounts for about 73% of total use of energy in SIC 33. With the inclusion of three other components (aluminum, iron foundries, and copper), about 92% of total use of SIC 33 energy is accounted for. Although each of the 13 components was analyzed individually to estimate its energy conservation potential (4), emphasis was placed on the four highest ranking components.

This was justified by sensitivity analyses which show, for example, that an error in estimation of 1% throughout the nine lowest-ranking components combined has only one-ninth the effect on the target of a 1% error in the steel plant component. The energy conservation potential of the nine lowest ranking components can be understated or overstated by about 10% before the SIC 33 target is changed by 1%. Each of the four highest-ranking components is involved in a voluntary reporting system.

The steel plant component reports voluntarily through the American Iron and Steel Institute (AISI). Each company reports its total use of energy by fuels and electricity consumed, using actual values for the heat content of fuels and 3,413 Btu per kwh for electricity. For 1972, the AISI report covered about 90% of steel production, and this increased to about 94% in 1975.

Using 1972 as the base year, the AISI reported an energy-efficiency improvement per ton of steel products of 2.7% in 1973, and an additional 0.1% in 1974. For 1975, however, energy use per ton of steel increased to 2.9% over the base year. Study and analysis have shown that this reversal is mostly a function of the drop in production in 1975.

On a raw-steel basis, in 1975 steel production dropped to about 81% of the average of the prior three years. During such a drop, considerable energy is required to keep the equipment in operating condition, even though the throughput is lower than normal. The AISI presently is working to take this important factor into consideration in its reports.

Analysis of the data presented suggests that in the steel industry a drop of 20% in production is accompanied by an increase of about 10% in the amount of energy consumed per ton of steel. Other factors that tended to increase the use of energy per ton of steel in 1975 included (1) an increase in the proportion of alloy steels which require more energy per ton than carbon steels, (2) conversions of some equipment from natural gas to less efficient heating with coal or oil, and (3) increasing use of energy to conform to environmental controls.

The aluminum component reports voluntarily through the Aluminum Association. Each company reports its total use of fuels and electricity, broken down by type and source. Also, the report includes details on energy use divided into the five operational categories of: (1) bauxite; (2) alumina; (3) hot metal; (4) hold, cast, and melt; and (5) fabrication. Participation has been increased from 14 to 39 companies, so that the latest report represents over 95% of the energy used in the domestic aluminum industry.

Using the Aluminum Association method of calculation, the aluminum component has reported an energy-efficiency improvement over the base year of 1972 in the amount of 2.1% for 1973, 5.7% for 1974, 4.0% for the first half of 1975, and 5.0% for the second half of 1975. This represents an indicated achievement of 50% toward their voluntary 1980 goal of an overall 10% improvement.

Although consistent within itself, the Aluminum Association reporting system uses different conversion factors for electricity than are used in other industries as noted by Battelle in their report to FEA (4). Purchased and self-generated hydropower are converted at 3,413 Btu/kwh. Purchased thermal electricity is converted at 10,500 Btu/kwh. Self-generated thermal electricity is converted at actual value which was 11,960 Btu/kwh in 1972. This detailed treatment of different sources for electricity is a side effect of the component's high use of electricity.

Because all other industries and components have standardized on 3,412 Btu per kilowatt hour for all electricity, regardless of source, the effect of the Aluminum Associations conversion factor is to overstate their energy consumption per unit of output when compared with other components. A sensitivity analysis shows that their reported 1972 base consumption of 935×10^{12} Btu becomes 588×10^{12} Btu when all electricity is converted at 3,412 Btu per kilowatt hour.

When analyzing the energy use of the aluminum component in detail, and estimating the component's conservation potential, the Aluminum Association's conversion factors were used by Battelle in their study for FEA (4). However, when integrating this component with other components for setting the SIC 33 target, use was made of the same conversion factors as used for the other components.

The iron foundry component reports voluntarily through the American Foundrymen's Society (AFS). At its present level of development, the AFS system is not keyed to SIC classifications and does not appear to have a representative sampling of both small and large foundries. Some of the larger foundries, for example, have achieved energy savings in the range of 15 to 25% per ton of castings since 1972. These savings, however, are not representative of the iron foundry component as a whole. Steel foundries and nonferrous foundries also report voluntarily through the AFS.

The copper component reports through the American Mining Congress (AMC), which made its first report early in 1976 using data supplied by the American Bureau of Metal Statistics. The sample includes about 70% of the primary metal produced from domestic ore and about all of the copper smelted and refined domestically. The report includes beneficiating energy which belongs in SIC 10, not SIC 33.

Using the AMC values, the energy consumed per ton of recoverable refined copper increased by 20% from 1972 to the first half of 1975. The increase is attributed by the AMC to about: (1) 7% additional because of decreasing grade of ore (not within the province of SIC 33), (2) 3 to 5% additional because of a decrease in production of about 10%, (3) 6 to 10% additional because of increased energy for environmental control, and because of other factors such as increased shipping distances for ore and colder than average weather.

The present voluntary reporting systems for the major components of SIC 33 are being improved according to Battelle (4) so as to allow for changes in energy consumption caused by factors other than energy conservation per se, and thus to allow more detailed tracking of energy conservation attainments.

PROCESS TECHNOLOGY INVOLVED

The process technology involved in each of the divisions of SIC 33 will be discussed under the specific division in question.

MAJOR ENERGY CONSERVATION OPTIONS TO 1980

The specific conservation options will be discussed under the particular industry divisions. Some of these options were reviewed earlier in Table 4.

However, a number of important trends and situations in SIC 33 have affected the industry's energy conservation implementation since 1972, and will continue to do so to 1980. These are discussed in the Battelle report to FEA (4), have been considered in target-setting, and are summarized here.

High use of coal and coke made from coal is a characteristic of SIC 33. For example, the steel plant component and the iron foundry component together account for over 75% of the total energy used in SIC 33. The steel plant component obtains about 70% of its energy from coal and coke and the iron foundry component about 30%. Among the ten most energy intensive 2-digit SIC industries, no other industry approaches SIC 33 in high use of coal as an energy source.

Energy intensiveness of operations in SIC 33 ranks in decreasing order as follows: (1) reduction of ores, as in smelting and electrolysis; (2) melting of metals; (3) heating of metals; and (4) rolling, forging, extrusion, and other mechanical working. Even a small percentage lowering of energy in the first or second of these types of operations yields much larger Btu savings than large percentages of improvement in the lower-ranking types of operations.

Housekeeping changes (the avoidance of obvious waste through methods involving little capital expenditure) likely have accounted for most of the conservation to date, and are expected to provide a large fraction of further conservation to 1980. Since 1972 there have been few important major process changes to save energy, and few are expected to 1980, but some of the few may be highly energy-conservative when viewed as single installations.

Conversions from natural gas to oil or electricity for heating are being made and will continue to be made. Relatively few are being made to coal because of expense and environmental problems. However, in the steel industry some boilers are being converted from coke-oven gas to coal, and the coke-oven gas so released is being used elsewhere in the plant to replace natural gas. Usually, conversions from natural gas to another fuel do not save Btu.

Therefore, because they involve an investment cost, most are unprofitable dollarwise when the investment is evaluated solely on the basis of the saving in Btu. Some establishments in SIC 33 have set a goal of becoming independent of a need for natural gas for heating by 1980. This goal is set not to conserve energy but to avoid the uncertainties of future supply of natural gas.

Yield of finished product as a percentage of metal produced or metal melted is a factor of great importance in determining energy efficiency in SIC 33. Improvement in yield offers one of the greatest opportunities for conservation of energy. For example, the yield of steel products is presently about 70% of the weight of raw steel produced. The other 30% recycles in the steelmaking system, and must be remelted and rerefined. In metal casting, the yield of shipped castings is in the range from 40 to 60% of the weight of metal melted. Again, the product not shipped is recycled through the energy-intensive melting step. Improvement in yield rarely involves capital-intensive measures. It depends mostly upon the application of technological know-how, quality control, and market factors.

Availability of capital restricts new investments in most SIC 33 companies. Availability is lowered and cost of capital increased by the historically low return on equity for SIC 33 companies generally when compared with the return possible from investor's equities in companies in other industries.

The capital that does become available to a company then must be allocated to competing projects. Most commonly, the list of projects competing for investment capital represents a total "demand" larger than the "supply." Thus a priority system comes into play, and is implemented by each company's capital-budgeting process.

For most SIC 33 establishments, at least three categories of projects usually have precedence over energy-conservation projects. These are (1) installation of new capacity to maintain or improve share of the market and to take advantage of technological improvements, (2) repair and modernization of existing facilities to maintain production capacity and/or to implement new technology, and (3) investments required to meet statutory or regulatory requirements of government on issues such as environmental control, health, and safety. In competition with such high priorities, investments to conserve energy usually fare poorly, even when "cost effective" and, lacking new incentives, will continue to fare poorly.

Desire to maintain production in a plant often dictates whether an operation will be shut down to make an equipment change or a process change. Major changes usually require that some production equipment be shut down, with corresponding loss of output. Because of the large scale and interconnected nature of many operations in SIC 33, the shutdown of one unit in a series adversely affects the production output of other units.

Also, major changes in SIC 33 often require weeks to months of shutdown to accomplish the change. Therefore, even though cost studies might show that a major investment in an energy-conservative device or technique has an acceptable rate of return, on balance the company might decide to defer the installation so as to avoid a loss in production and income.

Increased use of scrap is a means for conserving energy in SIC 33. Because of the large amount of energy needed to reduce ores to metal, whenever recycled scrap can be substituted for primary metal made from ore, a large saving in energy results. For this reason, most components in SIC 33 already make extensive use of recycled scrap, both self-generated (home) scrap and scrap purchased on the open market.

For example, in 1972 when the steel component shipped about 92 million tons of steel products and the ferrous foundries shipped about 18 million tons of castings, they jointly purchased on the open market and recycled about 42 million tons of ferrous scrap (in addition to the recycling of home scrap). Thus, scrap recycled from the public and from other industries provided about 38% of the weight of products shipped. In 1972, to augment the about 1.9 million tons of primary copper produced from ore, about 0.5 million tons of copper were produced by recycling of old scrap.

Of the metal that goes into the mill processing of aluminum, about 45% comes from primary metal (from ore), about 45% from recycled home scrap, and about 10% from scrap purchased on the open market. SIC 33 companies are well aware of the energy-conservation effects of the use of scrap, and of the associated environmental advantages of such recycling.

Therefore, the industry generally is continuously investigating methods to re-cycle even larger amounts of scrap collected from outside the industry. One technological constraint on the amount of scrap that can be used is concerned with product quality and characteristics. For many high-quality products, in-cluding deep-drawing materials, the amount of scrap that can be used is limited because of impurities in the scrap. Another constraint is that the demand for metals is larger than the supply of scrap, so that support of the growth of the national economy requires ultimately that a substantial fraction of the growing demand for metals be provided from ores, or from sources outside the country.

Environmental problems are high on the list of situations that determine the strategy of SIC 33 establishments. Operations such as coke plants, sintering plants, nonferrous smelters, and foundries have been high on the list of obvious major contributors to industrial pollution of air and water. Correction of this situation has required much attention by management, has strained the techno-logical state of the art, and has drained off large amounts of capital that would have been used for other purposes.

In general, although the environmental-control measures have required substantial increases in use of energy, the amount of energy used for environmental control generally has been relatively small in comparison to the total use of energy by SIC 33. From an energy standpoint, a major factor now is that each incremental improvement in air quality requires a much higher use of energy.

For example, one published report states that Stage 1 collection equipment of two electric steelmaking furnaces would collect about 88% of air-borne emissions at an energy cost of 0.12 kwh per pound of dust collected. A Stage 2 installa-tion would collect 97.3% of the dust with an energy cost of 7.7 kwh per addi-tional pound of dust. A Stage 3 installation would increase collection efficiency to 97.4% at an energy cost of 632 kwh per additional pound of dust.

The range in energy consumption per pound of dust from Stage 1 to Stage 3 is over 5,000 to 1. Although the numbers are different in different situations, the behavior is general, additional small amounts of benefit can have very high costs for energy. Presently the steel industry in particular is confronted with the possibility of having to control so-called "fugitive emissions."

If forced to this level of control additional requirements for energy for this purpose will be several times those already in use to control to the present levels.

Availability of electricity after 1977-78 is beginning to worry many establishments in SIC 33 and many utilities supplying these establishments. The trend to increased use of electricity is being supported strongly by conversions from natural gas. Many establishments take the view that by converting to electricity they move the fuel-availability and environmental problems back to the utility, thus easing the establishment's immediate problem.

It is not clear whether the utilities will be able to supply the increased demand, but most SIC 33 establishments are making the hopeful presumption that they will. The utilities in some cases seem less confident.

Credibility is low in many companies about the existence of a real "energy crisis" today. Managements of many companies, especially smaller companies, state or suggest the following opinion. The future availability and prices of energy represent a business risk that must be planned for. One option is to make substantial investments today to conserve energy in the future. The risk is that energy will not be so scarce or expensive as planned for.

Another option is to more or less ignore the situation (except for housekeeping measures) and hope that Doomsday on availability and price does not materialize, or that this can be dealt with when it does materialize. Meanwhile, the company has a higher percentage of its capital available for projects that it rates higher than energy conservation. The second option seems to be more attractive to many companies than the first. Simply stated, a general sense of urgency is lacking. However, some individual companies do view the situation with urgency.

Government incentives to encourage companies to invest in energy-conservative equipment and techniques are being proposed both from within SIC 33 and within government. The Battelle study for FEA (4) is based on the assumption that no new government incentives for conservation investments will be provided.

Other incentives for lower consumption of Btu per unit of product in SIC 33 are of two types. The first derives from the usual business practice of trying to minimize the consumption of expensive ingredients. This is the major incentive now operational in SIC 33. The second incentive is the federal government's activity in analyzing the industrial use of energy, and uncertainty on the part of companies as to where this activity will lead.

There is a third incentive, but of a different class, not specifically energy-conservative when evaluated in terms of Btu. This is the incentive to be able to stay in business and to maintain production even though some source of energy becomes available. This incentive drives the conversions from one fuel to another or to electricity.

GOAL YEAR (1980) ENERGY USE TARGET

The base year is 1972 in the Battelle study for FEA (4). The target year is defined as a year throughout which the industry makes its products with the operating procedures and technologies in place as of January 1, 1980. The apparent consumption of energy by SIC 33 in 1972 was about $4,246 \times 10^{12}$ Btu. If the industry operated in the target year at the same energy efficiency (units of output per Btu) as in 1972, because of anticipated growth in units of output the energy consumption in the target year would be about $5{,}167 \times 10^{12}$ Btu.

Based on the data and information available, it is judged that, based solely on technological feasibility and economic practicability (without the influence of special circumstances), SIC 33 could lower its consumption of energy in the target year to $4{,}456 \times 10^{12}$ Btu. This represents a net conservation of 14% in energy use per unit of output from 1972 to the target year.

Special circumstances are defined as those beyond the control of establishments within SIC 33. The most significant is the need to comply with environmental control regulations. Another is the declining grade of certain ores.

After consideration of special circumstances over which the industry has no control, it is judged that SIC 33 could lower its consumption of energy in the target year to $4{,}674 \times 10^{12}$ Btu. This represents a net conservation of 10% in energy use per unit of output from 1972 to the target year.

The SIC 33 conservation targets expressed above were determined by aggregating the conservation potential for each of 13 components that make up SIC 33.

These components include all 26 SIC 4-digit classifications that comprise SIC 33. The identification of the components and their conservation potential are given in Table 8.

TABLE 8: CONSERVATION POTENTIAL OF COMPONENTS OF SIC 33

| Component | . . . . .10^{12} Btu in Target Year . . . . | | |
	At 1972 Efficiency	Technologically Feasible and Economically Practicable	With Consideration of Special Circumstances
Steel plants	3,753	3,245	3,413
Aluminum	741	639	646
Iron foundries	170	125	136
Copper	92	80	102
Ferroalloys	90	82	86
Nonferrous foundries	44	40	41
Steel foundries	53	43	45
Other primary nonferrous	62	55	56
Nonferrous processing	43	41	41
Miscellaneous metal products	42	39	39
Secondary nonferrous smelting and refining	39	33	33
Primary zinc	29	26	28
Primary lead	9	8	8
Total SIC 33	5,167	4,456	4,674

Source: Reference (4)

Aggregation of the technologically feasible potential for energy conservation to the target period for each component of SIC 33 yields a technologically feasible level of 4,142 x 10^{12} Btu per year for SIC 33. This is derived from the summary in Table 9. As shown in Table 8, operations of SIC 33 during the target period but at 1972 efficiencies would require about 5,167 x 10^{12} Btu per year. Therefore, the technologically feasible potential for energy conservation by 1980 is 5,167 – 4,142 = 1,025 x 10^{12} Btu per year, which is about 19.8% of the amount of energy which would be required in 1980 at 1972 efficiencies.

The summation of the economically practicable potentials for all the individual components in SIC 33 gives the economically practicable potential for SIC 33 by the target period. This sum is 4,456 x 10^{12} Btu annually. It is shown in Table 8. By definition, this potential is technologically feasible, because only technologically feasible changes were included in the economic analyses. This potential is identified as "gross" potential, because it is computed before the effects of special circumstances. Of the energy saving judged to be technologically feasible by the target period, an average of about 69% is judged to be also economically practicable.

1980 energy use at 1972 efficiency = 5,167 x 10^{12} Btu*
Technologically feasible by 1980 = 4,142 x 10^{12} Btu**
Technologically feasible saving = 1,025 x 10^{12} Btu
Economically feasible by 1980 = 4,456 x 10^{12} Btu
Economically feasible saving = 711 x 10^{12} Btu

$$\frac{711}{1,025} \times 100 = 69.4\%$$

*See Table 8 for derivation.
**See Table 9 for derivation.

TABLE 9: AGGREGATION OF TECHNOLOGICALLY FEASIBLE POTENTIAL FOR ENERGY CONSERVATION TO 1980 FOR COMPONENTS OF SIC 33

Component	1972 (10^6 Btu/ton)	...Technologically Feasible by 1980...		
		1980 (10^6 Btu/ton)	Projected Production (10^6 tons)	1980 Potential (10^{12} Btu)
Steel plants	29.9	24.0	125.5	3,012
Aluminum	112.6	89.5	6.6	591
Iron foundries	16.0	10.3	10.6	109
Copper	51.2	43.2	1.8	78
Ferroalloys	31.1	25.2	2.8	71
Nonferrous foundries	34.1	28.9	1.3	38
Steel foundries	28.0	21.8	1.9	41
Other primary non-ferrous	149.1	138.0	0.4	55
Nonferrous processing	16.1	15.0	2.7	41
Miscellaneous metal products	26.1	24.4	1.6	39
Secondary nonferrous smelting and refining	17.7	14.9	2.2	33
Primary zinc	33.6	24.2	1.1	27
Primary lead	13.0	10.5	0.7	7
				4,142

STEEL INDUSTRY

This component consists of establishments engaged in the production of iron and steel and the manufacture of steel wire, sheet, strip, pipe, and tubes. The component corresponds to those establishments in the four SIC codes 3312, 3315, 3316, and 3317. This analysis is primarily directed toward SIC 3312—Blast Furnaces (Including Coke Ovens), Steel Works, and Rolling Mills—which represents the basic iron and steel industry. SIC 3315, 3316, and 3317 are only briefly reviewed because of their relatively low production volume and low energy consumption.

In the segments represented by SIC 3315, 3316, and 3317, basic steel shapes (e.g., rod, bar, strip) are purchased from the steel-producing companies (SIC 3312) and then processed into final products (e.g., wire, nails, cold-rolled sheet, pipe). The processing usually is relatively simple consisting of such steps as cold drawing, cold rolling, welding, annealing, and heat treatment. Most of the establishments are relatively small and usually produce a limited range of types of products.

PROCESS TECHNOLOGY INVOLVED

In SIC 3312, steel is produced in either integrated plants or in so-called "cold metal" plants. In the integrated plants, the primary source of iron units is iron ore either in the form of lump ore, sinter, or pellets. The flow of steelmaking operations for an integrated plant is shown in Figure 2.

In a cold metal shop, the basic raw materials are iron and steel scrap which are melted and refined in an electric-arc furnace. The steel is then cast and rolled into the final shape (usually bar or rod form). Most cold metal shops have relatively low capacity as compared to integrated plants. It should be noted that many integrated plants also have electric-arc furnaces for making steel from scrap.

A brief description of the major processes used in the production of finished steel shapes is presented in the following review of integrated operations.

20

FIGURE 2: FLOWSHEET OF INTEGRATED STEELMAKING OPERATIONS

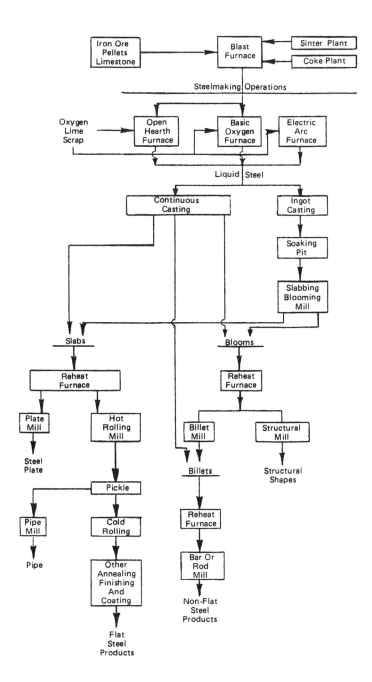

Source: Reference (4)

The primary source of iron units is "hot metal" obtained from a blast furnace. It is obtained from such basic raw materials as iron ore in one or more of the forms previously mentioned and coke which is produced from coal at the steel plant. Only a small percentage of coke used is purchased from outside producers. When hot metal is cast into small bars it is known as pig iron.

Presently in the United States there are three types of facilities used to produce liquid steel:

> Open-hearth furnaces
> Basic-oxygen furnaces (BOF)
> Electric-arc furnaces

In 1972, 26% of the steel was produced in open-hearth furnaces, 56% in BOF's, and 18% in electric-arc furnaces. In both the open-hearth and BOF processes, steel scrap is used as part of the charge to the furnace, in addition to hot metal. A rapid replacement of old open-hearth furnaces with new BOF facilities has been taking place over the last 15 years. In 1959, total BOF raw steel production was about 1.9 million net tons; in 1972 it was almost 75 million tons or 56% of total raw steel production of 133 million tons.

The steel industry not only remelts the scrap generated within the steel plants (home scrap) but also purchases a large tonnage of scrap. In 1972 it produced about 45 million tons of home scrap but also received an additional 31 million tons.

The liquid steel from any one of the three steelmaking processes is cast by either one of two processes. The metal may be "ingot cast," in which case the liquid steel is poured into iron molds and allowed to solidify. The other process is known as "continuous casting," in which the liquid steel is cast into an oscillating water-cooled mold and such semifinished shapes as slabs, blooms, or billets are produced as a continuous strand.

In order that ingots may be converted into semifinished forms, they are heated to a temperature of 2200° to 2400°F in refractory chambers known as "soaking pits." The hot ingots are then rolled into forms such as slabs, blooms, or billets.

After removal of surface defects, the slabs, blooms, or billets are then brought up to a temperature of about 2400°F in reheat furnaces. They are then hot rolled into various forms such as sheet, strip, structurals, bar, rod, etc.

Various additional finishing operations such as those listed below are carried out to obtain some final steel products: pickling, cold rolling, annealing, tin plating and zinc coating.

Figure 3 shows the primary steps in the steel manufacturing process using the basic oxygen furnace. The major energy consumption operations are coking of coal; agglomerating of iron ore; ironmaking; steelmaking; soaking of ingots; reheating of blooms, billets, and slabs; and heat-treating or forging. These operations account for more than 80% of the total energy consumption.

Energy consumption in the process is highly dependent on the ratio of scrap-to-blast-furnace iron that is fed to the furnace. The energy values used are based on a scrap-to-pig iron feed ratio of 1:2, i.e., approximately the average of the overall steel industry.

FIGURE 3: STEEL ENERGY CONSUMPTION DIAGRAM

1972 USA shipments: 83.5 x 10^9 kg (184 x 10^9 lb)
1972 energy consumption (primarily coal and natural gas): 110,000 Mw (3,300 x 10^{12} Btu)

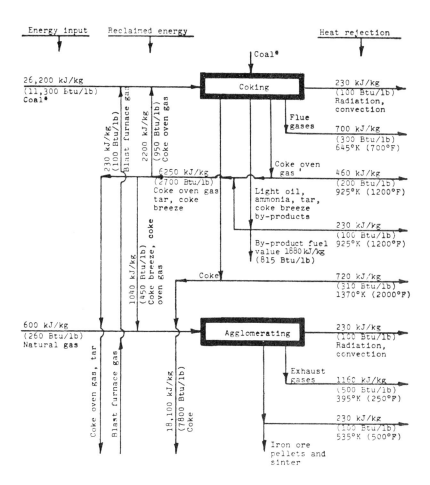

*Coal is feed and energy source in this operation.

(continued)

FIGURE 3: (continued)

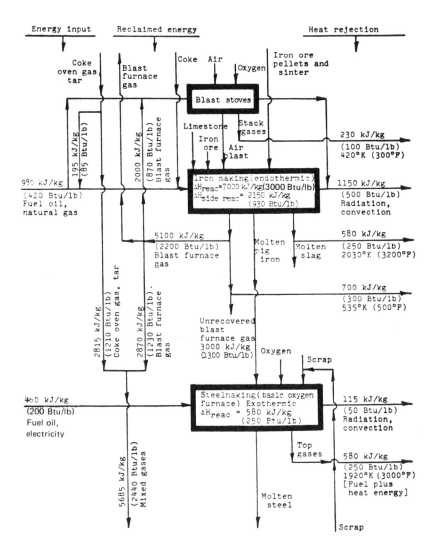

(continued)

FIGURE 3: (continued)

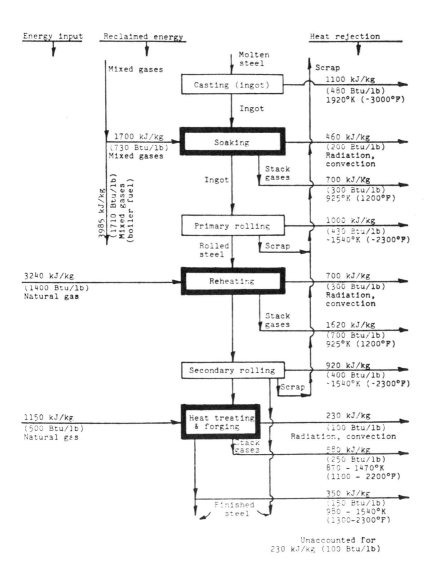

Figure 4 shows the coking operation. Coal is distilled at approximately 1370°K (2000°F) using the combustion of coke oven gas and blast furnace gas as the heat source. The coke oven is a rather complicated piece of equipment. Coal is located in narrow slots typically 40 feet long by 20 feet high by 18 inches wide.

Hot combustion gases pass between slots containing coal, down through brick checkerwork, and then out waste heat flues. Combustion air is preheated by passing through previously heated brick checkerwork. Flow of combustion gases is reversed periodically so that each half of the brick checkerwork (regenerators) is being heated half the time and cooled half the time.

Products from the coke oven include coke, coke oven gas, tar, light oil, ammonia solution, and coke breeze (small pieces of coke which pass through ½-inch screen). The coke oven gas, coke breeze, and part of the tar are used as fuel in other portions of the steel manufacturing process.

Figures 5 and 6 show two agglomerating operations that are commonly used to improve iron ore permeability and improve gas-solid contact in the blast furnace. Agglomeration also decreases the amount of fine material that is blown out of the blast furnace. The agglomerating operations are sintering and pelletizing.

Figure 5 shows that the sintering operation occurs on a traveling grate that conveys a bed of ore fines, limestone fines, and coke breeze. The bed (coke breeze) is ignited by gas burners and, as the mixture moves along the grate, air is pulled down through the mixture to keep the breeze burning. The heat sinters the mixture at 1640°K (2500°F) into pea- to baseball-size lumps. Approximately one-third of the iron ore burden in a typical blast furnace is sintered.

Figure 6 shows that the pelletizing operation also occurs on a traveling grate. Pellets are formed from iron ore, bentonite, and moisture. The pellets are coated with coal which is ignited on the traveling grate. Recuperated hot air from the cooling hood is used to dry and preheat the pellets. Natural gas is used to ignite the pulverized coal or coke breeze fuel.

Some heat is sometimes obtained from the oxidation of magnetite to hematite. The pelletizing operation occurs at 1370° to 1590°K (2000° to 2400°F). Approximately 50% of the iron ore burden in a typical blast furnace is pelletized.

Figure 7 shows the ironmaking portion of the steel manufacturing process. The reduction of iron ore to iron takes place in the blast furnace. The reducing agent is coke which not only reduces the iron ore but also provides heat to melt the iron.

Additional heat is also provided by the hot blast which is a mixture of air and oxygen that has been heated in the blast stoves. The fuel for the blast stoves is blast furnace gas plus natural gas. Some hydrocarbon fuels are also generally injected into the blast furnace. The temperature in the lower part of the blast furnace is 1750°K (2700°F). The blast furnace gas coming off the top of the furnace is used as a fuel throughout the steel process.

Figure 8 shows another type of flow diagram for the ironmaking process (7) which indicates both material and energy balances.

FIGURE 4: EQUIPMENT DIAGRAM—COKE OVEN

Rejected heat:
 Radiation—230 kJ/kg (100 Btu/lb)
 Flue gases—700 kJ/kg (300 Btu/lb) at 645°K (700°F)
 Coke oven gas—460 kJ/kg (200 Btu/lb) at 925°K (1200°F)
 Tar, light oil, ammonia water stream—230 kJ/kg (100 Btu/lb) at
  925°K (1200°F)
 Coke—720 kJ/kg (310 Btu/lb) at 1370°K (2000°F)

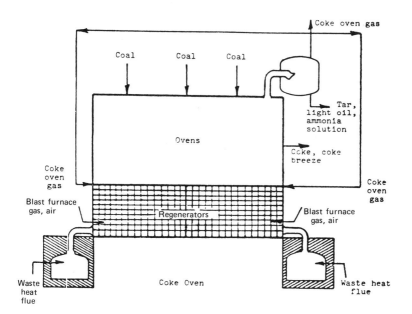

Energy input		
Coal	26,200 kJ/kg	(11,300 Btu/lb)
Gases	2,440 kJ/kg	(1,050 Btu/lb)
Energy output		
Coke	18,100 kJ/kg	(7,800 Btu/lb)
Coke oven gas	5,560 kJ/kg	(2,400 Btu/lb)
Tar	1,270 kJ/kg	(550 Btu/lb)
Light oils	325 kJ/kg	(140 Btu/lb)
Coke breeze	970 kJ/kg	(420 Btu/lb)
Losses	2,440 kJ/kg	(1,050 Btu/lb)

Note: All energy is expressed in terms of energy per unit weight of finished steel.

Source: Reference (2)

FIGURE 5: EQUIPMENT DIAGRAM—SINTER OPERATION

Rejected heat:
   Radiation—230 kJ/kg (100 Btu/lb)
   Exhaust gases—1,160 kJ/kg (500 Btu/lb) at 395°K (250°F)
   Hot sinter or pellets—230 kJ/kg (100 Btu/lb) at 535°K (500°F)*

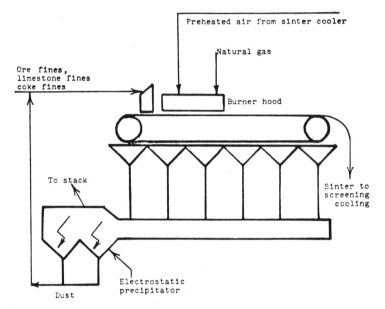

*Rejected heat quantities are totals for sintering and pelletizing operations.
Note: All energy is expressed in terms of energy per unit weight of finished steel.

FIGURE 6: EQUIPMENT DIAGRAM—PELLETIZING*

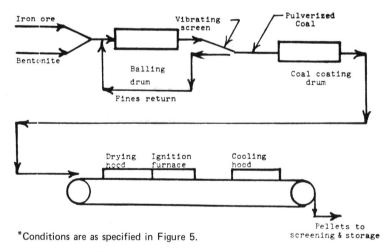

*Conditions are as specified in Figure 5.

Source: Reference (2)

FIGURE 7: EQUIPMENT DIAGRAM—BLAST FURNACE AND BLAST STOVES

Rejected heat:
   Radiation, other—1,150 kJ/kg (500 Btu/lb)
   Sensible heat in blast furnace gas—700 kJ/kg (300 Btu/lb) at 535°K (500°F)
   Stack gas from blast stoves—230 kJ/kg (100 Btu/lb) at 420°K (300°F)
   Molten slag—580 kJ/kg (250 Btu/lb) at 2030°K (3200°F)
   Molten iron transfer—70 kJ/kg (30 Btu/lb) at 2030°K (3200°F)
Lost fuel:
   Blast furnace gas—3,000 kJ/kg (1,300 Btu/lb)

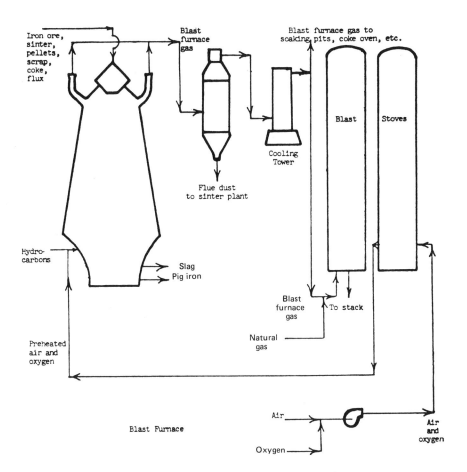

Note: All energy is expressed in terms of energy per unit weight of finished steel.

Source: Reference (2)

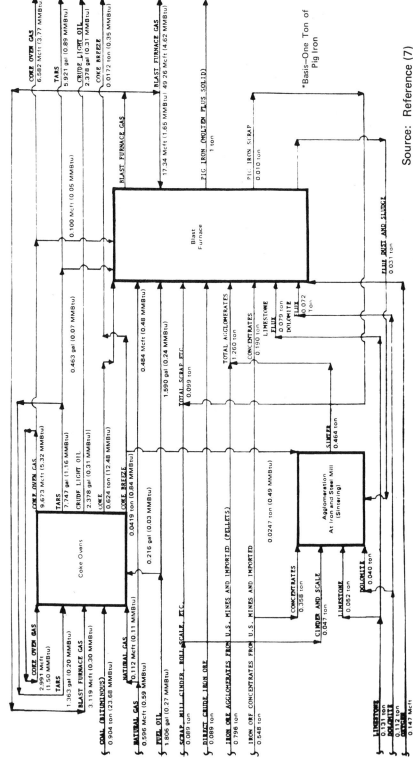

FIGURE 8: MATERIALS AND ENERGY CONSUMPTION DETAIL FOR THE BLAST FURNACE AREA IN 1970*

Source: Reference (7)

*Basis—One Ton of Pig Iron

Figure 9 shows the steelmaking operation using the basic oxygen furnace. Pig iron from the blast furnace generally contains excessive amounts of carbon, silicon, manganese, and phosphorus. These impurities are quickly oxidized by oxygen which is blown onto the molten metal. These oxidation reactions liberate heat so that very little additional heating is needed to keep the metal molten.

FIGURE 9: EQUIPMENT DIAGRAM—STEELMAKING, BASIC OXYGEN FURNACE

Rejected heat:
    Radiation—115 kJ/kg (50 Btu/lb)
    Top gases (fuel value plus sensible heat)—580 kJ/kg (250 Btu/lb) at $1920°K$ $(3000°F)$

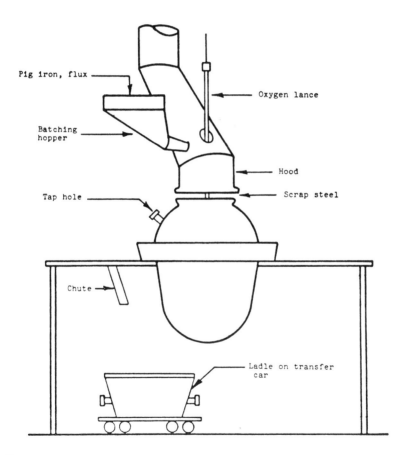

Note: All energy is expressed in terms of energy per unit weight of finished steel.

Source: Reference (2)

Figure 10 shows the steelmaking operation using the open hearth furnace. Air for combustion passes through brick checkerwork regenerators in one direction for 15 to 20 minutes and then is directed in the opposite direction. The air provides oxygen to oxidize impurities in the steel and to burn fuel for heat. The combustion gases pass through the brick checkerwork regenerators and then to boilers.

The open hearth furnace is declining in importance because of its slowness. It also requires more energy than the now dominant basic oxygen furnace. The use of an oxygen lance increases the speed of this operation and also decreases energy consumption. The decrease in energy consumption is due primarily to a decrease in the amount of heat-absorbing nitrogen which passes through the system.

FIGURE 10: EQUIPMENT DIAGRAM—STEELMAKING, OPEN HEARTH FURNACE WITH OXYGEN INJECTION

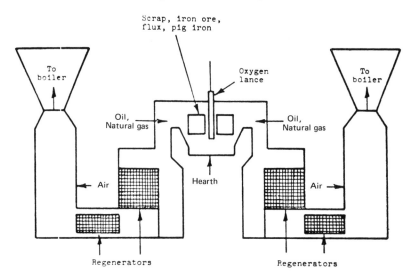

Rejected heat:
   Radiation—1,350 kJ/kg (580 Btu/lb)
   Stack gas heat—840 kJ/kg (360 Btu/lb) at 480°K (400°F)
Heat used for steam generation: 700 kJ/kg (300 Btu/lb)

Note: All energy is expressed in terms of energy per unit weight of finished steel produced using the open hearth furnace.

Source: Reference (2)

Figure 11 shows the steelmaking operation using a direct-arc electric furnace. Electricity is used to generate heat to melt scrap steel. Preheating of the charge can be used to reduce energy requirements in melting scrap. Oxygen lancing speeds oxidation of pig iron and results in energy savings. This method of producing steel is increasing in importance.

FIGURE 11: EQUIPMENT DIAGRAM—STEELMAKING, ELECTRIC-ARC FURNACE

Rejected heat:
Radiation, conduction—700 kJ/kg (300 Btu/lb)

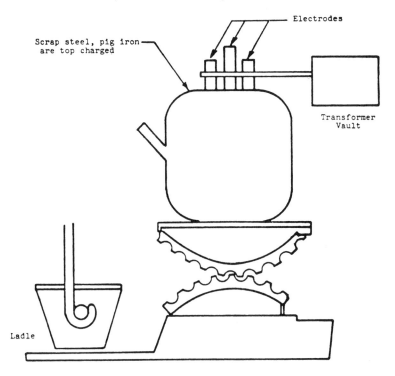

Note: All energy is expressed in terms of energy per unit
weight of finished steel produced using an electric-arc
furnace.

Source: Reference (2)

Figure 12 shows a block flow diagram (7) for the three steelmaking processes just described indicating both material and energy balances.

Figure 13 shows the soaking pit operation using a two-way gas-fired soaking pit. Often coke oven gas and blast furnace gas are used to heat solidified ingots to approximately 1600°K so that they can be rolled into blooms, billets, and slabs.

Recuperators allow some of the heat from combustion gases to be retained in the furnace by transferring this heat to incoming combustion air.

FIGURE 12: MATERIALS AND ENERGY CONSUMPTION DETAIL FOR THE RAW STEEL PRODUCTION AREA IN 1970*

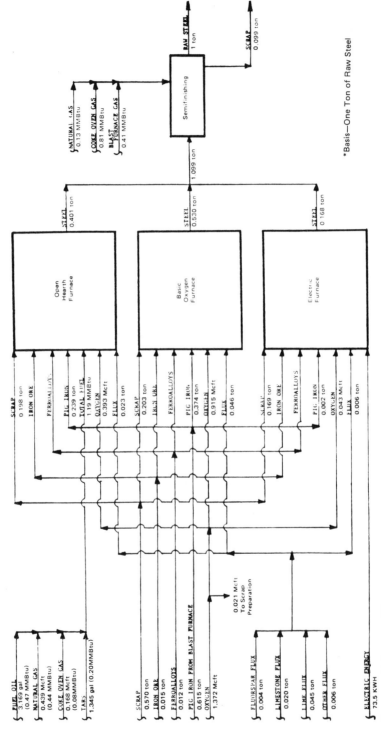

*Basis—One Ton of Raw Steel

Source: Reference (7)

FIGURE 13: EQUIPMENT DIAGRAM—SOAKING PIT

Rejected heat:
Radiation, convection—460 kJ/kg (200 Btu/lb)
Stack gases—700 kJ/kg (300 Btu/lb) at 925°K (1200°F)

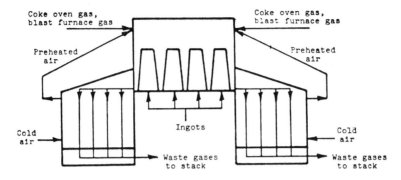

Two-way fired soaking pit

Note: All energy is expressed in terms of energy per unit weight of finished steel.

Source: Reference (2)

Figure 14 shows the reheating operation using a five-burner, countercurrent-fired, pusher-type, continuously reheating furnace. Natural gas can be used as fuel in this operation which heats slabs, blooms, and billets to approximately 1530°K so that they can be further rolled or milled into finished products. Recuperators allow some of the heat from combustion gases to be retained in the furnace by transferring this heat to incoming combustion air.

FIGURE 14: EQUIPMENT DIAGRAM—REHEATING FURNACE, COUNTER-CURRENT PUSHER-TYPE CONTINUOUS

Rejected heat:
Radiation convection—700 kJ/kg (300 Btu/lb)
Stack gases—1,620 kJ/kg (700 Btu/lb) at 925°K (1200°F)

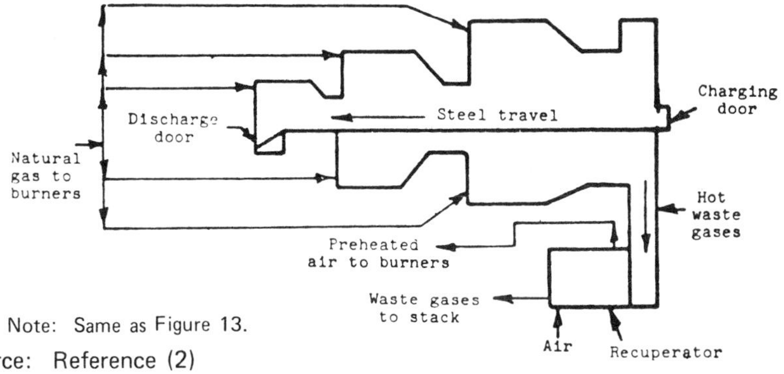

Note: Same as Figure 13.

Source: Reference (2)

Figure 15 shows a radiant-type annealing furnace. Approximately 25% of finished steel products are given an annealing treatment at 920° to 1090°K (1200° to 1500°F) to relieve stresses in the steel. Another 15% is processed at 1450° to 1510°K (2150° to 2250°F) in forging furnaces. Natural gas is often used to provide energy for these operations.

FIGURE 15: EQUIPMENT DIAGRAM—ANNEALING OVENS

Rejected heat*:
 Radiation, conduction, other—230 kJ/kg (100 Btu/lb)
 Exhaust combustion gases: 580 kJ/kg (250 Btu/lb) at 870° to 1470°K
  (1100° to 2200°F)

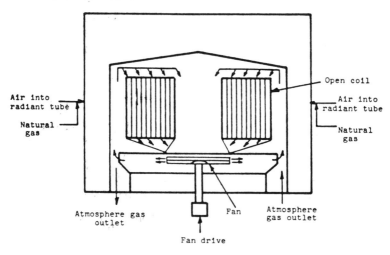

Annealing furnace for coiled strip

*Includes annealing and forging oven energy.
Note: All energy is expressed in terms of energy per unit weight of
 finished steel.

Source: Reference (2)

The products from any one of the three steelmaking processes may be converted into steel slabs by one of two routes. [1] They may be cast into ingots, soaked (see Figure 13), and then reheated (see Figure 14) before being rolled into slabs. The energy requirements of this sequence are summarized in Figure 16. [2] They may be continuously cast into slabs. The energy requirements for this mode of operation are shown in Figure 17.

The weighted average energy consumption of the various steelmaking processes in 1973 is shown in Figure 18. The energy values for the various unit operations are given in the boxes in units that represent millions (10⁶) of Btu.

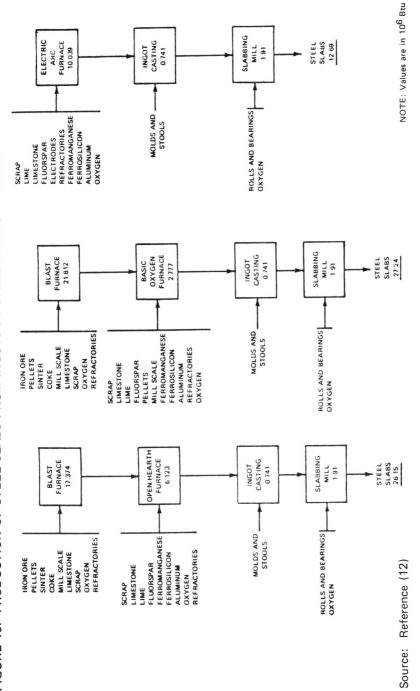

FIGURE 16: PRODUCTION OF STEEL SLABS FROM INGOTS BY STEELMAKING PROCESS

NOTE: Values are in 10^6 Btu

Source: Reference (12)

FIGURE 17: PRODUCTION OF CONTINUOUS-CAST STEEL SLABS BY STEELMAKING PROCESS

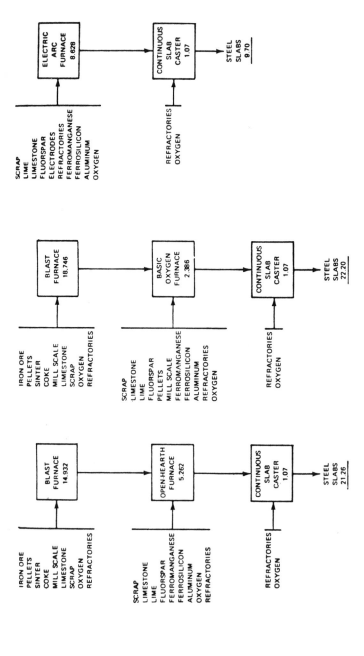

NOTE: Values are in 10⁶ Btu

Source: Reference (12)

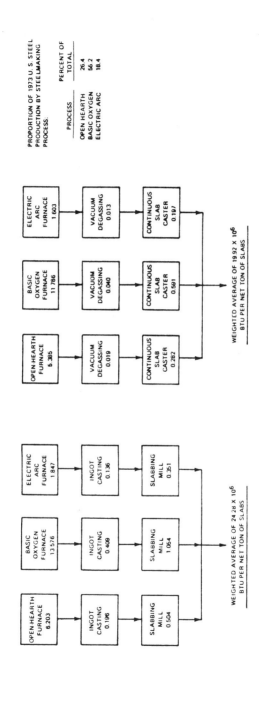

FIGURE 18: WEIGHTED-AVERAGE ENERGY CONSUMPTION OF STEELMAKING PROCESSES IN THE UNITED STATES FOR 1973—INGOT CASTING VERSUS CONTINUOUS CASTING TO PRODUCE SLABS

Source: Reference (12)

MAJOR ENERGY CONSERVATION OPTIONS TO 1980

The energy consumption situation and the potentials for energy conservation in the steel industry have been the subject of a number of studies including:

A Gordian Associates study of the potential for energy conservation (7)(11).

A 1975 Battelle study for FEA (8).

A Resource Planning Associates study for the Department of Commerce (9). This report emphasized energy requirements for environmental control.

An FEA study of energy conservation in manufacturing (10) made as a part of the overall Project Independence and looking ahead as far as 1990.

A 1975 Battelle study for the U.S. Bureau of Mines (12).

A 1974 Report to the Energy Policy Project of the Ford Foundation (14).

A 1975 British report on the energy conservation experiences of several large firms (15) including British Steel Corp.

A 1974 British report on energy conservation in industry (16) including steelmaking.

The proceedings of a 1975 British conference on energy recovery in process plants (17).

Table 10 shows the causes of energy losses in the operations of the steel process. It also gives an estimate of the magnitude of the losses and some possible energy conservation approaches.

TABLE 10: STEEL ENERGY CONSERVATION APPROACHES

Causes of Energy Losses	Approximate Magnitude of Losses	Energy Conservation Approaches
(1) Coke ovens		
(a) Radiation and convection	230 kJ/kg (100 Btu/lb)	Insulation Maintenance
(b) Partial nonrecovery of sensible heat of flue gases	700 kJ/kg (300 Btu/lb)	Design modification (waste heat recovery)
(c) Nonrecovery of sensible heat of coke	720 kJ/kg (310 Btu/lb)	Design modification (dry quench with heat recovery)
(d) Nonrecovery of heat in by-products stream	230 kJ/kg (100 Btu/lb)	–
(e) Nonrecovery of sensible heat of coke oven gas	460 kJ/kg (200 Btu/lb)	–
(f) Wastage of coke oven gas	100 kJ/kg (40 Btu/lb)	Waste utilization
(2) Agglomeration		
(a) Radiation and convection	230 kJ/kg (100 Btu/lb)	–
(b) Exhausting of hot gases from sintering or pelletizing machines and from coolers	1,160 kJ/kg (500 Btu/lb)	Design modification (waste heat recovery)

(continued)

TABLE 10: (continued)

Causes of Energy Losses	Approximate Magnitude of Losses	Energy Conservation Approaches
(c) Heat in product sinter and pellets	230 kJ/kg (100 Btu/lb)	Design modification (feed hot sinter and pellets to blast furnace)
(3) Blast furnace		
(a) Radiation, convection, other	1,150 kJ/kg (500 Btu/lb)	Insulation Maintenance
(b) Partial nonrecovery of sensible heat of blast furnace	700 kJ/kg (300 Btu/lb)	Design modification (waste heat recovery)
(c) Nonrecovery of sensible heat of slag	580 kJ/kg (250 Btu/lb)	–
(d) Wastage of blast furnace gas	3,000 kJ/kg (1,300 Btu/lb)	Waste utilization
(4) Steelmaking furnace (basic oxygen)		
(a) Radiation and convection	115 kJ/kg (50 Btu/lb)	Insulation Maintenance
(b) Sensible heat in top gases	580 kJ/kg (250 Btu/lb)	Design modification (waste heat recovery)
(5) Soaking pit		
(a) Partial nonrecovery of sensible heat of combustion gases	700 kJ/kg (300 Btu/lb)	Design modification (waste heat recovery)
(b) Radiation and convection, other	460 kJ/kg (200 Btu/lb)	Insulation Maintenance
(6) Reheating furnace		
(a) Partial nonrecovery of sensible heat of combustion gases	1,620 kJ/kg (700 Btu/lb)	Design modification (waste heat recovery)
(b) Radiation and convection, other	700 kJ/kg (300 Btu/lb)	Insulation Maintenance
(7) Annealing or forging furnace		
(a) Partial nonrecovery of sensible heat of combustion gases	580 kJ/kg (250 Btu/lb)	Design modification (waste heat recovery)
(b) Radiation and convection	230 kJ/kg (100 Btu/lb)	Insulation Maintenance
(8) Overall process		
(a) Higher energy requirement for sintering as compared to pelletizing	460 kJ/kg (200 Btu/lb)	Process modification (increase use of pellets)
(b) Higher energy requirement for open hearth furnace	700 kJ/kg (300 Btu/lb)	Process modification (replace open hearth with basic oxygen furnace)
(c) Formation of scrap throughout the process	2,090 kJ/kg (900 Btu/lb)	Process modification (increase use of continuous casting)
(d) Loss of sensible heat of ingots between casting and soaking	1,700 kJ/kg (730 Btu/lb)	Process modification (use continuous casting)

Note: All energy values in this table are expressed in terms of energy per unit weight of finished steel.

Source: Reference (2)

The following paragraphs describe some possible energy conservation measures as outlined by Battelle in a report to FEA (4).

Increased Pellet Usage in Blast Furnaces: The use of pellets in the blast furnace has been responsible for an increase in productivity and a decrease in coke rate. Expansion and construction of new domestic pellet plants which have been firmly committed will provide an additional capacity of 25 to 30 million gross tons between now and 1980.

During the last 10 years, pellet consumption (as a percentage of total blast furnace ore consumption) has increased at an average annual rate of 2.4%. Assuming that pellet consumption continues to increase at the same rate for the 1972 to 1980 period, it is estimated that the energy saving potential by 1980 is 1.09 $\times 10^6$ Btu per ton of steel product. Although this saving in energy accrues to SIC 33, an increased energy requirement to make the larger amount of pellets would occur in SIC 10.

The following indicates the calculations from which the energy savings were estimated for increased pellet usage in blast furnaces. Based on data reported on a 10-year study of pellet usage in blast furnaces, a conservative estimate indicates that 4 pounds of coke can be saved per net ton of hot metal (NTHM) for each 1% increase in pellets in the blast furnace burden. The energy for coke is taken at 30.7 $\times 10^6$ Btu per net ton which includes the energy required to convert coal to coke in the coking process. For the base year 88.942 $\times 10^6$ net tons of hot metal and 84.05 $\times 10^6$ net tons of steel product were produced.

$$\frac{(2.4\%/\text{year})(4 \text{ lb}/\%/\text{NTHM})}{2,000 \text{ lb/ton}} (30.7 \times 10^6 \text{ Btu/ton}) = 0.147 \times 10^6 \text{ Btu/NTHM/year}$$

$$(88.942 \times 10^6 \text{ NTHM}) (0.147 \text{ Btu/NTHM/year}) = 13.07 \times 10^{12} \text{ Btu/year}$$

$$\frac{13.07 \times 10^{12} \text{ Btu/year}}{84.05 \times 10^6 \text{ tons of steel product}} = 0.155 \times 10^6 \text{ Btu/ton steel product year}$$

For the target period:

$$7 \text{ years} \times (0.155 \times 10^6) = 1.09 \times 10^6 \text{ Btu/ton of steel product}$$

The same energy saving potential is considered both technologically feasible and economically practicable, because no significant increase in pellet capacity could be achieved by 1980 other than present firm commitments for new and expanded facilities.

Increase in Coke Ash: Energy penalties accrue in steelmaking because the quality of coking coal with regard to ash and sulfur content has been deteriorating and this trend is expected to continue.

The blast furnace coke rate increases by about 30 pounds per net ton of hot metal for each 1% increase in the ash content of the coke. During the period 1970 to 1975 a typical increase in ash in coal was about 0.5%. About 1.5 tons of coal are required per ton of coke. Therefore, for this 5-year period the ash in coke increased by about 0.75%, or 0.15% per year.

$$(0.15)(30) = 4.5 \text{ lb coke/NTHM/year}$$

$$\frac{4.5}{2,000} \times (30.7 \times 10^6 \text{ Btu/ton of coke}) = 0.069 \times 10^6 \text{ Btu/NTHM/year}$$

In 1972, 88.942×10^6 NTHM produced:

$$(88.942 \times 10^6) \times (0.069 \times 10^6) = 6.14 \times 10^{12} \text{ Btu/year}$$

In 1972, 84.05×10^6 net tons of steel produced:

$$\frac{6.14 \times 10^{12}}{84.054 \times 10^6} = 0.07 \times 10^6 \text{ Btu/tons of steel product}$$

For the 7-year period to 1980:

$$7 \text{ years} \times (0.07) \times 10^6) = 0.49 \times 10^6 \text{ Btu increase in energy consumption}$$
per net ton of steel produced

Increase in Coke Sulfur: The coke rate increases about 60 pounds per net ton of hot metal for each 1% increase in sulfur in coke. For the period 1970 to 1975, the increase in sulfur content in coal was about 0.02%. For 1.5 tons of coal per ton of coke, the increase in sulfur content of coke was about 0.03%, or 0.006% per year.

$$(0.006)(60) = 0.36 \text{ lb coke/NTHM/year}$$

$$\frac{0.36}{2,000} \times (30.7 \times 10^6 \text{ Btu/ton of coke}) = 0.0055 \times 10^6 \text{ Btu/NTHM/year}$$

For 88.942×10^6 NTHM:

$$(88.942 \times 10^6) \times (0.0055 \times 10^6) = 0.49 \times 10^{12} \text{ Btu/year}$$

For 84.05×10^6 NT steel:

$$\frac{0.49 \times 10^{12}}{84.05 \times 10^6} = 0.0058 \times 10^6 \text{ Btu/ton of steel product}$$

For the 7-year period to 1980:

$$7 \text{ years} \times (0.0058 \times 10^6) = 0.04 \times 10^6 \text{ Btu increase in energy consumption}$$
per net ton of steel produced

This is a technological factor of such a nature that its effect on energy consumption is the same on an economic basis as on a technological basis.

Increased Recovery of Blast Furnace Gas: All blast furnace gas is not presently being recovered. An increase in recovery of about 6% (during the 1972 to 1980 period) of the total blast furnace gas generated would give an energy saving of 0.39×10^6 Btu per ton of steel product made by this component.

The following are the calculations upon which Battelle (4) based their estimate of energy savings. About 65,000 cubic feet of blast furnace gas are produced per net ton of hot metal. The heating value of the gas is 95 Btu per cubic foot.

The total gas produced in 1972 was:

$$88.942 \times 10^6 \text{ NTHM} \times 65 \times 10^3 \text{ cu ft /NTHM} = 5.78 \times 10^{12} \text{ cu ft}$$

In 1972, in-plant consumption of BF gas was 4.24×10^{12} cu ft (5).

Therefore, percent recovered and used was:

$$\frac{4.24 \times 10^{12}}{5.78 \times 10^{12}} \times 100 = 73\% \text{ in 1972}$$

If 10% is considered to be an inevitable loss then 17% is the maximum additional amount which can be recovered.

$$(0.17)(65,000 \text{ cu ft/NTHM})(95 \text{ Btu/cu ft}) = 1,050,000 \text{ Btu/NTHM}$$

Assume that during the 10 year period after the base year (1972) operational and design improvements can be made which will permit the practical recovery of half of this maximum potential, or 525,000 Btu/NTHM.

For 88.942×10^6 NTHM:

$$(88.942 \times 10^6) \times (0.525 \times 10^6) = 46.7 \times 10^{12} \text{ Btu}$$

For the 7-year period to 1980, the energy saving would be:

$$(0.7)(46.7 \times 10^{12}) = 32.69 \times 10^{12} \text{ Btu}$$

For 84.05×10^6 NT steel:

$$\frac{32.69 \times 10^{12}}{84.05 \times 10^6} = 0.39 \times 10^6 \text{ Btu per net ton of steel product}$$

The capital cost for increased blast furnace gas recovery is relatively low, because little major equipment would be required. Action would involve the prevention of leakage and the recovery and use of gas which heretofore has been flared. Thus the energy saving potential which is technologically feasible is taken to be economically practicable.

New and Modernized Blast Furnaces: It is expected that from 1972 to 1980 about five or six new blast furnaces, ranging in capacity from 5,000 to 8,000 tons per day, will be in operation. These are expected to give improved energy efficiency, and, coupled with modernization of some existing furnaces, are expected to provide an energy saving averaging about 0.13×10^6 Btu per ton of steel product made in this component. The following gives some of the details of how this estimate was arrived at by Battelle in their report to FEA (4).

Modernization programs of several steel companies indicate that new and modernized blast furnaces will total about 14×10^6 tons per year of capacity by 1980. Such new and larger furnaces will provide the means to achieve lower coke rates and higher productivity. No quantitative data are available as to coke rate as a function of furnace size (all other variables remaining constant). For the purpose of this study, it is assumed that the operation of such furnaces will provide a decrease in coke consumption of 50 pounds per net ton of hot metal.

With coke at an energy value of 30.7×10^6 Btu per net ton, the energy saving would be:

$$\frac{50}{2,000} \times (30.7 \times 10^6) \times (14.0 \times 10^6) = 10.74 \times 10^{12}$$

For 84.05×10^6 NT steel:

$$\frac{10.74 \times 10^{12}}{84.05 \times 10^6} = 0.13 \times 10^6 \text{ Btu per ton of steel product}$$

Because of the long time for design, construction, installation, and shakedown operations, it is not considered technologically feasible to install additional blast furnaces by 1980, other than those presently committed. The economics of new large blast furnaces are evaluated on a host of factors, of which energy saving is just one factor. Other cost factors (e.g., labor) play a significant part in lowering the overall cost per net ton of hot metal. The fact that the construction of several new blast furnaces is underway indicates that these specific furnaces are economically practicable.

Dry Quenching of Coke: At present, hot coke (about 1850°F) is pushed from the coke oven into a quenching car which is delivered to a quenching station where the coke is sprayed with recirculated water. This quenching is done to prevent combustion of the coke. It produces a large quantity of steam which is vented to the atmosphere. Thus, no attempt is made to utilize the sensible heat of the hot coke.

Dry quenching involves the use of an inert gas as the coke-cooling agent and as a heat-transfer medium. The gas is continuously recirculated in a closed system, removing and transferring a major portion of the sensible heat in the coke to a waste-heat boiler or other type of heat exchanger. Most of the existing dry-quenching systems are used to generate high-pressure steam.

Dry quenching of coke is not new; it was introduced in the European coking industry in the 1920's. To date there is no commercial facility in operation in the United States. The estimated energy saving is about 1.2 million Btu per ton of coke. By 1980 it is estimated that the technologically feasible energy saving resulting from the dry quenching of 10×10^6 tons of coke annually would be 0.14×10^6 Btu per ton of steel product made by this component. The following gives the calculation upon which this estimate was based.

At an energy saving of 1.2×10^6 Btu/ton of coke the total would be:

$$(10 \times 10^6) \times (1.2 \times 10^6) = 12 \times 10^{12} \text{ Btu}$$

$$\frac{12 \times 10^{12} \text{ Btu}}{84.05 \times 10^6 \text{ tons of steel product (in 1972)}} = 0.14 \times 10^6 \text{ Btu/ton of steel product}$$

However, taking economic factors into consideration, there is no energy saving potential by 1980. Recently an economic analysis was made based on the following considerations.

Energy saving per ton of coke (as steam): 1.2×10^6 Btu per ton of coke
2,000 tons of coke/day x 350 days/yr: 0.7×10^6 tons of coke/yr
Steam value per 1,000 pounds: $2.25
Savings per ton of coke: $2.00
Gross savings per year: $1,460,000
Net savings/yr (after operating costs): $820,000
Total capital investment: $15,000,000

The investment per million Btu saved annually is:

$$\frac{\$15 \times 10^6}{(1.2 \times 10^6)(0.7 \times 10^6)} = \$18 \text{ per million Btu}$$

This figure substantially exceeds the investment of $10 per million Btu saved annually which is in this study considered to be the maximum practicable investment. It should also be noted that the payback period is about 18 years, which any company would consider unfavorable.

A dry quenching process has been described by R. Kemmetmueller (36) and is one in which an inert gas used for dry-quenching is also used directly for drying and preheating coal prior to delivery of the coal in the dry preheated condition to a coke oven. Figures 19 and 20 show the first and second portions of a pollution-free coal preheating plant which uses waste heat from dry coke quenching.

Referring first to Figure 20, there is schematically illustrated at the right thereof a dry-quenching bunker **10** which is charged with hot coke pushed out of a coke oven. For this purpose the hot coke is charged into the bunker **10** through the top end thereof which in a known way is provided with a closure assembly **12**. Thus, this assembly **12** closes off the interior of the bunker **10** from the outer atmosphere, and the closure assembly **12** is capable of being opened in a known way to receive a charge of hot coke directly from a coke oven, these operations preferably being carried out in such a way that there is no escape of pollutants to the outer atmosphere.

After the coke is cooled in the dry-quenching bunker **10** it is discharged out of the lower discharge end **14** thereof also in a known way. At the lower end **14** the bunker **10** is provided with suitable gates which are known and which control the discharge of the cooled coke from the coke oven, this coke being screened and delivered to suitable bins where it is available for use in blast furnaces, for example. When the coke is discharged from the dry-quenching bunker it has a temperature on the order of 50° to 70°C, since at this temperature the coke has been cooled sufficiently so that it will not burn belt conveyors. The interior of the dry-quenching bunker **10** communicates with a pressure release valve **16** for safety purposes.

It will be apparent from the description which follows that during the dry-quenching there is no generation of water gas in the dry-quenching bunker. This is one of the important advantages achieved with this process.

The inert gas which flows upwardly through the hot coke in order to cool that latter while extracting heat therefrom is preferably nitrogen or a mixture of nitrogen and carbon dioxide. This inert gas is fed in a perfectly dry condition to the lower end region of the bunker **10** by way of a recirculating fan **18** which drives the inert gas through a conduit **20** into the interior of the bunker **10**.

FIGURE 19: FIRST PORTION OF PLANT

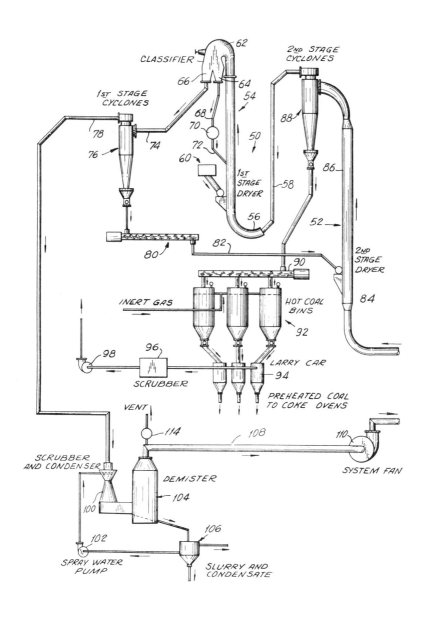

Source: Reference (36)

FIGURE 20: SECOND PORTION OF PLANT

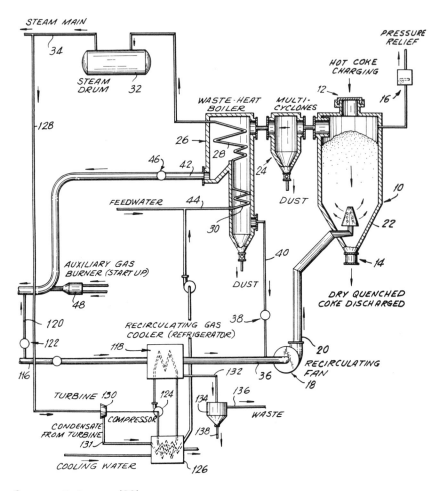

Source: Reference (36)

As is well known at the interior of the bunker **10** the latter carries a discharge head **22** in the form of a hollow component of substantially conical configuration into the interior of which the inert gas is delivered with this distributor head being formed with openings through which the inert gas escapes to flow up through the body of hot coke in the bunker **10**.

The inert gas flows out of the bunker **10** at a temperature on the order of 800° to 850°C, and immediately after discharging from the upper region of the bunker **10** the inert gas is directed through the multicyclone unit **24** where dust is removed from the inert gas. This dust is a valuable combustible product and may be used at any desired location.

The inert gas which is in this way cleaned by the multicyclone unit and which has the above temperature on the order of 800° to 850°C then is received by a steam generator means **26** in the form of a waste-heat boiler having upper coils **28** and lower coils **30** as illustrated. The lower coils communicate with the upper coils and receive feed water from any suitable supply as schematically illustrated in Figure 20. This water is converted to steam which is received in a steam drum **32** from which the steam is delivered through suitable steam main **34** to any desired location where use will be made of the steam which is generated in this way.

Part of the inert gas, which is still in a perfectly dry condition, is returned from the steam generating means to a conduit **36** which communicates with the suction end of the recirculating fan, and through the conduit **36** the perfectly dry inert gas is returned to the fan to be delivered thereby to the bunker as described above. The proportion of inert gas from the steam generating means which is mixed with the inert gas flowing along the conduit **36** can be controlled by a valve **38** in the conduit **40** which communicates with the conduit **36** and with the lower end region of the steam generator.

It will be noted that additional dust is delivered out of the steam generator. This dust also is combustible and forms a valuable product. Generally the inert gas flowing along the conduit **40** to the conduit **36** will have a temperature on the order of 200° to 250°C and approximately 30% of the gas delivered by the fan to the bunker will be derived from the conduit **40**, the remaining 70% being derived through the conduit **36** which in a manner described below delivers a cool dry inert gas to the fan, this latter gas which flows along the conduit **36** to join with the gas from the conduit **40** having a temperature on the order of 0° to 10°C. In this way the temperature of the gas flowing into the bunker can be controlled.

In other words by operating the damper or the valve it is possible to control the proportions of the inert gas derived from the conduits **36** and **40** so as to regulate the temperature of the inert gas which is delivered to the bunker. At the same time it is possible to control the temperature of the inert gas discharging out of the steam generator means through the conduit **42** by controlling the flow of cool boiler feed water into the steam generator through the supply conduit **44**. In this way it is possible to maintain a relatively constant predetermined temperature for the dry inert gas which is used for drying and preheating coal as described below.

The flow of the inert gas from the steam generator means along the conduit **42** is controlled by a valve or damper **46**, and the arrangement is such that a completely dry gas having a temperature on the order of 600° to 650°C flows out of the steam generator means along the conduit **42**. Of the inert gas which flows to the waste-heat boiler **26** from the bunker, approximately two-thirds of the gas will flow out of the boiler along the conduit **42** while the remaining inert gas will flow along the conduit **40** to be recirculated back to the bunker by way of the fan together with the remaining cool inert gas derived from the conduit **36** upstream of its connection with the conduit **40**.

The conduit **42** may be provided with an auxiliary gas burner **48** for starting purposes, this unit being used only at the beginning of an operating cycle, and the initial heating provided by way of the unit **48** will serve to generate a certain amount of inert gas. Part of the dust discharged out of the units **24** and **26** may be used in the auxiliary gas burner **48**.

Referring now to Figure 19, it will be seen that the perfectly dry inert gas which in the above example is at a temperature on the order of 600° to 650°C reaches a coal-drying means **50**. In the illustrated example the coal-drying means is a two-stage dryer, and the hot, dry inert gas first flows through the second stage **52** of the coal-drying means before reaching the first stage **54** thereof. With the illustrated coal-drying means, the coal is dried in a countercurrent manner with respect to the flow of inert gas in the sense that the second stage **52** is situated upstream of first stage, but in each stage the coal flows concurrently with the gas so that concurrent drying takes place in each stage of the coal-drying means while the stages thereof are arranged in a countercurrent manner with respect to the flow of inert gas.

It is to be understood that although a two-stage dryer is illustrated in Figure 19 and described in detail below, the process can equally well be used with a single stage dryer or with a fluidized bed type of coal dryer installation.

The first stage dryer unit is in the form of an elongated tubular structure which has an upstream end region **56** communicating with a conduit **58** through which the heated inert gas flows into the tubular structure of the first stage unit in order to dry coal therein. The wet coal is fed in a crushed particulate form, after passing through suitable crushing mills as is well known, into the tubular structure of the dryer unit at the region of the upstream end thereof. Thus Figure 19 shows schematically a wet coal feed means **60** which serves to feed the wet coal in particulate form into the tubular structure of the first stage unit.

The particles of wet coal become suspended in the flowing stream of inert gas to be carried along with the gas, while being dried thereby, toward the downstream end region **62** of the unit **54**. This end region **62** is provided with a pair of classifier units **64** and **66**. Thus, as the particles of coal suspended in the inert gas flow with the latter toward and through the curved end region **62** of the unit **54**, the larger and therefore heavier coal particles will discharge out of the classifier outlet **64** while the smaller and lighter coal particles will discharge out of the subsequent classifier outlet **66**. In this way a classification is carried out to separate particles which are larger than a given size from coal particles which are smaller than a given size.

The particles larger than a given size which discharge out of the classifier outlet **64** are carried along with part of the inert gas through a conduit **68** to a hammer mill **70** which is situated outside the path of flow of the inert gas and the coal suspended therein, and in the hammer mill the size of the particles is further reduced after which the particles of reduced size are returned to the tubular structure of the first stage dryer unit through a conduit **72**, as schematically represented in Figure 19.

Thus the process provides outside of the path of flow of the inert gas and the coal suspended therein a means for reducing the size of coal particles which are greater than a given size, with the particles which have their size thus reduced being returned to the tubular structure of the coal dryer to again flow along with the inert gas toward the downstream end region where the classifier means **64, 66** is situated. In this way the particles of coal which flow with the inert gas beyond the first stage of the dryer means **50** along the conduit **74** will necessarily be no larger than a given grain size. The inert gas with the coal particles suspended therein is received by a first-stage cyclone means **76** which includes a number of cyclones in which the particles of coal dried at the first stage are separated from the inert gas. The inert gas continues to flow beyond the cyclone separator means **76** along a conduit **78** where the inert gas has a temperature on the order of 110°C.

The particles of coal which are discharged from the cyclone means **76** are received by a conveyor means **80** such as a suitable screw conveyer, as schematically illustrated, and these particles of coal which are dried in the first-stage unit are delivered by the conveyer **80** through a conduit **82** into an upstream end region **84** of the tubular structure **86** which forms the second-stage unit of the coal-drying means. Of course in this second-stage unit the temperature of the inert gas is higher than in the first-stage unit, the temperature of the gas at the second stage being on the order of 600°C, as pointed out above.

In the second-stage unit the coal particles flow together with the inert gas toward the upstream end of the conduit **58** which interconnects the first and second stages, and in this second stage unit the coal not only has been completely dried but is preheated to a temperature well above the ambient temperature so that coal in completely dry and preheated condition is discharged out of the second stage of the coal-drying means. The downstream end of the second-stage unit communicates also with a cyclone means **88** which include a plurality of second-stage cyclones in which the inert gas is separated from the dry and preheated coal. This inert gas which is thus separated from the particles of coal at the second stage then flows through the conduit **58** to the first stage.

The dry and preheated coal is received from the second-stage cyclone means **88** by a conveyer means **90** which may also take the form of a suitable screw conveyer. It is to be noted that during drying and preheating the coal has been maintained entirely out of contact with oxygen.

The conveyer means delivers the dry preheated coal to a plurality of hot coal bins **92** which form a bin means in which the dry, preheated coal is temporarily stored prior to delivery to a larry car installation **94** which serves in a known way to deliver the dry, preheated coal to the coke ovens as schematically illustrated in Figure 19. The gas which discharges from the larry car unit is delivered to a scrubber means **96** where the gas is cleaned before being discharged to the outer atmosphere so that pollution is in this way avoided, a suitable blower **98**, or the like, being provided to draw the gases out of the larry car unit and through the scrubber means before discharging the clean gas to the outer atmosphere.

In the event that the system used for transfer of the hot coal from bin means **92** to the coke ovens is relatively tight with respect to the outer atmosphere, the gas discharged after the scrubber by the fan **98** will still for the most part be in the form of an inert gas, and in this case the inert gas can be returned to any desired part of the closed path of flow along which the inert gas circulates in accordance with the process, as will be apparent from the further description below.

In accordance with one of the further features of the process, for safety purposes an inert gas derived from any suitable source such as a nitrogen tank or the like is delivered to the interior of the bins **92** into upper regions thereof to form an inert gas cushion over the dry preheated coal which is temporarily stored in these bins. In this way an extremely safe installation is provided free of any possibllity of undesirable explosions. The inert gas which thus forms the gas cushions over the dry, preheated coal in the bins **92** is capable of escaping through the conveyer means into the second stage cyclone means to be combined with inert gas flowing through the second stage dryer unit, so that in this way the inert gas which forms the gas cushions above the preheated coal in the coal bins also serves to make up any losses in the continuously circulating inert gas which flows first through the bunker and then through the coal-drying means.

The inert gas which flows along the conduit 78 from the first stage cyclones is delivered to a scrubber and condenser means 100. The scrubber means 100 thus receives the cooled inert gas which has a temperature on the order of 100°C together with water vapor which has been removed from the coal during the drying thereof. In the scrubber and condenser means the water is quenched out of the gas. The scrubber means may be a single or two-stage scrubber and spray water is pumped into the scrubber means by way of a suitable pump 102.

After traveling through the scrubber the gas flows through a demister unit 104 which serves further to remove any moisture from the gas, with the liquid which is removed being delivered to a clarifier means 106 which discharges the condensate and slurry. Make-up water if required is delivered to the clarifier with this water being delivered to the spray water pump. At the same time the gas which has been dried flows from the demister unit along a conduit 108 to the inlet of the system fan 110 with part of the gas being discharged to the outer atmosphere through a suitable vent unit controlled by a valve or damper 114, so that in this way balanced operating conditions are maintained.

The gas in conduit 108 is a saturated gas at the dew point temperature of 30° to 40°C. This gas has on the order of 0.02 gram of water in each cubic meter of gas.

The system fan delivers the inert gas to a conduit 116 which serves to deliver the inert gas to a cooling means 118. A conduit 120 may be provided if desired to interconnect the conduits 116 and 42. This conduit 120 is optional. It may be used to deliver part of the inert gas directly from the fan to the conduit 42 in order to be combined with the hot inert gas from the steam-generating means before reaching the second stage dryer unit. If the connecting conduit 120 is provided, it will be equipped with a valve or damper 122 enabling the proportion of gas which flows from the conduit 116 to the conduit 42 to be regulated.

As has been indicated above the inert gas may be a mixture of nitrogen and carbon dioxide and this inert gas flows from the demister unit to the system fan at a temperature on the order of 30° to 40°C. In addition desulfurization, if necessary, is carried out at the demister unit. In this way a relatively cool, relatively dry, clean gas is delivered to the recirculating gas cooler means 118.

This cooling means forms part of a refrigerating unit in which through suitable coils a refrigerant is circulated as by a compressor means 124, suitable cooling water being directed through a section 126 of the cooling means as illustrated in Figure 20. The cooling installation will include suitable compressors 124 which must be driven, and for this purpose, steam is tapped from the main 34 and delivered by way of a conduit 128 to the steam turbine unit 130 of the cooling means so as to drive the compressors 124 thereof. Condensate from turbine 130 is delivered by conduit 131 to unit 126 together with the cooling water and is then fed by a pump and conduit to feedwater supply 44.

As a result of the cooling action taking place at the cooling means, the temperature of the gas is lowered to the order of 0.5°C, and all moisture in the gas is condensed out of the gas at the cooling means, with the condensate being received by the conduit **132** which serves to deliver the condensate to the condensate treating unit **134** from which gas is discharged by the conduit **136** while the condensate is discharged by the outlet **138**.

In this way a perfectly dry, chilled inert gas is delivered along the conduit **36** where it is combined with part of the gas flowing along the conduit **40** from the steam generating means, as described above, enabling in this way the temperature of the dry inert gas which is delivered to the bunker to be precisely regulated.

It is apparent, therefore, that a closed path of flow is provided for the inert gas, this closed path of flow having the body of hot coke through which the inert gas flows, at one part of the closed path, and a coal-drying means, to which the wet coal is delivered, at another part of the closed path.

It will be noted that an extremely efficient operation is achieved, making full use of the waste heat from the dry coke-quenching. Furthermore it will be noted that there is no generation of water gas as a result of the flow of the inert gas through the hot coke in the bin **10**. Full use is made of the energy of the waste heat from the dry coke-quenching by utilizing this heat not only for drying the wet coal initially but also for preheating the coal so that the coal is in a preheated as well as dry condition when delivered to the coke ovens.

The situation of the hammer mill at the exterior is also of advantage since in this way the hammer mill is not influenced by the heat and vapors in the dryer itself. Moreover it will be seen that throughout the entire system there is never a location where any gas with pollutants therein can escape to the outer atmosphere so that there is no problem in connection with pollution of the atmosphere. Furthermore, the installation does not include any expensive heat exchanging units and is for the most part relatively inexpensive as well as highly efficient in its operation.

Injection of Coal Into Blast Furnaces: This technology was proven several years ago and has been in continuous use by one steel company in two of its blast furnaces. By 1980 it appears feasible that coal-injection facilities could be installed for the production of 15×10^6 tons of hot metal annually. This would provide an energy saving of 0.09×10^6 Btu/ton of steel product made by this component. The following are calculations upon which this estimate is based.

Assuming that coal replaces coke on a 1 to 1 basis, a 20% coke replacement, coal at 26×10^6 Btu per net ton, coke at 30.7×10^6 Btu per net ton, and 1,200 pounds of coke per net ton of hot metal, the energy saving potential is about 0.5×10^6 Btu per net ton of hot metal.

By 1980, it is considered technologically feasible to install coal injection equipment on facilities producing 15×10^6 net tons per year of hot metal. The total energy saving would be

$$(15 \times 10^6) \times (0.5 \times 10^6) = 7.5 \times 10^{12} \text{ Btu}$$

For 84.05 x 10^6 NT steel:

$$\frac{7.5 \times 10^{12}}{84.05 \times 10^6} = 0.09 \times 10^6 \text{ Btu per net ton of steel product}$$

However, a recent economic appraisal indicated that the cost to install coal injection facilities on a furnace producing 1.225 x 10^6 net tons of hot metal per year would be about $8.6 million. From an energy point of view, the investment per million Btu saved annually is

$$\frac{\$8.6 \times 10^6}{(1.225 \times 10^6)\,(0.5 \times 10^6)} = \$14 \text{ per million Btu}$$

This exceeds the maximum practicable value of $10 million Btu saved annually. It has been indicated by representatives of the steel industry that an investment in coal injection facilities which would replace existing liquid fuel injection systems is economically unattractive. It should also be noted that liquid fuel injection systems are considered necessary at certain establishments where a large quantity of tar is generated in cokemaking and must be consumed if no outside market is available. Some of the new blast furnaces after 1980 will undoubtedly be fitted for coal injection.

External Desulfurization of Hot Metal: External desulfurization refers to those processes which are used to lower the sulfur content of hot metal external to the blast furnace (i.e., after tapping the hot metal from the blast furnace). Up to the present, the sulfur content of the hot metal is controlled through blast-furnace practice.

External desulfurization is being practiced commercially in several steel plants in Europe and Japan. As yet there is no routine commercial operation in the United States. The primary driving force for external desulfurization is the decreasing availability and increasing cost of low-sulfur charge materials.

External desulfurization provides a means for markedly lowering blast-furnace slag weight with a consequent decrease in the coke rate. One analysis of the energy saving indicated a coke decrease equivalent to 0.9 x 10^6 Btu per ton of hot metal. It is estimated that by 1980 the use of external desulfurization could be applied to 15 x 10^6 tons of hot metal annually to provide an energy saving of 0.16 x 10^6 Btu per ton of steel product made by this component. The following calculations show the derivation of this energy conservation estimate.

By 1980, it is considered technologically feasible to install external desulfurizing facilities to treat 15 x 10^6 tons of hot metal per year.

The total energy saved would be:

$$(15 \times 10^6) \times (0.9 \times 10^6) = 13.5 \times 10^{12} \text{ Btu}$$

For 84.05 x 10^6 NT steel:

$$\frac{13.5 \times 10^{12}}{84.05 \times 10^6} = 0.16 \times 10^6 \text{ Btu per ton of steel product}$$

No actual cost figures are available for such facilities. Factors which are considered constraints by some of the steel companies include the possibility of explosions from waste slags containing calcium carbide and space limitations around old blast furnaces. However, an estimate has been made that the capital cost for a facility to treat 1 million tons of hot metal per year by the Mag-Coke process would cost about $1.4 million. From an energy point of view, the investment per million Btu saved annually is:

$$\frac{\$1.4 \times 10^6}{(1.0 \times 10^6)\,(0.9 \times 10^6)} = \$1.6 \text{ per million Btu}$$

This is considerably below the maximum practicable value of $10 per million Btu saved annually. Thus the energy saving of 0.16×10^6 Btu per ton of steel product is considered both technologically feasible and economically practicable.

Substitution of Basic-Oxygen Furnaces for Open-Hearth Furnaces: Some analysts of the steel industry conclude that it takes substantially less energy to make steel in a BOF as compared to the open-hearth furnace. This is true if one looks only at the fuels and electrical energy consumed in the two processes.

However, the BOF requires substantially more hot metal (less scrap) than the open hearth. When the energy of the hot metal is considered, it is found that in total the BOF takes about 0.4×10^6 Btu more per ton of steel than does the open hearth. Thus with the historical and continued replacement of open-hearth furnaces with new BOF facilities, it is expected that by 1980 the energy increase over the base year will be 0.07×10^6 Btu per ton of steel product made by this component.

Over the past 20 years, the basic-oxygen (BOF) steelmaking process has been replacing the older open-hearth process to the extent that by 1974, about 56% of total domestic steel was made in basic-oxygen furnaces (BOF). In recent years the average annual increase in BOF capacity has been about 2.25×10^6 tons per year. It is assumed that this annual increase will be applicable for the 1972-1980 period.

If the energy in hot metal is included, the total energy to produce BOF steel is about 0.4×10^6 Btu higher per net ton of steel than for open-hearth steel. Therefore, the energy penalty per year is:

$$(2.25 \times 10^6) \times 0.4 \times 10^6 = 0.9 \times 10^{12} \text{ Btu/year}$$

For 84.05×10^6 NT steel:

$$\frac{0.9 \times 10^{12}}{84.05 \times 10^6} = 0.01 \times 10^6 \text{ Btu/ton of product/year}$$

For the seven-year target period, the energy increase is:

$$7 \times (0.01 \times 10^6) = 0.07 \times 10^6 \text{ Btu/ton of product}$$

Thus, it can be seen that from an energy point of view the installation of BOF furnaces causes an increase in energy consumption. The economic practicability of BOF furnaces is not justified by energy considerations. The justification is that steel can be produced at lower cost because it can be produced in one-fifth to one-tenth the time it takes to make open-hearth steel. New BOF facilities have been committed to 1980. Thus, this energy penalty will prevail under both technological and economic considerations.

Recovery of BOF Offgas: During the oxygen-blowing period of a BOF, a large volume of gas is generated. This gas contains up to 90% carbon monoxide, and can be used as a fuel if collected. In several foreign countries, the gas is collected by means of a closed hood over the furnace and a suitable gas-handling system.

In the United States, some of the sensible heat in the gas is recovered with waste-heat boilers. Only a few closed-hood BOF systems are operational in the United States, and in none of the installations is the gas collected and used for its fuel value. The energy saving is about 0.5×10^6 Btu per ton of steel. It is estimated that by 1980 an energy saving of 0.04×10^6 Btu per ton of steel product made by this component is technically feasible by collecting and recovering 10% of the BOF offgas. The following are the calculations upon which this estimate was based by Battelle in their report to FEA (4).

In 1972, 74.6×10^6 net tons of BOF steel were produced. It is considered technologically feasible by 1980 to retrofit or install facilities which would recover the offgas for use as a fuel equivalent to 10% of 1972 BOF production. The energy saving would be:

$$0.10 \ (74.6 \times 10^6) \times (0.5 \times 10^6) = 3.73 \times 10^{12} \ \text{Btu}$$

For 84.05×10^6 NT steel:

$$\frac{3.73 \times 10^{12}}{84.05 \times 10^6} = 0.04 \times 10^6 \ \text{Btu/net ton of steel product}$$

Data on capital costs for such gas recovery facilities is not readily available, but one estimate cited a retrofit capital cost of $10.2 million for a BOF installation with an annual capacity of 2 million tons. The figure used for energy saving in this estimate was 0.422×10^6 Btu per net ton. The investment per million Btu saved annually is:

$$\frac{\$10.2 \times 10^6}{(0.422 \times 10^6) \ (2.0 \times 10^6)} = \$12 \ \text{per million Btu}$$

This exceeds the maximum practicable value of $10 per million Btu saved annually. Thus, this saving is not considered economically practicable by 1980. After that time, with anticipated higher fuel costs, steel companies may be able to justify economically the recovery of offgas and its use as fuel.

Preheating Scrap for Basic-Oxygen Furnace with Oxygen-Fuel Burners: Only a few companies in the United States preheat scrap in the basic-oxygen furnace, but it has been shown to be technologically feasible. Some steel plants which have insufficient blast furnace capacity preheat the scrap charge so they can use more scrap and less hot metal per ton of steel produced.

The energy saving derives mainly from the decreased amount of hot metal (which has a high energy content) per ton of steel produced. It is estimated that by 1980 an energy saving of 0.12×10^6 Btu per ton of steel product made by this component is technologically feasible by preheating the scrap for 10% of BOF steel at an energy saving of 1.4×10^6 Btu per ton of BOF steel using preheated scrap. The following calculations furnish the basis for this estimate.

In 1972, 74.6×10^6 net tons of BOF steel were produced. It is considered technologically feasible by 1980 to install scrap preheating facilities equivalent to 10% of 1972 BOF steel production. The energy saving would be:

$$0.10 \ (74.6 \times 10^6) \times (1.4 \times 10^6) = 10.44 \times 10^{12} \text{ Btu}$$

For 84.05×10^6 NT steel:

$$\frac{10.44 \times 10^{12}}{84.05 \times 10^6} = 0.12 \times 10^6 \text{ Btu/net ton of steel product}$$

No specific economic data are available concerning the capital investment for such a scrap preheating facility, but steel industry representatives indicate that it has severe economic handicaps, unrelated to energy considerations, because preheating scrap in the furnace takes a substantial amount of time and thus daily steel production from a given furnace is markedly lowered and the cost per ton of steel increases. Thus, for this reason no steel plant with sufficient hot metal capacity will install fuel-oxygen preheating facilities and the technologically potential energy saving, from a practical overall cost point of view, is taken to be not economically practicable.

Increased Electric-Furnace Capacity: Steel scrap is the major metallic component in the charge to the electric furnace. In the Battelle report to FEA (4), steel scrap is taken to have a zero energy value. Consequently, with electrical energy at 3,412 Btu per kwh, the energy required to make steel in an electric furnace is only about 2×10^6 Btu per ton of steel. This compares with an average energy value for BOF and open-hearth steel of about 16.4×10^6 Btu per ton of liquid steel. Thus there is a considerable energy saving by melting scrap in an electric furnace.

For the 1964-1974 period, electric furnace production, as a percent of total, increased at an average of 1% per year. With an average raw steel production of about 130×10^6 tons per year over the same period, the annual increase for electric furnace steel was 1.3×10^6 tons per year. On the basis of economic practicability the historical rate of growth is assumed to prevail to 1980.

It is considered technically feasible to install sufficient electric-furnace capacity by 1980 to achieve an energy saving of 2.01×10^6 Btu per ton of steel product made by this component. Of this potential saving, about 78% is accounted for by the historic growth rate of 0.69% per year, and 22% by an assumed doubling of growth rate in 2 of the 7 years. The calculations leading to the energy conservation estimate are as follows:

$$(1.3 \times 10^6) \ [(16.4 \times 10^6) - (2.0 \times 10^6)] = 18.7 \times 10^{12} \text{ Btu per year}$$

For 84.05×10^6 NT steel, the energy saving is

$$\frac{18.7 \times 10^{12}}{84.05 \times 10^6} = 0.2225 \times 10^6 \text{ Btu/ton of steel product}$$

For the 7 years of the target period, the energy saving is:

$$(0.2225 \times 10^6) \times 7 = 1.56 \times 10^6 \text{ Btu/ton of steel product}$$

This is considered economically feasible because the steel industry has followed this pattern of increasing electric furnace capacity, and it is assumed that they are not doing it if it is not financially advantageous.

From the point of view of technological feasibility, it is considered possible to increase electric furnace capacity by two times the historical rate for each of the last two years of the 1972-1980 period. Thus, the technologically feasible energy saving is taken to be:

$$(1.56 \times 10^6) + 2(0.2225 \times 10^6) = 2.01 \times 10^6 \text{ Btu/ton of steel product}$$

Increased Continuous Casting Capacity: Continuously cast steel can be produced with less energy consumption than ingot cast steel because of the higher yield of semifinished steel such as slabs, blooms, or billets. In recent years, the installation of continuous casting facilities has been at an average rate of about 2.0×10^6 net tons per year. Taking into consideration the yield factors and the energy content for electric furnace steel versus BOF and open-hearth steel, the average energy saving for continuously cast steel is 3.4×10^6 Btu per ton. Based on the historical growth of continuous casting, it is assumed that it is economically practicable and that the same average growth will prevail during the 1972-1980 period. Therefore, the energy saving will be:

$$(2 \times 10^6) \times (3.4 \times 10^6) = 6.8 \times 10^{12} \text{ Btu per year}$$

For 84.05 NT steel the energy saving is:

$$\frac{6.8 \times 10^{12}}{84.05 \times 10^6} = 0.081 \times 10^6 \text{ Btu per year per ton of steel product}$$

For the target period, the energy saving is:

$$7 \times (0.081 \times 10^6) = 0.57 \times 10^6 \text{ Btu/ton of steel product}$$

From the point of view of technological feasibility, it is considered possible to increase continuous casting capacity by two times the historical rate for each of the last two years of the 1972-1980 period. Thus, the technologically feasible energy saving is taken to be:

$$(0.57 \times 10^6) + 2(0.081 \times 10^6) = 0.73 \times 10^6 \text{ Btu/ton of steel product}$$

Increased Use of Induction Heating of Steel Slabs: One domestic steel company has been using electric induction heating of steel slabs for several years, whereas the usual practice is to use fuel-fired reheat furnaces which in many cases are thermally inefficient.

A recent analysis of electrical induction heating versus fossil-fuel heating of steel slabs indicated an energy saving with induction heating of 1.3×10^6 Btu per ton. In this same analysis, the economics of induction heating were examined on the basis of the following factors.

Annual capacity, net tons   2,400,000
Capital investment-induction system $79,600,000
Capital investment-fossil fuel $66,600,000
Differential capital investment $13,000,000

To replace an existing fossil-fuel furnace with an induction system, the investment per million Btu saved annually is:

$$\frac{\$79.6 \times 10^6}{(1.3 \times 10^6)\,(2.4 \times 10^6)} = \$26 \text{ per million Btu}$$

This exceeds the maximum value of $10 per million Btu saved annually, and, therefore, is not considered economically practicable. However, for a new facility the differential capital investment is much lower and the investment per million Btu saved annually is:

$$\frac{\$13.0 \times 10^6}{(1.3 \times 10^6)\,(2.4 \times 10^6)} = \$4.2 \text{ per million Btu}$$

This makes such a choice economically practicable. Based on the assumption that by 1980 new induction heating facilities for additional capacity of 5.0×10^6 tons per year are economically practicable, the energy saving would be:

$$(5.0 \times 10^6) \times (1.3 \times 10^6) = 6.5 \times 10^{12} \text{ Btu}$$

For 84.05×10^6 NT steel, the saving would be:

$$\frac{6.5 \times 10^{12}}{84.05 \times 10^6} = 0.08 \times 10^6 \text{ Btu/ton of steel product}$$

For technological feasibility, the major constraint is the limitation of the builders of induction furnaces to design, produce, and install such equipment. It is considered technologically feasible that by 1980 induction heating equipment to process 15×10^6 tons per year could be installed. In such a case the energy saving would be:

$$\frac{15}{5} \times (0.08 \times 10^6) = 0.24 \times 10^6 \text{ Btu/per ton of steel products}$$

Improvements in Soaking Pits, Reheat, Annealing, and Heat-Treating Facilities:
Steel ingots are reheated in soaking pits and (after being formed into slabs, blooms, or billets) are again heated in reheat furnaces for hot rolling. After subsequent processing steps, much of the steel is either annealed or heat treated. These improvements have potential for energy savings, such as:

Improved skid insulation in reheat furnaces to lower energy losses in cooling water.
Improved reheat-furnace control as related to rolling-mill schedules.
Installation of recuperators on those reheat and annealing furnaces not presently fitted with such heat-recovery devices.
Conversion of batch-type annealing furnaces from radiant tubes to direct firing.
Increased use of fibrous refractory insulation in annealing and heat-treating furnaces.
Lower leakage from furnaces.
Improved burner design and maintenance.

There is no quantitative data available on energy efficiency economics for the above types of reheating operations covering the total steel industry. Obviously, some of these energy-saving innovations are economically practicable, because many establishments have been making such changes over the years up to the present. It should also be noted that some steel products go through all four of the above types of heating operations at different stages in their production.

For the purpose of this study, it is assumed that on an economically practicable basis, during the 1972-1980 period, the steel plant subcomponent can achieve energy improvements equivalent to the processing of 50×10^6 tons of steel at an estimated energy saving of about 0.75×10^6 Btu per ton of steel. Thus the energy saving would be:

$$(0.75 \times 10^6) \times (50 \times 10^6) = 37.5 \times 10^{12} \text{ Btu}$$

For 84.05×10^6 NT of steel, the energy saving is:

$$\frac{37.5 \times 10^{12}}{84.05 \times 10^6} = 0.45 \times 10^6 \text{ Btu/ton of steel product}$$

It is estimated that it is technologically feasible (without any financial constraints) to additionally replace inefficient facilities with a capacity of 25 million tons. Therefore, the total energy saving technologically feasible by 1980 is:

$$\frac{(50 + 25)}{50} (0.45 \times 10^6) = 0.68 \times 10^6 \text{ Btu/ton of steel product}$$

Housekeeping: The only available figure for energy saving through improved house-keeping is the estimate by the AISI that a 2% decrease could be achieved in the 1972-1980 period. All housekeeping items are considered technologically and economically practicable. In 1972 the average energy consumption to produce a ton of steel product was 32.61×10^6 Btu. Thus the energy saving which is expected to be technologically feasible and economically practicable by 1980 would be:

$$0.02 \times (32.61 \times 10^6) = 0.65 \times 10^6 \text{ Btu/ton of steel product}$$

Energy Penalties: As noted in Table 11, there is an energy penalty for increased sulfur and ash in coke due to a gradual decrease in the quality of coking coal.

GOAL YEAR (1980) ENERGY USE TARGET

The potential for energy conservation by 1980 with economic constraints includes most of the innovations previously discussed as technologically feasible. Specifics are listed in Table 11. As will be noted in the table, some of the items which were considered technologically feasible are not considered to be economically practicable. For instance, increased injection of coal to blast furnaces by 1980 is not considered economically practicable because most of the present furnaces are fitted for injection of hydrocarbon liquids. However, after 1980 new blast furnaces will be built, and it is likely that these will incorporate coal-injection facilities.

As will be seen in Table 11 for certain line items, the energy saving that is economically practicable is less than that listed as technologically feasible. For instance, for the item covering increased electric-furnace capacity, the saving that is considered economically feasible represents the historical 10-year average rate of increase in electric-furnace capacity, whereas a greater rate is considered technologically feasible by the target period.

TABLE 11: ENERGY EFFICIENCY IMPROVEMENT POTENTIAL BY 1980; STEEL PLANT SUBCOMPONENT (SIC 3312)

Technique	Potential Saving, 10^6 Btu per Ton of . . . . .Finished Steel*	
	Technologically Feasible	Technologically Feasible and Economically Practicable
Increased pellet usage in blast furnace	+1.09	+1.09
Increase in coke ash	-0.49	-0.49
Increase in coke sulfur	-0.04	-0.04
Increase blast furnace gas recovery	+0.39	+0.39
New and modernized blast furnaces	+0.13	+0.13
Dry quenching of coke	+0.14	0.0
Blast furnace coal injection	+0.09	0.0
External desulfurization	+0.16	+0.16
Substitution of basic oxygen furnaces for open hearth furnaces	-0.07	-0.07
Recovery of BOF offgas	+0.04	0.0
Preheating scrap for BOF with oxygen-fuel burner	+0.12	0.0
Increased electric furnace capacity	+2.01	+1.56
Increased continuous casting	+0.73	+0.57
Increased use of induction heating of steel slabs	+0.24	+0.08
Improvements in soaking pit, reheat, annealing and heat-treating facilities	+0.68	+0.45
Houskeeping	+0.65	+0.65
Total	5.9**	4.4**

*+ indicates a saving, – indicates more energy required.
**Rounded totals.

Source: Reference (4)

Economic practicability must be seasoned with a certain measure of judgement and not related strictly to the potential for energy conservation. For instance, up to 1980 there will continually be new basic-oxygen furnaces installed, because this process is economically better than the open-hearth process. However, the economics are almost wholly related to the production costs.

The total energy required to make BOF steel is about the same as the total for open-hearth steel. This is so because of the relatively high amount of hot metal required for the BOF. Therefore, the BOF process is known to be economically attractive for investment, but the feasibility is not really related to energy considerations.

The future economic constraints on the steel plant subcomponent (SIC 3312) are severe because it is a capital-intensive industry which will require large investments to meet pollution-control standards. A recent estimate for a new greenfield integrated steel plant was $3.2 billion for a facility with a capacity of 4 million tons. This amounts to $800 per annual ton of capacity. Even though a new modern plant could operate with up to 40% less energy per ton than old facilities, it is not likely that many new greenfield plants will be built.

To increase capacity to meet future demand for steel and to install the required facilities for pollution control, it has been estimated that between now and 1983 the steel industry will have to spend an average of about $5 billion per year. This is more than 2.5 times the average annual rate of capital investment in the industry over the last 10 years. Thus, there is a very real question as to whether the required financing will be available.

Over the two decades prior to the energy crisis, the American steel industry decreased its average consumption of energy per ton of steel shipped by about 16%, as shown in Table 12. During this period, coal and coke accounted for about 70% of the energy. As a percentage of all energy used, the major trends involved a decrease of about 50% in the importance of oil as an energy source and doubling of the importance of purchased electricity.

In comparison to foreign steel plants, in 1973 the American companies used on the average to make 1 ton of steel about the same amount of energy as United Kingdom plants, about 13% more than West German plants, and about 24% more than Japanese plants.

In Germany and Japan, plants generally are newer, and in Japan the average size of major steel plants is substantially larger than in the United States. Also, in these countries, energy historically has been more expensive than in the United States. Therefore, they have had more incentive to conserve energy.

TABLE 12: HISTORICAL TRENDS IN USE OF ENERGY BY U.S. STEEL PLANTS

Year	Energy Used per Ton of Steel Shipped, Index	. . . . .Percentage of Total Btu				
		Coal, Coke	Oil, LPG	Natural Gas	Purchased Electricity	Total
1950	100	72.7	10.1	15.6	1.6	100
1960	89	77.4	7.8	12.6	2.2	100
1970	84	73.5	5.4	17.4	3.7	100

Source: Reference (4)

Various reports have been issued in recent years which analyze the energy consumption to produce steel such as (12). These reports were based on one or more of the following conditions:

> Electrical energy was based on a figure (e.g., 10,500 Btu per kwh) which included the efficiency of the electrical utility in conversion of coal to electricity.
> The energy value for steel was determined in terms of Btu per net ton of raw steel and not for the finished steel product mix.
> Energy was included for such items as mining, transportation, and consumable items (e.g., oxygen and electrodes).

However, in the 1976 Battelle report to FEA (4), the basis is for a ton of finished steel product, with electrical energy at 3,412 Btu per kwh, no inclusion of energy consumed outside the plant boundaries, and no energy for consumable items other than fuels and electrical energy.

Data from the 1972 Census of Manufactures were analyzed, but it was found that the values given were based solely on purchased fuels and energy and not on total direct energy consumption. The data were deficient mainly in the reported tonnage of coal consumed; 5,085,000 net tons versus an actual consumption of 79,450,000 net tons.

The American Iron and Steel Institute (AISI) publishes an Annual Statistical Report which gives some data on quantities of fuels and energy consumed. However, because the Institute recognized that there were deficiencies in these data, it formed an Energy Committee composed of members from steel companies.

Data gathered quarterly from the steel companies on their energy consumption starting with the year 1972 is aggregated by AISI. The basis for AISI reporting, which is a continuing procedure, is as follows:

> Electrical energy is at 3,412 Btu per kwh.
> Each company reports its energy consumption for each particular item
> (e.g., coal, natural gas, fuel oil, etc.) in terms of British thermal units.
> In this way each item reflects the actual energy value for the fuel consumed at a particular plant.
> The energy per ton of finished steel product is derived from tonnage of shipments adjusted for inventory changes.

The AISI energy data were used as the basis for the analysis in the Battelle report (4) because of the breadth and depth of the coverage and because the factors used for the data were the same as the ground rules selected.

A further breakdown of the 1972 base-year energy of 32.6×10^6 Btu per net ton of finished steel produced is given in Table 13 to show the appropriate energy distribution between major unit operations. It should be noted that the blast-furnace operation, because of the large coke consumption, requires about 60% of the total energy.

Table 14 summarizes the data reported by the AISI for the 4 years 1972 through 1975. There was an increase in energy consumed in 1975 as compared to either the base year or the previous year (1974). The increase amounts to 2.9% or almost 1 million more Btu per ton than in the 1972 base year. This increase in energy is attributed primarily to the fact that 1974 was a low production year for the steel industry, during which it produced less steel than in any of the previous 10 years.

It was pointed out above that the steel industry (SIC 3312) a few years ago recognized the need for better energy data and began a voluntary reporting program. As is shown in Table 14, improvements in energy consumption were made in 1973 and 1974 as compared to the 1972 base year. However, due to a sharp downturn in steel production in 1975, the energy efficiency decreased by 2.9% over the base year. The effect of plant size and level of production on energy consumption are reviewed in the Battelle report to FEA (4).

TABLE 13: ENERGY CONSUMPTION FOR MAJOR STEELMAKING UNIT
OPERATIONS—1972 BASE YEAR

	10^6 Btu per Net Ton of Finished Steel Produced	Percent of Total Energy Used
Coke ovens	3.7	11.3
Blast furnace	19.7	60.5
Steel furnaces	2.2	6.7
Soaking pits, reheat furnace, annealing and heat treatment	4.8	14.7
Rolling and finishing operations	2.2	6.8
Total	32.6	100.0

Source: Reference (4)

TABLE 14: ENERGY USE IN THE STEEL INDUSTRY (SIC 3312) 1972
THROUGH 1975 AS REPORTED BY AISI(*)

	1972	1973	1974	1975
Finished steel production**, 10^6 net tons	84.05	95.52	96.92	78.40
Total energy consumption, 10^{12} Btu	2,742	3,031	3,073	2,633
Energy per ton of steel, 10^6 Btu	32.61	31.73	31.71	33.58
Percent improvement	Base year	2.7	2.8	-2.9

*Survey companies accounted for 89.5% of the total industry production in 1972,
  89.3% in 1973, 88.1% in 1974, and 94.0% in 1975.
**Annual shipments of finished product adjusted for inventory changes.

Source: Reference (4)

The formation of the AISI Energy Committee, which is composed of steel-company representatives, was a positive step in appraising the performance of the industry with regard to energy.

Just recently another AISI Committee completed a Handbook of Energy Conservation Technology which consisted of 15 chapters, each covering a specific subject. Both energy and economic factors were considered. The subjects covered included, among others, dry quenching of coke, coal injection into the blast furnace, utilization of BOF off gas, external desulfurization of hot metal and induction heating of slabs.

Primary technological barriers to major increases in energy conservation in the steel industry are that a considerable part of the steel plant facilities are old, are costly to replace, and are inherently limited with regard to major change in their energy-consuming characteristics over a short period, such as the period 1976 to 1980.

Control of pollution of air and water is a special circumstance that affects the steel industry and has a significant effect on energy consumption. Large amounts of capital are required for the installation of facilities which will meet 1977 and 1983 standards.

One investigation directed primarily to the economic factors related to environmental control for the steel industry covered the period 1972 to 1983. Energy requirements for pollution control were only briefly considered by these investigators, but they estimated the total energy requirement out to 1983 for full compliance would be about 141×10^{12} Btu per year. Adjusting back to the first of 1980 (which is the target period for the Battelle study), the energy required would be about 110×10^{12} Btu per year. This translates into an increased energy consumption for pollution control from 1972 to 1980 amounting to 4.0% of total energy.

In another analysis done by an American Iron and Steel Institute Committee, the total energy consumed by 1983 for all air and water pollution control, including that required for control of "fugitive" dust emissions, would amount to about 6.4 percent of total energy consumption. Adjusting back to the first of 1980 gives an additional need for about 4.8% of total energy for the 1972 to 1980 period.

Still another study has indicated that by 1983 the energy consumption for pollution control would amount to 10% of total energy consumed by the steel industry. However, it should be noted that this analysis was based on the following factors:

> It included preplant and postplant energy requirements. If these are excluded the energy requirement for pollution control is about 9% of total energy.
> The air-pollution control standards used were the most stringent visible-emission limitation levels currently being enforced by state and local agencies. These are more stringent than existing numerical standards (process weight or concentration limits) which are based on actual measurements of air pollutants.
> In 1983 water-pollution control standards were used because they are the most stringent currently established.
> Air-flow rates which were higher than other studies to facilitate compliance with Occupational Safety and Health Act (OSHA) regulations within a shop or cast house.
> For sinter-plant and basic-oxygen facilities, energy-intensive scrubbers were used for emission control.

These investigators also analyzed the impact of revised (less stringent) air and water pollution control standards. Such revisions indicated that air-pollution-control energy requirements could be cut 50% and water-pollution energy requirements by two-thirds.

Thus, it can be seen that an additional 9% (excluding preplant and postplant energy) of total energy for air and water pollution control is the extreme case. The steel industry estimates that the present energy consumption for air and water pollution control is from 2.0 to 2.5% of total energy. It should also be noted that about half the existing sinter plants and basic-oxygen furnaces are electrostatic precipitators instead of high-energy scrubbers, and that the precipitators require considerably less energy than the scrubbers.

If one assumes a linear increase in energy for pollution control and adjusts the

9% back to the first of 1980, the energy requirement for pollution control would be, under the very stringent standards, about 6% of total energy.

There is an element of uncertainty concerning the state of compliance that will be reached by the first of 1980, and the type of equipment and standards of control which at present vary for different geographic locations. For the purposes of the Battelle report to FEA (4), and after consideration of the data available, the additional amount of energy required for pollution control for 1972 to 1980 is taken at 4.5% of total energy. This amounts to 1.47 x 10⁶ Btu per ton of steel product (for SIC 3312).

For SIC 3315, 3316, and 3317, additional energy for pollution control is taken at 2% of total energy or 0.1 x 10⁶ (rounded value) Btu per net ton of steel product. Table 15 gives a summary of the energy conservation potential and projected 1980 energy per unit, including the requirements for environmental control ("Net" Column 5) for the four SIC codes which make up the steel-plant component. Applying these estimated energy requirements for environmental controls to the aggregated component and projecting these to 1980 gives a total estimated net potential of 3413 x 10¹² Btu for this component.

TABLE 15: ENERGY CONSERVATION AND PROJECTED 1980 ENERGY POTENTIAL

Code	(1) 1972 Energy/Unit 10^6 Btu per Net Ton	(2) Projected 1980 Production, 10^6 Net Tons	(3) 1980 Energy Consumption Using 1972 Efficiency, 10^{12} Btu	(4) Gross	(5) Net	(6) Gross	(7) Net
				Projected 1980 Energy/Unit 10^6 Btu per ...Net Ton...		Projected 1980 Energy Consumption, $..10^{12}$ Btu...	
SIC 3312(a)	32.61	113.6(b)	3,704.5	28.21(c)	29.68(d)	3,204.7	3,371.6
SIC 3315(e)	4.5	3.8	17.1	3.8(f)	3.9(g)	14.4	14.8
SIC 3316(h)	3.4	4.5	15.3	2.9(f)	3.0(g)	13.0	13.2
SIC 3317(i)	4.0	3.9	15.6	3.4(f)	3.5(g)	13.3	13.7
Total	29.9(j)	125.5(k)	3,752.5	25.86(l)	27.20(m)	3,245.4	3,413.3

(a)   SIC title: Blast furnaces and steel mills.
(b)   Adjusted for 100% of industry production based on AISI data for 1972 which covers 89.5% of the total industry.
(c)   Based on a gross energy conservation potential of 4.4 x 10⁶ Btu/ton over the base year.
(d)   Includes an energy cost of 1.47 x 10⁶ Btu/ton over the base year for the operation of air and water pollution control facilities.
(e)   SIC title: Steel wire and related products.
(f)   Based on a gross energy conservation potential of 15% over the base year. This estimated energy saving is assigned primarily to improved housekeeping and reduced energy consumption in annealing and heat-treating furnaces.
(g)   Includes an energy cost of 0.1 x 10⁶ Btu/ton for air and water pollution control.
(h)   SIC title: Cold finishing of steel shapes.
(i)   SIC title: Steel pipe and tubes.
(j)   Determined by dividing total Btu consumed in 1972 (3,103 x 10¹² Btu) by total tons of product (103.9 x 10⁶ tons).
(k)   Determined by dividing total of Column 3 by total of Column 1.
(l)   Determined by dividing total of Column 6 by total of Column 2.
(m)   Determined by dividing total of Column 7 by total of Column 2.

Source: Reference (4)

Energy consumption in SIC codes 3315, 3316, and 3317 is low compared to SIC 3312. In total it amounts to slightly over 1% of the energy required for the basic steel industry (SIC 3312). The energy consumption used for these three segments was derived from Census of Manufactures data and is summarized in Table 16.

TABLE 16: ENERGY USE IN SIC 3315, 3316, and 3317 in 1972

	3315	3316	3317
Production, 10^6 net tons	3.1	3.7*	3.2*
Total energy consumption, 10^{12} Btu	13.8	12.7	12.8
Energy per ton of steel, 10^6 Btu	4.5	3.4	4.0

*Estimated from dollar value in census data.

Source: Reference (4)

The energy conservation for these three SIC components is assumed at a potential of 15% reduction from the base year 1972. Such an energy saving is expected to be achieved primarily by improved housekeeping and lower energy consumption in annealing and heat-treating furnaces. This level of improvement is expected because it has been found in recent years that relatively small metalworking operations can make substantial energy savings by various housekeeping improvements, including use of better insulation. Such improvements also can be made at low cost. Therefore, the energy savings potentials for these three SIC components are taken to be both technologically feasible and economically practicable, and are as follows:

SIC Code	1972 Energy per Unit, 10^6 Btu/net ton	Energy Saving Potential at 15% of 1972 Energy, 10^6 Btu/net ton (rounded)
3315	4.5	0.7
3316	3.4	0.5
3317	4.0	0.6

The energy conservation potential for SIC 3315 (wire drawing, nails, and spikes), 3316 (cold-rolled steel), and 3317 (pipe and tubes) up to 1980 which is considered economically practicable is the same as that taken to be technologically feasible.

Applying all of the economically practicable savings to the aggregated steel-plant component and projecting these to 1980 gives a total estimated gross potential of 3,245 x 10^{12} Btu. (See Table 15.)

SOME PROJECTIONS BEYOND 1980 to 1990

The estimates for energy conservation in the manufacturing sector made as a part of the Project Independence Blueprint in 1974 (10) made some estimates for the steel industry.

The projections shown in Table 17 are based on the following critical assumptions: (1) Pig iron production grows commensurately with steel output at slightly less than 2.5% per year and (2) Average coke rates decline to 0.56 in 1980 and to 0.52 by 1990. Current energy usage per ton of pig iron is 16.5 MM Btu. This was calculated on the basis of 5.0 MM Btu/ton of coke for coke ovens times the coke rate of 0.6 plus 13.5 MM Btu/ton of pig iron consumed in the blast furnace. The forecasts show a 2.7% improvement in energy use per ton of pig iron by 1980 and a 7.1% improvement in 1990 compared to current usage.

TABLE 17: ENERGY USE PROJECTIONS FOR COKE OVENS AND BLAST FURNACES*

Measure / Year	Change 1973-1980 (MM Btu/ton pig)	Change 1973-1990 (MM Btu/ton pig)
Increased agglomeration	-0.37 (-0.017)	-0.74 (-0.034)
Increased fuel injection	-0.5 (-0.022)	-0.7 (-0.032)
New ore burden improvements	-	-0.4 (-0.018)
Savings from reduced coke production	-0.2	-0.42
Losses as coke rate falls		
Equals: Gross Savings	-1.07	-2.26
Less:		
More fuel injection	+0.5	+0.7
Lost blast furnace gas**	+0.13	+0.39
Equals: Net Savings	-0.44	-1.17

*1973 base is 16.5 MM Btu/ton pig iron for both stages. Numbers in parentheses refer to tons of coke at 22.4 Btu/ton.
**Assumes 8.3 MM Btu/ton of coke charged to blast furnace less 10% by-product gas losses.

Source: Reference (10)

The assumptions for steelmaking taking into account the hot metal factor, growth rate and process efficiency improvements are summarized in Table 18.

TABLE 18: STEELMAKING ENERGY USE PROJECTIONS*

Year	Process	Percent of Raw Steel Output	Energy use (MM Btu/ton raw steel)
1973	Open hearth	26.4%	
	BOF	55.2%	1.41
	Electric	18.4%	
1980	Open hearth	15.6%	
	BOF	63.1%	0.88
	Electric	21.2%	
1990	Open hearth	4.3%	
	BOF	71.7%	0.46
	Electric	23.9%	

*Several industry experts were more pessimistic about possible energy gains in this area based on the belief that the open hearths will still comprise 10% of output by 1990 with little or no improvements in energy use per ton of steel within each production process.

Source: Reference (10)

The rate of adoption of continuous casting by the industry is a critical parameter in the overall energy use projections. At present continuous casting technology is suited for only a fairly narrow product range. Quality problems are also a constraint on its use. Steel industry sources have indicated that continuous casting is presently applicable to roughly 70% of the current product mix. Therefore, through 1980 it is assumed that 70% of the new capacity through 1980 will be in the form of continuous casters.

Traditional equipment will not be replaced but is assumed to provide backup and supply capability over a wider range of products. It was also assumed (10) that the technology would improve sufficiently by 1980 to allow it to cover 90% of the product mix between 1980-1990 and that it will begin to substitute (at the same percentage) for soaking pit capacity which is assumed to be retired at a rate of 1.5% of capacity per year in the 1980's. The results of this projection are shown in Table 19.

The assumed rate of introduction of continuous casting also improved the yield per ton of output from 86% in 1973 to 89% by 1990. Replacing one unit of primary rolled output of ingot pouring-soaking pit method with a continuous caster saves 0.123 ton of raw steel.

TABLE 19: ENERGY USE PROJECTIONS FOR SOAKING PITS AND CONTINUOUS CASTING

Year	Process	Output (percent)	Weighted Average (MM Btu/ton)	Improvement Over Base Year (percent)
1973	Soaking pits	93%	1.7	–
	Continuous casting	7%		
1980	Soaking pits	83%	1.6	6%
	Continuous casting	17%		
1990	Soaking pits	58%	1.3	24%
	Continuous casting	42%		

Source: Reference (10)

The energy use forecasts derived here (10) are premised on a rate of growth of net shipments produced domestically of 2.5% per year, with imports maintaining their recent share of the U.S. market. This rate of growth has been used in several reports and speeches by the steel industry.

It should be emphasized that this growth rate will be difficult to achieve in the near-term even if market demand is strong and investment funds are available. As mentioned previously, steel capacity has not expanded significantly for a long time.

Equipment manufacturers which catered to this industry have folded or shifted their expertise to meet the expansion requirements of other sectors. Two recent expansions by the industry had to be accomplished through a joint arrangement with European expertise and U.S. construction firms. Thus, this industry is likely to be under severe construction limitations resulting in long lead times to significantly expand new capacity. To the extent that actual growth takes place at a slower pace than assumed here, energy consumption per unit output will tend to be higher than forcasted below but, of course, total energy use will be lower due to any slower growth in domestic output.

Any forecast of energy use per unit output can be broken into three elements:

(1)   Shifts in the output mix and process changes that are likely to occur;
(2)   Improved energy efficiency of new capacity compared to the present average efficiency;
(3)   Further improvements in energy utilization through better operating practices and capital investments designed to reduce energy losses in present installations.

Table 20 presents energy/output projections to 1990 based only on the first factor, shifts in output mix and the introduction of new processes. These energy savings are the result of the following factors listed roughly in the order of their importance:

Further improvements in ore preparation, leading to a continued decline in the coke rate;
The expanded utilization of continuous casting which saves energy in the primary rolling phase of production and also reduces the raw steel requirement per ton of shipments;
A continuation of the present trend to replace the open hearth with oxygen converters and electric furnaces;
Increasing reliance on imported agglomerated iron ore, thus reducing domestic energy requirements for ore processing;
Continued trend to continuous processing in reheat furnaces due to shifts in the product mix.

These developments could reduce energy use per unit of output by 0.9% per annum.

TABLE 20: ENERGY USE FORECASTS* (MM Btu/ton of net shipments)

Process Step	1973 (MM Btu)	. . . . 1980		. . . . . . 1990	
		(MM Btu)	Total Savings,%	(MM Btu)	Total Savings,%
Agglomeration	1.4	1.3	5	1.2	4.9
Coke ovens	2.85	2.6	12.2	2.4	11.4
Blast furnace	12.8	12.43	20.6	11.7	26.7
Steel making	2.0	1.2	42.2	0.6	31.9
Soaking pit	2.55	2.4	10.0	1.9	15.8
Reheat furnaces	3.9	3.8	6.7	3.7	5.6
Heat treating and forging	1.0	0.95	2.8	0.9	2.3
Internal generation of electricity	0.8	0.79	1.0	0.77	0.7
Other uses	4.3	4.3	0	4.3	0
Total energy use	31.7	29.8		27.3	
Percent reduction per unit output from 1973 level		6.0		13.9	
Average percent reduction per unit output per annum		0.9		0.9	

*Due to process changes and shifts in the output mix; no allowance is included for housekeeping or short-term, quick capital fixes to save energy.

Source: Reference (10)

Other possible energy-conserving technological process changes on the horizon include: recovery of top gases in the basic oxygen process, introduction of the dry quenching process in coke production, more rapid displacement of energy intensive steel processing steps by technologies tied to the use of continuous casting and greater reliance on oxygen injection into the blast furnace. These measures could reduce energy consumption per ton of shipments by at least 1.0 to 1.5 million Btu, equivalent to a roughly 4% further decline in energy consumption for new installations. Although there are still some unresolved technical problems associated with these technologies, the major obstacle to adoption is economics. The industry appears unwilling to apply these technologies unless fuel costs rise to $3 or more per million Btu.

In addition to the energy savings through process changes, improved operating procedures and waste energy recovery on both new and existing installations can cut energy consumption. Since such opportunities have not yet been systematically evaluated by the steel companies, it is not possible to provide an estimate of what a "package" of these measures could achieve based on experience. However, discussions with industry personnel suggest that significant savings are possible through the cumulative impact of a series of relatively small energy conservation measures.

In addition to the savings through process change, there was assumed (10) a further reduction in energy/output levels of at least 5% by 1980 and another 5% by 1990 if a high priority management program is established. Energy savings of this magnitude would be roughly equal to the expected savings as a result of shifts in the output mix and new processes, raising the rate of decline in energy consumed per unit output to 1.7% per annum through 1990.

ALUMINUM INDUSTRY

The aluminum component as considered here consists of three major segments
(1) the production and refining of alumina from bauxite by the Bayer process,
(2) the reduction of alumina to aluminum metal by the Hall-Heroult electrolytic
process, and (3) the preparation and fabrication of aluminum and aluminum alloy
shapes such as sheet, plate, extrusions, and bar. The first two segments correspond to SIC 3334 while the third covers establishments corresponding to
SIC 3353, 3354, and 3355.

About 1.93 tons of alumina are required to produce 1 ton of aluminum. Thus,
the total alumina requirement of the aluminum component when operating at
full capacity is 9.69×10^6 tons. This is 1.93×10^6 tons more alumina than can
be produced domestically. Therefore, at full capacity, about 20% of the alumina
used must be imported, primarily from the Caribbean, northern South America,
and Australia.

It is unlikely (4) that any new Bayer alumina process plants will be built
within the U.S. to produce the alumina that is presently imported, particularly
because the bauxite-producing nations have come together in an OPEC-like organization and are pressuring the U.S. aluminum companies to locate their alumina-
producing operations near the source of the bauxite. Other incentives for converting the bauxite to alumina at the source include a lowering of the freight
charges, higher availability of financing, and lower cost for labor in the bauxite-
producing countries. Also, most of the bauxite-producing countries are relatively
underdeveloped and as such are able to secure loans from organizations such as
The World Bank for the transportation and infrastructure requirements of alu-
mina-production facilities. For these reasons, the only current construction of
alumina plants is overseas.

PROCESS TECHNOLOGY INVOLVED

A simplified flowsheet for the production of aluminum metal is shown in Figure
21. The manufacturing process starts with bauxite, the principal raw material,
which may contain from 40 to 60% alumina.

FIGURE 21: PRODUCTION OF ALUMINUM FROM BAUXITE

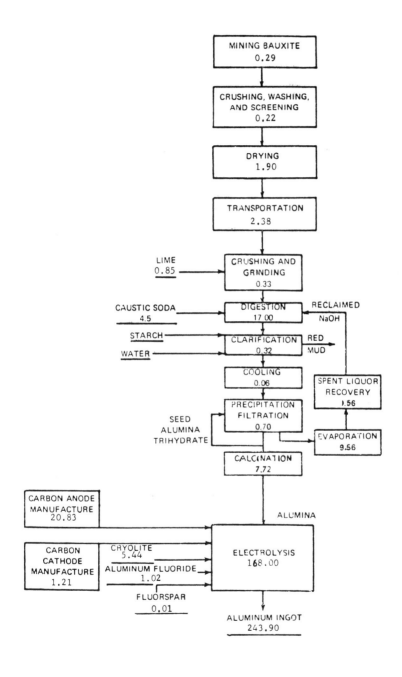

Source: Reference (12)

The method of separating the alumina from impurities in the bauxite is the Bayer process. In this process, the bauxite is mixed with caustic soda and digested under heat and pressure so that the alumina in the bauxite is dissolved in the caustic soda as sodium aluminate, leaving behind the impurities as insoluble solids which are filtered out in a series of tanks and filter presses. The solution containing alumina is cooled and is seeded with fine crystalline alumina hydrate to precipitate alumina trihydrate which is then filtered, washed, and heated in kilns to drive off chemically attached water and yield commercially pure alumina. The dry alumina is then shipped to the aluminum smelter.

At the smelter, the alumina is reduced to aluminum metal using the Hall-Heroult process. The electrolytic reduction cell (or pot as it is known in the trade) breaks alumina down into its components, aluminum and oxygen. The alumina is dissolved in a molten cryolite (sodium-aluminum fluoride) electrolyte. High-amperage, low-voltage direct current passes through a suspended anode of carbon located at the top of the pot and deposits aluminum on the bottom lining of carbon which, covered with a molten layer of aluminum, acts as a cathode. As the aluminum builds up in depth, it is siphoned off. The separated oxygen plates out on the anode where it combines with carbon and is released as carbon dioxide.

The energy values in Figure 21 for various unit operations are given in boxes in units that represent millions (10^6) of Btu. Products, feed materials and waste products are underlined with energy values assigned where appropriate. No fuels or electrical energy consumptions are itemized on this flowsheet.

After smelting, a portion of the refined metal is cast into ingots. However, a substantial proportion of the liquid metal is also delivered to both on-site casting houses within a given plant and to outside users. Such liquid-metal transfers are routinely made within a 10-mile radius and occasionally for much longer distances. Brooks has estimated that the present energy for melting might be 25 to 40% greater in the absence of such liquid-metal deliveries.

At the casting sites, the molten primary aluminum is charged into holding furnaces, which are typically of the gas-fired reverberatory type, along with scrap and alloying additions. A substantial amount of scrap is recycled in the melting operation. An average of 55 tons of scrap is added for each 43 tons of primary metal and 2 tons of alloying additions to form a melt of 100 tons.

Initial processing of cast ingots is usually conducted at about 800° to 900°F. This is either a hot-rolling (blooming) or extrusion operation. For some products, intermediate conditioning and continued hot rolling is necessary. Most mill products are cold rolled or drawn to finish dimensions.

Figure 22 shows the principal steps in the Bayer-Hall aluminum manufacturing process. The major energy consumption operations are the steam digestion of bauxite, the evaporation of water from used caustic solution, the calcining of aluminum trihydrate, and the remelting and heat treating of aluminum.

These operations account for over 90% of the total energy consumption in the aluminum manufacturing process.

FIGURE 22: ALUMINUM ENERGY CONSUMPTION DIAGRAM

1972 USA production: 3.74×10^9 kg (8.24×10^9 lb)
1972 energy consumption (electricity, carbon, natural gas): 15,000 Mw (450×10^{12} Btu)*

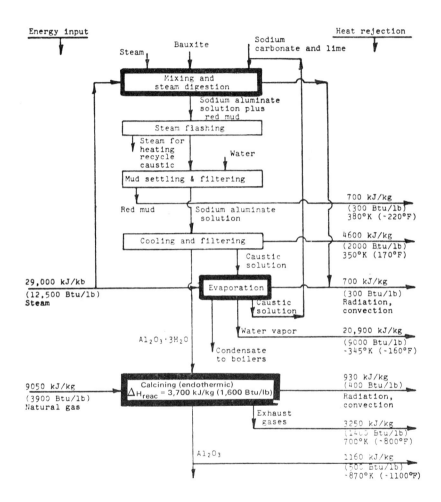

*Electricity is counted as 3,600 kJ/kwh (3,413 Btu/kwh).

(continued)

FIGURE 22: (continued)

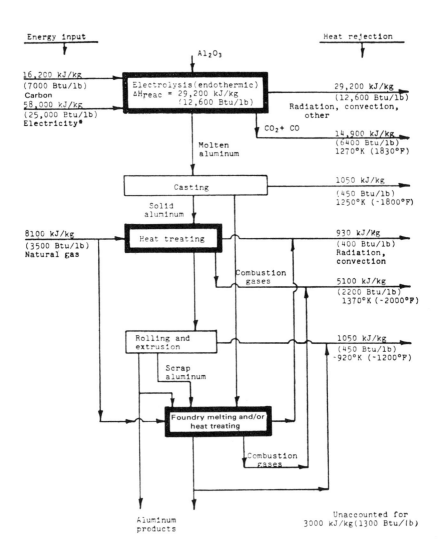

*Electricity is counted as 3,600 kJ/kwh (3,413 Btu/kwh).

Source: Reference (2)

Figure 23 shows the steam digestion of bauxite and evaporation of water from caustic. The alumina in bauxite is dissolved in caustic at elevated temperature and pressure. Iron oxide, titania, and silicates do not dissolve but form a red mud which is later separated from the sodium aluminate solution. Later in the process, steam is used to boil water from a dilute caustic solution. Aluminum trihydrate has previously been filtered out of the caustic solution. A set of multi-effect evaporators is used to boil water from the caustic.

FIGURE 23: EQUIPMENT DIAGRAM—STEAM DIGESTION OF BAUXITE AND EVAPORATION OF WATER FROM CAUSTIC

Rejected heat:
  Radiation, convection—700 kJ/kg (300 Btu/lb)
  Heat in red mud—700 kJ/kg (300 Btu/lb) at $380°K$ ($220°F$)
  Heat in water vapor from evaporators—20,900 kJ/kg
    (9,000 Btu/lb) at $345°K$ ($160°F$)
  Heat in sodium aluminate solution—4,600 kJ/kg (2,000 Btu/lb)
    at $350°K$ ($170°F$)

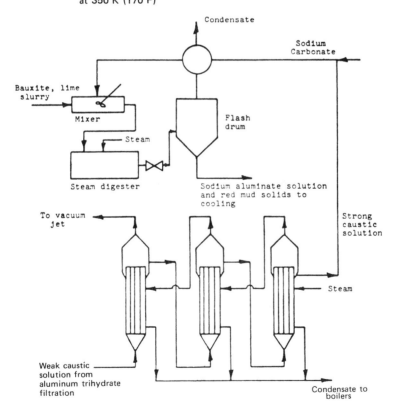

Multi-effect evaporators

Source: Reference (2)

According to the conventional method, the discharge of waste steam and the separation of sand are carried out independently of each other. Lacking effective use of the steam, this method inevitably involves heavy consumption of steam. It also requires a special device designed exclusively for the removal of sand. This, accordingly, necessitates installation of pumps to effect the transfer of the slurry to these separate devices. Slurry pumps are highly liable to develop mechanical trouble and demand troublesome maintenance. It is, therefore, desirable to minimize use of slurry pumps.

A process described by C. Sato et al (37)(38) is one in which the overflow from the red mud washing thickener is introduced into a column in which the flashing of slurry down to the atmospheric pressure is effected. Inside this column, the overflow liquid is brought into counterflow contact with the steam generated by the flashing so that the overflow liquid is heated by the heat recovered from the steam. The heated liquid is then refluxed to be used as the wash water in the washing thickener. In the meantime, the speed at which the slurry is discharged the bottom of the column is relatively below the speed at which the sand in the slurry settles so as to separate the sand from the slurry. The sand thus separated is discharged through the outlet provided at the bottom.

A production system was actually operated using this device and the amount of steam released into the atmospheric air was about 1 ton/hour, while in the operation excluding the use of this device, the amount of released steam was about 13 tons/hour. This represents a saving of about 0.15 ton of steam per ton of alumina produced.

Figure 24 shows the use of a rotary kiln to calcine aluminum trihydrate to alumina. The combustion of natural gas in preheated air provides heat to remove water of hydration at $1370°K$ ($2000°F$).

FIGURE 24: EQUIPMENT DIAGRAM—ROTARY KILN

Rejected heat:
Radiation—930 kJ/kg (400 Btu/lb)
Combustion gases—3,250 kJ/kg (1400 Btu/lb) at 700°K (800°F)
Heat in alumina—1,160 kJ/kg (500 Btu/lb) at 870°K (1100°F)

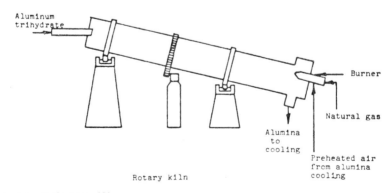

Rotary kiln

Source: Reference (2)

Figure 25 shows the electrolytic separation of alumina into aluminum and oxygen (which then reacts with carbon to form carbon dioxide or carbon monoxide). Electricity supplies energy to keep the molten salt bath and molten aluminum at 1250°K (1790°F). Electricity and carbon oxidation also provide energy to dissociate the alumina.

FIGURE 25: EQUIPMENT DIAGRAM—ELECTROLYTIC CELL

Rejected heat:
 Radiation, convection, other—29,200 kJ/kg (12,600 Btu/lb)
 Exit gases—14,900 kJ/kg (6,400 Btu/lb) at 1270°K (1830°F)

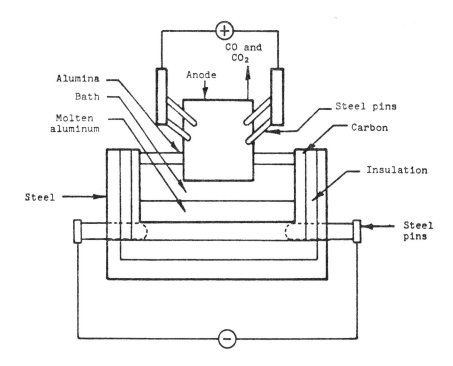

Alumina electrolysis cell

Source: Reference (2)

Figure 26 shows the melting of aluminum in a crucible furnace. Approximately one-third of aluminum castings are remelted to obtain finished products. Natural gas is burned to provide heat for melting and for heat treating.

FIGURE 26: EQUIPMENT DIAGRAM—MELTING FURNACE

Rejected heat*:
Radiation, other—930 kJ/kg (400 Btu/lb)
Combustion gases—5,100 kJ/kg (2,200 Btu/lb) at –1370°K (–2000°F)
Heat in products—1,050 kJ/kg (450 Btu/lb) at –920°K (–1200°F)

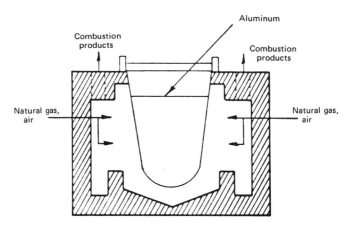

Crucible furnace

*Heat quantities include melting and heat treating operations.

Source: Reference (2)

MAJOR ENERGY CONSERVATION OPTIONS TO 1980

In making unit energy comparisons on a per ton basis, the physical properties
of metals are important. For example, aluminum alloys are of lower density
than are steel or copper products. The energy required to produce 1 cubic inch
of the common metals through mill processing is as follows:

Metal	10^3 Btu per in^3
Aluminum	9.8
Copper	16.0
Steel	5.7

Thus, even though aluminum may appear as the most energy intensive of these
three metals on a unit-weight basis, copper is the most energy consumptive on
a unit-volume basis.

The distribution of types of energy used in 1972 for the various processes comprising the aluminum component is given in Table 21. The Btu percentage totals
of 64.7, 24.3, and 11.0 for electricity, natural gas, and all other fuels, respectively,
may be compared with the analogous percentage totals of 69.5, 27.5, and 3 reported for these same fuel types in the 1971 Census of Manufactures energy totals.

TABLE 21: 1972 ENERGY USE DISTRIBUTION FOR ALUMINUM
COMPONENT

| Process | . . . . .Btu, percent. . . . . | | | |
	Electricity	Natural Gas	Other Fuels	Totals
Alumina	12.1	80.4	7.5	100.0
Molten metal	85.2	3.2	11.6	100.0
Hold, cast, melt	7.4	80.0	12.6	100.0
Fabrication	38.1	50.8	11.1	100.0
Total for component	64.7	24.3	11.0	100.0

Source: Reference (4)

The most energy intensive step by far in this entire component is the operation
of the Hall-Heroult electrolysis cells. These require about 85% of the total Btu
to produce molten metal from alumina and accounted for about 85% of the
total energy consumed by this component in 1972.

Within the past few years, numerous investigators have reported detailed analysis
of energy consumption in the production of aluminum. Table 22 compares the
results obtained from six of these studies, including Battelle (12), Bravard (39),
Chapman (40), Gordian (11), Kellogg (41), and Smith (42) with data reported
by the Aluminum Association for the base year 1972. In order to permit comparison, the data from all six sources were reworked slightly to the constant baseline described in the footnotes to Table 22. This includes the use of averaged
conversion factors of 10,072 and 8,021 Btu/kwh for the electricity consumed
in the production of alumina and aluminum, respectively, which are based on
the specific mixes of hydroelectric, purchased thermal electric, and self-generated
electricity that were reportedly used to produce these materials in 1972.

The reported energy required to produce a ton of alumina ranged from a low of
14.9×10^6 Btu (Gordian) to a high of 23.8×10^6 Btu (Chapman). Battelle (4)
prefers the 19.6×10^6 Btu per ton value of the Aluminum Association on the
basis that (a) this value is based on data obtained from 39 companies and represents over 95% of the energy used in the domestic aluminum industry, and
(b) these data are specifically related to the base year of 1972. In 1972, 6.22×10^6
tons of alumina were produced domestically for refining to aluminum. Multiplication of this value by 19.6×10^6 Btu per ton gives the 1972 total energy
consumption for producing alumina as 122×10^{12} Btu.

Other data on energy requirements for various aluminum ore processing steps
appear in reference (13).

Similarly, from Table 22, the reported values for smelting one ton of aluminum
ranged from a low of 145.7×10^6 Btu (Smith) to a high of 166.7×10^6 Btu
(Kellogg). Battelle (4) prefers the Aluminum Association value of 153.6×10^6
per ton of aluminum for the same reasons cited above. In 1972, 4.12×10^6 tons
of primary aluminum were produced in the U.S. The product of this value and
153.6×10^6 Btu per ton gives the 1972 calculated energy consumption for primary aluminum metal of 633×10^{12} Btu. The 1972 Census of Manufactures
(M72SR-6) reported that the total energy consumed by primary aluminum
SIC 3334 in 1971 was 53.7×10^9 kwh plus 140×10^{12} Btu.

TABLE 22: SUMMARY OF ESTIMATED ENERGY CONSUMPTION IN ALUMINUM PROCESSING*

Production Process	Aluminum Association	Battelle (12)	Bravard (39)	Chapman (40)	Gordian (11)	Kellogg (41)	Smith (42)
	Energy Consumption, 10^6 Btu per Ton of Alumina						
Refining							
Steam generation	–	14.2	13.1	10.7	6.7	11.0	–
Calcining	–	4.0	4.0	3.6	3.9	4.9	–
Caustic replacement	–	2.3	–	17.3	1.5	0.6	–
Other	–	1.1	0.3	2.2	2.8	1.5	–
Subtotal	19.6	21.6	17.4	23.8	14.9	18.0	15.9
	Energy Consumption, 10^6 Btu per Ton of Aluminum						
Reduction							
Electrolysis	–	128.3	148.0	109.5	125.8	120.3	124.1
Energy content of anode	–	18.9	–	22.5	31.7	13.8	15.9
Anode manufacture	–	2.7	12.0	11.3	0.8	32.6	2.1
Other	–	6.5	–	3.2	1.5	–	3.6
Subtotal	153.6	156.4	160.0	146.5	159.8	166.7	145.7

*Data from all sources reworked to a constant baseline as follows:

Material flow was adjusted to 1.93 parts of alumina to 1 part of aluminum (excepting Aluminum Association). Unit inputs of materials consumed in aluminum production (e.g., the amounts of anode carbon, caustic, or electricity) were not changed from those reported in the references.

All energy data were converted either to 10^6 Btu per ton of alumina or 10^6 Btu per ton of aluminum.

Electrical energy was converted to 10^6 Btu using the same conversion factors used by the Aluminum Association in 1972:

Source of Electricity	Conversion Factor (Btu/kwh)	Percent of Use Refining	Reduction
Hydropower	3,413	17.1	38.9
Purchased thermal	10,500	17.1	37.8
Self-generated thermal	11,690	65.8	23.3
		100.0	100.0

Thus, weighted averages used for conversion were 10,072 Btu/kwh for refining and 8,021 Btu/kwh for reduction.

Source: Reference (4)

Converting these kwh to Btu using the Aluminum Association factors and adjusting the total to the reported 1972 total primary metal production gives an estimated total energy consumption of 602 x 10^{12} Btu. This is appreciably less than the derived total value of 755 x 10^{12} Btu for alumina and aluminum.

During the base year, 8.49 x 10^6 tons of metal were processed through the holding and melting furnaces into unalloyed and alloyed ingot and cast products. The total included 4.12 x 10^6 tons of primary aluminum with the balance consisting of recycled scrap, alloying additions, plus some imported primary metal. The total energy required was 58.7 x 10^{12} Btu giving an average of 6.91 x 10^6 Btu per ton of metal processed.

This includes data for some continuous casting which had been adopted by the industry prior to 1972. Specifically, much of the wire product intended for electrical conductors was and is being made by the Properzi process where liquid metal is continuously cast to a rod shape which emerges hot into a special rolling mill where $\frac{3}{8}$ inch diameter rod is produced at the rate of 7.5 tons per hour. Reportedly, the run-around scrap is about 8% for this process versus 10 to 15% for older methods.

In 1972, a total of 121.2 x 10^{12} Btu were required to fabricate 5.22 x 10^6 tons of aluminum and aluminum alloy mill products for an average energy consumption of 23.2 x 10^6 Btu per ton of product. Mill-product shipments in 1972 totaled 4.62 x 10^6 tons with the product breakdown as follows:

	% of Mill Shipments
Sheet, plate, and foil	58.9
Extruded shapes and tubing	25.4
Rod, bar, and wire	12.0
Powder	2.4
Forgings	1.3
Total	100.0

The 1972 Census of Manufactures reported the total 1971 energy consumption for the combined establishments of SIC 3353, 3354, and 3355 was 4.8 x 10^9 kwh plus 66 x 10^{12} Btu. Converting these kwh to Btu using the Aluminum Association factors and adjusting the total by the 1972:1971 ratio of reported mill-product shipments gives an estimated total energy of 120 x 10^{12} Btu for the 1972 fabrication of mill products. This compares favorably with the 121.2 x 10^{12} Btu that was derived using the Aluminum Association data.

In general, aluminum alloys exhibit greater resistance to plastic deformation at their hot working temperatures than do many other metals. For this reason, mill equipment (e.g., hot rolls, extrusion presses, et cetera) is constructed to be able to exert greater fabrication pressures than for other tonnage metals. This implies generally somewhat greater energy consumption during hot working of aluminum than for carbon steel or copper alloys. On the other hand, hot working of aluminum is conducted at relatively low temperatures (aluminum alloys melt at 1000° to 1250°F compared with 2400° to 2800°F for steels and 1750° to 2100°F for copper alloys), and the energy required to heat the material to hot-working temperatures is substantially less than for these other metals. Cold fabrication processes for aluminum are essentially the same as for other easy-to-work metals. Machine energy requirements are not much different for cold-fabricating aluminum alloys than for low-carbon steel or copper alloy products of equivalent size and shape. Process annealing and heat-treatment energy consumption are somewhat less, in general, for aluminum than for steel or copper alloys, again owing to lower temperature requirements.

Most of the major plants making up this component are energy conservation conscious and have active conservation programs. For example, several of the major companies in this component have designated an energy coordinator at every plant. Each plant reports quarterly on conservation efforts in a company wide energy conservation newsletter, and employee communication is an integral part of the program. Several of the major companies have also led the aluminum component in promoting the recycling of aluminum scrap.

This component, through the Aluminum Association, was among the first to cooperate with the FEA in establishing a voluntary 1980 energy conservation goal to lower energy consumption in Btu per pound of aluminum shipped to 10% below 1972 levels. Progress through the second half of 1975 has been as follows:

| | - - - - - - - - - - Btu x 10^{12} - - - - - - - - - | | |
Time Period	Energy Base, 1972 Efficiencies	Energy Consumed in Period	Percent Improvement
1972	935.3*	935.3*	–
1973	1039.4	1025.0	2.1*
1974	1118.2	1057.6	5.7*
1st half 1975	448.4	414.8	4.0*
2nd half 1975	428.9	386.7	5.0*

*Revised data

The percentage of improvement as tabulated above cannot be calculated from the other data in this tabulation because the Aluminum Association calculates a weighted improvement value from values for each process by the method described below.

The energy conservation analysis shows a 5.0% improvement in energy efficiency through the second half of 1975. The figure reported previously for the first half of 1975 (6.5%) has been revised to 4.0%. Industry analysis of the earlier method of energy efficiency calculation disclosed a distortion that reflected a change in the mix of hydropower and fossil fuel-generated electricity rather than changes in efficiency.

This mix is significant because, as reported by the aluminum industry, electricity generated by hydropower has a lower Btu value than that generated by fossil fuels. To remove this distortion, energy consumption was converted by the Aluminum Association to the 1972 electricity pattern. The converted values are as follows:

| | | | - - - - - - 1975 - - - - - - - | |
1972	1973	1974	1st Half	2nd Half
935.3*	1017.9	1054.1	430.3	407.3

*Revised data

The percent improvement for 1975 is the net effect of process-by-process improvement as reflected in the table below where values are given in Btu x 10^{12}.

| Process | 1972 | ------- -1975- - - - - - - - | |
		1st Half	2nd Half
Bauxite	0.3	0.2	0.1
Alumina	122.0*	54.6	51.3
Hot metal	633.1	308.5	285.3
Hold, cast and melt	58.7	23.3	24.8
Fabrication	121.2	43.7	45.7
Total	935.3*	430.3	407.2

*Revised data

The percent improvement for each process is determined as follows.

Step 1: $\dfrac{1972 \text{ Btu}}{1972 \text{ lb}}$ = 1972 Btu/lb

Step 2: 1972 Btu/lb x 1975 lb = 1975 Btu at 1972 efficiencies

Step 3: 1975 Btu at 1972 efficiences - 1975 Btu = 1975 Btu saved as a result of new conservation efforts

Step 4: 1975 Btu saved ÷ 1975 Btu at 1972 efficiencies x 100 = % improvement

Table 23 lists causes of energy losses in the aluminum manufacturing process. It also gives estimates of energy losses and possible conservation approaches.

TABLE 23: ALUMINUM ENERGY CONSERVATION APPROACHES

Causes of Energy Losses	Approximate Magnitude of Losses	Energy Conservation Approaches
(1) Digestion of bauxite and evaporation of water from caustic		
(a) Radiation and convection	700 kJ/kg (300 Btu/lb)	Insulation Maintenance
(b) Heat in red mud	700 kJ/kg (300 Btu/lb)	
(c) Heat removed in cooling of aluminate solution	4,600 kJ/kg (2,000 Btu/lb)	
(d) Heat in vapor leaving evaporators	20,900 kJ/kg (9,000 Btu/lb)	Design modification (optimize evaporation scheme) Operation modification (close control of wash water volume)
(2) Calcining		
(a) Heat in exit combustion gases	3,250 kJ/kg (1,400 Btu/lb)	Operation modification (control of air/fuel ratio) Design modification (more complete heat recuperation)
(b) Radiation and convection	930 kJ/kg (400 Btu/lb)	Insulation Maintenance

(continued)

TABLE 23: (continued)

Causes of Energy Losses	Approximate Magnitude of Losses	Energy Conservation Approaches
(c) Heat in alumina	1,160 kJ/kg (500 Btu/lb)	Operation modification (feed hot alumina to cells) Design modification (more complete heat recuperation)
(3) Electrolytic reduction		
(a) Anode overvoltage, resistance and electrical connection	4,650 kJ/kg (2,000 Btu/lb)	Operation modification (lower current density) Research and development (catalytic additive to anode)
(b) Cathode resistance	4,650 kJ/kg (2,000 Btu/lb)	Operation modification (lower current density) Research and development (alternative cathode materials)
(c) Electrolyte resistance	20,400 kJ/kg (8,800 Btu/lb)	Design modification (closer anode-cathode spacing) Operation modification (lower current density)
(d) Resistance between cells	1,850 kJ/kg (800 Btu/lb)	Operation modification (lower current density) Design modification
(e) Recombination of aluminum with oxygen	8,100 kJ/kg (3,500 Btu/lb)	Operation modification (closer control of cell operation)
(f) Excess carbon consumption	5,350 kJ/kg (2,300 Btu/lb)	
(4) Remelting and heat treating		
(a) Heat in combustion gases	5,100 kJ/kg (2,200 Btu/lb)	Design modification (waste heat recovery)
(b) Radiation and convection	930 kJ/kg (400 Btu/lb)	Insulation Maintenance
(5) Overall process losses		
(a) Lack of aluminum recycling	32,000 kJ/kg (14,000 Btu/lb)	Waste utilization (more aluminum recycling)
(b) Higher energy requirement of Hall process as compared to new Alcoa process	25,000 kJ/kg (11,000 Btu/lb)	Process modification (replacement of Hall process with new Alcoa process)
(c) Radiation convection from electrolysis cell	29,200 kJ/kg (12,600 Btu/lb)	Insulation
(d) Heat in exit gases from electrolysis cell	14,900 kJ/kg (6,400 Btu/lb)	Design modification (waste heat recovery)

Note: Electrolytic reduction losses are electrical. Overall process losses (a) and (b) are primarily electrical. The fuel equivalent for these losses would be approximately three times the listed values.

Source: Reference (2)

EPA has compared the energy requirements and solid-waste production for various control and treatment technologies in the production of primary aluminum. See Table 24.

TABLE 24: ENERGY REQUIREMENTS AND SOLID WASTE PRODUCTION FOR WATER EFFLUENT CONTROL AND TREATMENT TECHNOLOGIES

Process	Energy Use Electrical (kwh/ton)	Energy Use Thermal Equivalent (kwh/ton)	Sludge Production (kg/ton)
Dry scrubbing	233	0	0
Primary wet scrubbing with recycle, Process A	84	200	73
Secondary wet scrubbing with recycle	394	200	76
Primary wet scrubbing, once through, Process B	84	–	40
Process A plus bleed and filtrate treatment	85-395	200	77
Process B plus alum treatment	100	–	123
Process B plus activated alumina treatment	100	–	110
Process B plus hydroxyl-apatite treatment	100	–	–
Process B plus reverse osmosis treatment	546	–	60

Note: Ton = metric ton; values are 10% lower for short ton.

Source: Reference (4)

Spokesmen for this component have reported the following energy requirements for the EPA and OSHA effects over the period of 1972 to 1980.

Process Stage	Energy Requirement
Alumina preparation	32 Btu per lb of alumina
Molten metal (refining)	885 Btu per lb of aluminum
Hold, cast, melt	7 Btu per lb of aluminum
Fabrication	173 Btu per lb of aluminum

Applying these estimated requirements for environmental controls to the aggregated subcomponents and projecting these to 1980 gives a total estimated net potential of 646×10^{12} Btu.

The following rationales and calculations were used to derive the energy efficiency improvement potential values summarized subsequently in Table 28. Attention is directed to the fact that the calculations in this component are based on use of the Aluminum Association conversion factors. These factors yield an estimated total energy use of 935×10^{12} Btu for the component in 1972 for a production of 5.22×10^6 tons of fabricated products, which is an average energy use of 179.1×10^6 Btu per ton of fabricated products.

Conversion of Rotary Kilns: In 1972, an estimated 6.22×10^6 tons of alumina were produced using calciners which were mostly (95% of capacity) equipped with rotary kilns. By 1980, it is estimated that it will be technologically feasible to convert a total of 66% of the calciner capacity to the more energy efficient fluidized bed calciners. Substitution of a fluidized bed calciner for a rotary kiln can save 600 Btu/lb of alumina, which is equivalent to 1.2×10^6 Btu/ton. Because about 5% of the industry was already equipped with fluidized bed calciners in 1972, the calculation for the technologically feasible potential savings by 1980 is as follows: $(0.61) \times (6.22 \times 10^6) \times (1.2 \times 10^6) = 4.6 \times 10^{12}$ Btu. Dividing this value by the 5.22×10^6 tons of fabricated products produced in 1972 gives the technologically potential savings of 0.9×10^6 Btu/ton.

Economic consideration of the energy savings gained by replacing rotary kilns with fluidized bed calciners leads to the estimate that it will be economically practicable for the aluminum component to convert about 20% of its total calciner capacity to the fluidized bed calciners by 1980. The diversion of more capital to effect a greater degree of conversion is not likely. On this basis, the 1980 fluidized bed calciner capacity will increase only by 15% over the 1972 value. This represents $0.15 \times (1.2 \times 10^6) = 0.2 \times 10^6$ Btu/ton as the estimated economically practicable potential.

Improvements of Hall Cells: In 1972, the average Hall cell required 8.15 kwh to produce one pound of aluminum, and the refining capacity of the component was 4.77×10^6 tons. The cells currently in use require about 6.5 kwh/lb of aluminum, and all of the capacity coming on stream by January 1, 1980 is expected to operate with this higher efficiency. Estimated 1980 refining capacity is 5.22×10^6 tons, which represents a gain of 0.45×10^6 tons over 1972.

By 1980 it is estimated to be technologically feasible to improve the efficiency of 5% of the 1972 cell capacity by a level of 10%. This would give 5% of the cells an efficiency of $8.15 - (0.10 \times 8.15) = 7.33$ kwh/lb of aluminum. By 1980, the combined electrical capacity might, therefore, be as follows:

$$
\begin{array}{lll}
\text{New cells:} & (0.45 \times 10^6) \times (2 \times 10^3) \times 6.5 & = 5.85 \times 10^9 \text{ kwh} \\
\text{95\% of old:} & 0.95 \times (4.77 \times 10^6) \times (2 \times 10^3) \times 8.15 & = 73.83 \times 10^9 \text{ kwh} \\
\text{5\% of old:} & 0.05 \times (4.77 \times 10^6) \times (2 \times 10^3) \times 7.33 & = \underline{3.52 \times 10^9} \text{ kwh} \\
& \text{Total} & 83.20 \times 10^9 \text{ kwh}
\end{array}
$$

Dividing this value by the 1980 tonnage output capacity gives an average 1980 cell efficiency of 7.97 kwh/lb of aluminum. This is 0.18 kwh less than the average 1972 cell efficiency, so the potential technological potential is:

$$
0.18 \frac{\text{kwh}}{\text{lb}} \times 8,021 \frac{\text{Btu}}{\text{kwh}} \times (8,245 \times 10^6 \text{ lb}) = 11.9 \times 10^{12} \text{ Btu}
$$

The 1972 values of 8,021 Btu/kwh and $8,245 \times 10^6$ pounds were obtained from industry spokesmen. Dividing 11.9×10^{12} by 5.22×10^6 gives the estimated potential savings of 2.3×10^6 Btu/ton of fabricated products.

The implementation only of the equipment changes and procedures necessary to upgrade 5% of 1972 cell capacity by a level of 10% by 1980 would lead to a potential savings of about 2.6×10^{12} Btu. Considerations of the capital costs required to do this, plus the losses in production due to cell shutdown, leads to the conclusion that this technique is not economically feasible. However, an estimated savings of 9.3×10^{12} Btu by 1980 is economically practicable due to the installation of the more efficient cells since 1972. This is equivalent to $(9.3 \times 10^{12}$ Btu$) \div (5.22 \times 10^6$ tons$) = 1.8 \times 10^6$ Btu/ton of fabricated products.

Conversion to Silicon Rectifiers: According to industry spokesmen, in 1972 about 25% of the primary aluminum smelters were using mercury arc rectifiers to convert AC power to the DC power required for reducing the alumina, while 75% had converted to the more electrically efficient, solid-state silicon rectifiers. Typically, such conversion results in an energy saving of 2 to 3%. By 1980, it is estimated to be technologically feasible and economically practicable for the remainder of the industry to install the silicon rectifiers.

In 1972, the equivalent of approximately 540×10^{12} Btu of electrical energy was used in the production of primary aluminum. The annual potential energy saving from this equipment installation is computed as follows:

$$(0.25) \times (0.025) \times (540 \times 10^{12}) = 3.4 \times 10^{12} \text{ Btu}$$

Dividing this value by 5.22×10^6 tons gives the estimated potential saving of 0.6×10^6 Btu/ton of fabricated products as both technologically feasible and economically practicable over the target period.

Decreasing Melting Losses, Improving Fabrication Yield, Changing Scrap Recovery Rates: Because of the interdependence of these techniques, their effects were assessed simultaneously using the concept of an aluminum industry flow diagram. Figure 27 presents the basic diagram which, in this instance, depicts the balance of materials flow based on the best available information for 1972. Some of the key data were provided by industry spokesmen.

In order to assess the effects of changing various operations, each of the six major processing steps was assigned the basic energy values shown in Figure 27. The basic reference value of the energy required to ship one ton of aluminum product in 1972 was then computed using the procedure shown below.

Primary metal: $\quad \dfrac{48.5}{54.4} \times 183 \times 10^6 = 163.15 \times 10^6$ Btu/ton

Alloying: $\quad \dfrac{2}{54.4} \times 4 \times 10^6 = 0.15 \times 10^6$

Hold, cast, melt: $\quad \dfrac{100}{54.4} \times 5 \times 10^6 = 9.19 \times 10^6$

Fabrication: $\quad \dfrac{98}{54.4} \times 23.2 \times 10^6 = 41.79 \times 10^6$

Recovered mill scrap: $\dfrac{43.6}{54.4} \times 0.7 \times 10^6 = 0.56 \times 10^6$

Recovered old scrap: $\dfrac{5.9}{54.4} \times 4.2 \times 10^6 = 0.46 \times 10^6$

Total = 215.30×10^6 Btu/ton

FIGURE 27: U.S. ALUMINUM INDUSTRY FLOW DIAGRAM WITH VALUES
ASSIGNED AND DERIVED FOR 1972 BASE YEAR
(Based on Data from Industry Spokesmen)

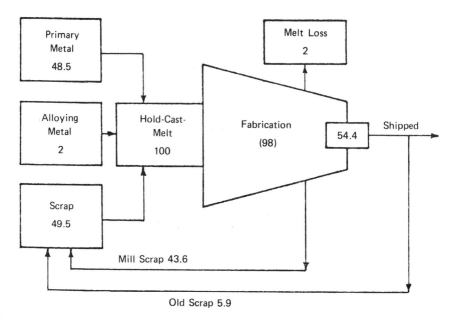

Item	Energy Value, 10^6 Btu/ton
Primary metal	183
Alloying metal	4.0
Hold, cast, melt	5.0
Fabrication	23.2
Recover mill scrap	0.7*
Recover old scrap	4.2**

*Estimated at one-eighth of old scrap recovery
**Btu value less Btu to melt = $(0.05 \times 183 \times 10^6) - (5 \times 10^6) = 4.2 \times 10^6$

Source: Reference (4)

Eight other hypothetical industrial cases were then evaluated in a similar manner
to explore the combined and individual effects of the variations listed in Table
25. Table 26 gives the results of the individual A values selected to depict a
given situation and the corresponding calculated B values which were obtained.
The total energy values obtained were then compared to the base year to quantify the potential savings.

In addition to Case 1 (representing the base year), the two other most important
cases derived in Tables 25 and 26 were Cases 2 and 6, because these represent
the specific combinations of techniques chosen to represent the technologically
feasible and economically practicable situations for the combined effects of decreasing melt losses, increased scrap recovery rates, and improved fabrication yield.

TABLE 25: HYPOTHETICAL CASES USED TO ISOLATE AND ASSESS THE EFFECTS OF DECREASING MELTING LOSSES, CHANGING SCRAP RECOVERY RATES, AND IMPROVING FABRICATION YIELD FOR ALUMINUM COMPONENT

Case Number	Objective or Purpose
1	Determine base year values using 1972 data for comparisons.
2	Assess the combined effects of (A) decreasing melt losses by 50%, (B) increasing old scrap recovery by 100%,* and (C) improving fabrication yield. Fabrication yield improvements included two components:
	(i) Increase yield of nonheat-treatable (NHT) sheet by 4% over base year (2.22 x 10^6 tons shipped) by installing larger, heavier rolling mills and ancillary equipment.
	(ii) Improving yield of other fabricated products by 0.5% using improved practices with existing equipment.
3	Isolate the effect of decreasing melt losses only in Case 2.
4	Isolate the effect of increasing scrap recovery only in Case 2.
5	Isolate the effect of improving fabrication yield only in Case 2.
6	Assess the combined effects of (A) decreasing melt losses by 25%, (B) increasing old scrap recovery by 100%,* and (C) improving fabrication yield by 1% for all fabricated products.
7	Isolate the effect of decreasing melt losses only in Case 6.
8	Isolate the effect of old scrap recovery only in Case 6.
9	Isolate the effect of improving fabrication yield only in Case 6.

*The values of old scrap used in computations for 1972 and 1980 were 0.5 x 10^6 and 1.0 x 10^6 tons, respectively.

Source: Reference (4)

Cases 3, 4, and 5 were computed to isolate the separate effects of these combined variables in Case 2 and to prorate these against the total value obtained from Case 2. Similarly, Cases 7, 8, and 9 were computed to isolate and prorate the separate effects of the variables combined in Case 6.

For example, the computations from Case 2 in Table 26 show this combination of techniques could save 12.91 x 10^6 Btu/ton of aluminum product shipped relative to the base year, for a potential saving of 6.0% of the total of 935 x 10^{12} Btu expended by the aluminum component in 1972. Dividing the value of 56.1 x 10^{12} Btu so obtained by 5.22 x 10^6 tons gives the combined total value of 1.6 + 0.4 + 8.8 = 10.8 x 10^6 Btu/ton shown for these three techniques in Table 28.

To isolate the individual effects of the three techniques, the computations for Cases 3, 4, and 5 were made to prorate the 56.1 x 10^{12} Btu as shown in Table 27.

While it may be technologically feasible to develop procedures for lowering the melt losses by 50% by 1980, on the basis of economic considerations, it is unreasonable to expect these to be fully implemented. Part of the technological potential will require new equipment and process changes which are expected to reach a point of vastly dimishing economic returns after savings of about 25% are achieved.

TABLE 26: CALCULATED ENERGY FOR SHIPPING ALUMINUM PRODUCTS ACCORDING TO NINE HYPOTHETICAL CASES USING THE FLOW DIAGRAM OF FIGURE 27

Percentage Breakdown (A) and Calculated Energy (B), 10^6 Btu per ton, to Ship One Ton of Aluminum Products for the Case Situations Listed Below*

	...1...		...2...		...3...		...4...		...5...		...6...		...7...		...8...		...9...	
	A	B	A	B	A	B	A	B	A	B	A	B	A	B	A	B	A	B
Primary metal	48.5	163.15	45.8	150.74	48.5	160.21	42.6	143.30	49.9	163.65	46	152.5	48.5	161.67	42.6	143.40	49.5	163.51
Alloying metal	2	0.15	2	0.14	2	0.14	2	0.14	2	0.14	2	0.14	2	0.15	2	0.14	2	0.14
Cast-melt-hold	100	9.19	100	8.99	100	9.02	100	9.19	100	8.96	100	9.06	100	9.11	100	9.19	100	9.03
Fabrication	98	41.79	99	41.31	99	41.46	98	41.79	98	40.75	98.5	41.40	98.5	41.62	98	41.79	98	41.04
Products shipped	54.4	-	55.6	-	55.4	-	54.4	-	55.8	-	55.2	-	54.9	-	54.4	-	55.4	-
Recovered mill scrap	43.6	0.56	43.4	0.55	43.6	0.55	43.6	0.56	42.2	0.53	43.2	0.55	43.6	0.56	43.6	0.56	42.6	0.54
Recovered old scrap	5.9	0.46	8.8	0.66	5.9	0.45	11.8	0.91	5.9	0.44	8.8	0.67	5.9	0.45	11.8	0.91	5.9	0.45
Total Btu/ton		215.30		202.39		211.83		195.89		214.47		204.32		213.56		195.89		214.71
Energy saving**																		
Percent		Base		6.0		1.6		9.0		0.39		5.1		0.81		9.0		0.27
10^{12} Btu		Base		56.1		15.0		84.2		3.6		47.7		7.6		84.2		2.5

*See Table 25 for a description of the nine situations listed here.
**Savings expressed as percent of 10^6 Btu per ton for base year and as this percentage applied to the total of 935×10^{12} Btu for the component in the base year.

Source: Reference (4)

TABLE 27: TECHNOLOGICALLY FEASIBLE POTENTIAL

	Cases 3, 4, and 5	Case 2, Prorated	Case 2, 10^6 Btu/ton of Product*
	. . . . .10^{12} Btu.		
Lower melt losses	15.0	8.19	1.6
Increase scrap recovery	84.2	45.95	8.8
Improve yield	3.6	1.96	0.4
Totals	102.8	56.10	

*Based on 1972 production of 5.22×10^6 tons of fabricated products.

Source: Reference (4)

Similarly, the potential savings which might have accrued through the installation of larger rolling mills were not deemed sufficient to justify the costs. On the other hand, it is assumed economically practicable to expect a 1% increase in the fabrication yield by 1980 by the expedient of such manufacturing process improvements as employing larger ingots, cutting edge trim, and rolling to wider widths.

Economic considerations also led to the conclusion that it will be practicable to double the amount of old scrap being recycled by the aluminum component from 1972 to 1980; i.e., from 0.5×10^6 to 1×10^6 tons. It is recognized that this is a formidable task. However, the aluminum component is keenly aware of the electric energy crisis and of impact that recycling has on decreasing their energy demands. One industry spokesman has already noted that between 1975 and 1980 his company expects to increase its level of recycled aluminum by 45%.

Nevertheless, to achieve and expand this goal for the entire component by 1980 it will clearly be necessary for the companies comprising the component to continue their intensive promotional efforts in the recycling of consumer products and for all branches of federal and municipal governments to continue their support of programs which are aimed at the collection of aluminum scrap from consumers, automotive scrap, and municipal wastes.

In summary, these technologically feasible and economically practicable savings were estimated simultaneously in Case 6 of Table 25, as noted earlier, and found to total 47.7×10^{12} Btu. Identical procedures were then used to prorate these savings among the melting loss decrease, improved fabrication yield, and increasing old scrap as described previously for the technologically feasible potential savings. The prorated values so obtained for the savings which were found to be economically practicable were:

Technique	Economically Practicable Potential	
	10^{12} Btu	10^6 Btu/ton of Products*
Cut melt losses	3.8	0.7
Increase scrap recovery	42.6	8.2
Improve yield	1.3	0.2
Total	47.7	

*Based on 1972 production of 5.22×10^6 tons of fabricated products.

Installation of Air Dampers: One of the major primary aluminum producers (representing 33% of the primary capacity in 1972) has announced that the installation of air dampers on their holding furnaces was under way, and that the expected energy savings were to be more than 0.25×10^{12} Btu/year. By 1980, it is estimated to be technologically feasible and economically practicable for the entire aluminum component to effect comparable savings. Therefore, the total potential savings equal 0.25×10^{12} divided by 0.33, or 0.76×10^{12} Btu. Division of this value by the 5.22×10^6 tons of fabricated products obtained in 1972 gives the potential technologically feasible and economically practicable savings of 0.2×10^6 Btu/ton of fabricated products.

Housekeeping: The potential savings from improved housekeeping practices were derived by considering the energy utilization in each of the four subcomponent operations, i.e., alumina, molten metal, hold-cast-melt, and fabrication.

It is estimated that, in 1972, 80×10^{12} Btu were used (a) to generate the steam used in digesting the alumina in the Bayer process, and (b) in the digestion process itself, i.e., in heating the caustic solution to dissolve the alumina. It is also estimated that, by 1980, 15% of the 80×10^{12} Btu, or 12×10^{12} Btu, could be saved by practices which include:

    (1)   Improving the digester heat exchanges
    (2)   Improvements in the heat transfer and process conditions
    (3)   Improved heat recuperation
    (4)   Insulation repair
    (5)   Improved combustion controls

It is estimated that, in 1972, 11×10^{12} Btu were used to manufacture the anodes and cathodes for use in the Hall cells. By 1980, it is estimated that 15% of this energy, or 1.6×10^{12} Btu, could be saved by good housekeeping practices which include:

    (1)   More extensive use of heat recuperation
    (2)   Increased use of economizers
    (3)   Improved process controls

It is estimated that, in 1972, 58.7×10^{12} Btu were used to either melt aluminum and aluminum alloys or to hold them in a molten state. It is also estimated that, by 1980, 20% of this energy, or 11.7×10^{12} Btu, might be saved by instituting practices which include:

    (1)   Improved combustion controls
    (2)   Installation of heat recuperators
    (3)   Combustion air preheating
    (4)   Improved process controls
    (5)   Insulation repairs
    (6)   Reuse of salvage oil

It is estimated that, in 1972, about 15% of the 121×10^{12} Btu consumed in the fabrication subcomponent, or 18.1×10^{12} Btu, might be saved by the widespread adoption of practices which include:

    (1)   Preheating of remelting furnace billets
    (2)   Installation of heat recuperators
    (3)   Direct firing of ingot and billet preheaters
    (4)   Insulation repair

Collectively, the aggregated potential of housekeeping savings is 43.4 x 10^{12} Btu. These were regarded as both technologically feasible and economically practicable. Dividing this value by 5.22 x 10^6 tons gives the component value of 8.3 x 10^6 Btu/ton of fabricated products.

GOAL YEAR (1980) ENERGY USE TARGET

Projected 1980 production figures for the elements of this component are as follows:

Operation	1980 Production, 10^6 tons
Alumina	7.0
Molten metal	5.22
Hold, cast, melt	11.35
Fabricated products	7.08

Solely on the basis of technological feasibility, it is estimated that this component could conserve an average of 23.1 x 10^6 Btu/ton of fabricated products produced by 1980 (see Table 28). To develop these energy-saving projections, the following factors have been presumed to apply:

> The component will be able to continue using existing bauxite sources.
>
> Continued availability of projected energy needs.
>
> Component continues assumed growth rate, product mix, and inventory practices.
>
> Realization of assumed post-consumer scrap recycling.
>
> No changes in quality or availability of key raw materials.
>
> Ability to operate at assumed production rate.

The high and relatively inefficient energy consumption in the Hall cell has resulted in a continuing conservation effort which has yielded good results. For example, prior to World War II, 12 kwh were required to produce 1 pound of aluminum. In 1972, the average smelter used 8.15 kwh, and the smelters being constructed today use 6.5 kwh/lb of aluminum.

One of the greatest potential energy savings for this component lies with the continuing development of the newest and most up-to-date process (Alcoa) which will decrease the electrical smelting requirements to 4.5 kwh/lb. A 30,000 ton-per-year pilot plant using this new process was scheduled to commence operations in 1976. Although its ultimate production is planned eventually to reach 300,000 tons per year, the lower initial output of this plant will not significantly affect this component's energy conservation potential before January 1, 1980.

Nevertheless, some improvement in the efficiency of the older Hall cells is technologically feasible. This, plus the added new capacity between 1972 and 1980, has the potential for lowering the energy in fabricated products by 2.3 x 10^6 Btu per ton. It will be noted from Table 28 that about 36% of the total projected savings for this component is accounted for by housekeeping improvements. These include, but are not limited to, some of the following major items:

> Improved digester heat exchange, heat transfer, and process conditions in Bayer process.
>
> Improved process controls and heat recuperation in generating the electrical energy used in the electrolysis operation.
>
> Installation of heat recuperators and improved combustion controls and more combustion-air preheaters in the melting, holding, and casting operations.
>
> Other activities as discussed under "Housekeeping".

TABLE 28: ENERGY EFFICIENCY IMPROVEMENT POTENTIAL FOR 1980: ALUMINUM COMPONENT

Technique	Potential Saving, 10^6 Btu/ton of . . . . Fabricated Products*	
	Technologically Feasible	Technologically Feasible and Economically Practicable
Convert 66% of rotary kilns for calcining alumina to fluidized bed calciners	0.9	0.2
Improved efficiency of Hall process	2.3	1.8
Complete 100% conversion to silicon rectifiers	0.6	0.6
Decrease melting losses by 50%	1.6	0.7
Improved fabrication yield	0.4	0.2
Double the amount of old scrap recycled	8.8	8.2
Install air dampers on holding furnaces	0.2	0.2
Housekeeping	8.3	8.3
Total	23.1	20.2

*Btu values in this table are based on use of Aluminum Association conversion factors that yield 935×10^{12} Btu as component energy use in 1972.

Source: Reference (4)

The technologically feasible potential energy savings for eight energy-conservation practices or techniques were listed in Table 28, which also summarizes the economically practicable savings that were estimated for these same practices. The estimated potential savings amount to a total of 20.2×10^6 Btu/ton of fabricated products. Applying all of these economically practicable savings to the aggregated component and projecting them to 1980 gives a total estimated gross potential of 639×10^{12} Btu.

As indicated in Table 28 above, use of the Aluminum Association's conversion factors for electrical energy gave 935×10^{12} Btu as the total 1972 energy consumption for the aluminum component. These conversion factors give a high value for energy when compared to the energy consumed by the other SIC 33 components. Accordingly, all of 1972 energy consumption data for aluminum were recalculated using the conversion factor of 3,412 Btu/kwh. This gave a revised total 1972 energy consumption of 588 x 10^{12} Btu. All of the potential 1980 savings which were deemed technically feasible and economically practicable were also recomputed using the same conversion factor of 3,412 Btu/kwh. These values are given in Table 29 along with the comparable Aluminum Association data.

TABLE 29: TECHNOLOGICALLY FEASIBLE AND ECONOMICALLY PRACTICABLE 1980 POTENTIAL ENERGY SAVINGS FOR ALUMINUM COMPONENT

| | ... Potential Saving, 10^{12} Btu .... | |
| | Aluminum Association | All Electricity at 3,412 Btu |
Technique	Conversion Factors	per kwh
Alumina (6.22 x 10^6 tons in 1972)		
Conversion of rotary kilns	1.1	1.1
Decreasing melting losses and improving scrap recovery and fabrication yield	27.9	25.7
Housekeeping	12.0	12.0
Subtotal	41.0	38.8
Molten metal (4.12 x 10^6 tons in 1972)		
Hall cell improvement	9.3	3.9
Conversion to silicon rectifiers	3.4	1.4
Decreasing melting losses, and improving scrap recovery and fabrication yield	18.4	9.4
Housekeeping	1.6	1.4
Subtotal	32.7	16.1
Hold-cast-melt (8.49 x 10^6 tons in 1972)		
Installation of air dampers	0.8	0.8
Decreasing melting losses, and improving scrap recovery and fabrication yield	0.6	0.6
Housekeeping	11.7	11.2
Subtotal	13.1	12.6
Fabrication (5.22 x 10^6 tons in 1972)		
Decreasing melting losses, and improving scrap recovery and fabrication yield	0.8	0.6
Housekeeping	18.1	15.1
Subtotal	18.9	15.7
Grand total	105.7	83.2

Source: Reference (4)

For purposes of the SIC 33 target definition, it was necessary to assign the potential energy savings among the four subcomponent divisions, i.e., alumina, molten metal, hold-cast-melt, and fabrication. As is evident from Table 25 above, the techniques of reducing melting loss, improving fabrication yield, and increasing scrap recovery are felt in all four subcomponent divisions. Accordingly, to prorate the energy values among the four processing operations, the computations were based on the energy differences between the A columns in Table 26 for the base year and the case in question. For Case 2, for example, the prorating of technologically feasible potential energy savings was accomplished as shown in Table 30.

TABLE 30: PRORATING OF TECHNOLOGICALLY FEASIBLE POTENTIAL ENERGY SAVINGS

Item	. . . . Potential Energy Savings		
	10^6 Btu/ton Δ Column A	10^{12} Btu, Prorated	Assigned Process Operation
Primary metal	12.41	53.93	Alumina and molten metal
Alloying metal	0.01	0.04	Hold-cast-melt
Hold-cast-melt	0.20	0.87	Hold-cast-melt
Fabrication	0.48	2.09	Fabrication
Recovered mill scrap	0.01	0.04	Fabrication
Recovered old scrap	−0.20	−0.87	Fabrication
Totals	12.91	56.10	

Source: Reference (4)

The saving ascribed to primary metal was then also divided between the alumina and molten metal operations according to the 1972 proportions of these materials processed by the component, i.e., 12.4 x 10^6 pounds and 8.2 x 10^6 pounds, respectively.

To assist in the SIC 33 energy target definition, the data of Table 29 were used along with the values of 1972 energy output and projected 1980 production for the four subcomponents to compute the projected 1980 energy efficiency improvement potential for the entire aluminum component, using both sets of conversion factors. The results of these computations are given in Table 31.

It should be noted that in order to derive a single total line representing the aluminum component from the seven columns with this methodology, it is mathematically necessary to carry only the real subcomponent totals from Columns 3, 6, and 7.

The total values from Columns 1 and 2 must then be chosen so their product equals the total of Column 3. For the aluminum component, the Column 1 totals were fixed using the procedure cited in Footnote (*).

This gave the Column 2 totals which were then used to calculate the Column 4 and 5 totals using the procedures cited in Footnotes (***) and (†). This synthetically derived totals line was then used to calculate the SIC 33 energy efficiency improvement targets.

TABLE 31: COMPUTATIONS TO DETERMINE PROJECTED 1980 ENERGY EFFICIENCY IMPROVEMENT POTENTIAL FOR ALUMINUM COMPONENT

Subcomponent	(1) 1972 Energy Output (10^6 Btu/ton)	(2) Projected 1980 Production (10^6 tons)	(3) 1980 Energy Consumption Using 1972 Efficiency (10^{12} Btu)	(4) Projected 1980 Energy Output (10^6 Btu/ton) Gross	(5) Net	(6) Projected 1980 Energy Consumption (10^{12} Btu) Gross	(7) Net
. . . . Aluminum Association Conversion Factors.							
Alumina	19.60	7.0	137.2	13.01	13.07	91.1	91.5
Molten metal	153.6	5.22	801.8	145.7	147.5	760.6	770.0
Hold-cast-melt	6.91	11.35	78.4	5.37	5.38	60.9	61.6
Fabrication	23.2	7.08	164.2	19.6	19.95	138.8	141.2
Total	179.1*	6.60**	1,181.6	159.3***	161.3†	1,051.4	1,064.3
. . . . All Electricity at 3,412 Btu/kwh.							
Alumina	18.1	7.0	126.7	11.86	11.92	83.0	83.4
Molten metal	78.3	5.22	408.7	74.4	75.3	388.4	393.1
Hold-cast-melt	6.62	11.35	75.1	5.14	5.15	58.3	58.5
Fabrication	18.4	7.08	130.3	15.39	15.67	109.0	110.9
Total	112.6*	6.58**	740.8	97.1***	98.2†	638.7	645.9

*Determined by dividing total 1972 energy consumption (935 x 10^{12} Btu for Aluminum Association and 588 x 10^{12} Btu for 3,412 Btu/kwh) by 1972 production of 5.22 x 10^6 tons of fabricated products.
**Determined by dividing total of Column (3) by total of Column (1).
***Determined by dividing total of Column (6) by total of Column (2).
†Determined by dividing total of Column (7) by total of Column (2).

Source: Reference (4)

SOME PROJECTIONS BEYOND 1980 TO 1990

The estimates for energy conservation in the manufacturing sector made as a part of the Project Independence Blueprint in 1974 (10) included some estimates for the aluminum industry. Table 32 presents projected energy/output coefficients through the year 1990.

TABLE 32: ENERGY/OUTPUT PROJECTIONS (10^{12} Btu)

Source	1971	1975	1977	1980	1985	1990
Purchased electricity	184.8	200.5	228.3	263.0	308.3	368.3
Internal generation	116.5	116.5	116.5	116.5	116.5	116.5
Purchased fuels	65.5	70.4	82.0	95.5	115.8	142.5
Total	366.8	387.4	426.8	475.0	540.6	627.3
Output (10^3 tons)	5,014	5,553	6,316	7,353	9,196	11,624.0
Energy/output (10^6 Btu/ton)	73.155	69.764	67.574	64.599	58.786	53.966

Source: Reference (10)

The aluminum industry has historically generated a large percentage of its electricity. It is assumed in this projection that this trend is discontinued. Most of the hydroelectric sites have already been exploited and the trend in recent years has been directed more at locating primary production facilities near marketing areas rather than in the Northwest or South where cheap power has always been available. This may be an overly restrictive assumption but recent literature and discussion with industry reinforce this belief.

The projections contained in Table 32 are presented on a net energy basis. That is, purchased electricity is valued at its delivered heat value, 3,412 Btu/kwh, while internally generated electricity is valued at a heat rate of 10,000 Btu/kwh. The heat rate for electricity generation with the aluminum industry is lower than for the U.S. as a whole because a disproportionate share is derived from hydroelectric power.

Based on projections contained in Table 32 the energy requirement per unit of output in the aluminum industry will decrease by roughly 1.6% per year. This represents a decrease in total energy requirements of roughly 223 trillion Btu of energy when compared to baseline projections using the 1971 energy/output coefficient. Overall energy requirements per unit of output are projected to decrease by 11.7% by 1980 and 26.2% by 1990.

Of considerable interest with respect to energy consumption in the aluminum industry is the Alcoa 30,000 ton per year pilot plant designed to produce aluminum by the electrolytic reduction of aluminum chloride. A flow sheet for this process is shown in Figure 28.

FIGURE 28: ALCOA SMELTING PROCESS FLOWSHEET

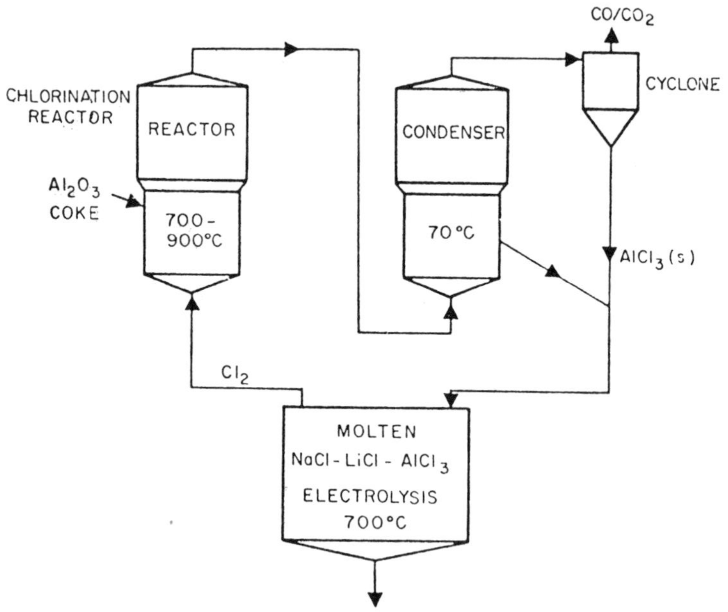

Source: Reference (8)

This process is reported to require 30% less electrical energy than the most efficient operating Hall-Heroult cell (this would indicate that energy consumption in the new process will be about 4.5 kwh/lb of aluminum). Since the process is fully enclosed, pollution problems will also presumably be reduced. As shown in Figure 28, the process starts with refined alumina. If pilot plant operations are encouraging, it is possible that a major portion of new reduction capacity would use this process.

The adaptation of this process to some degree has apparently been included in the preparation of Table 32 but it is not apparent from the Project Independence Blueprint (10) what contribution has been assigned to this specific process improvement.

IRON FOUNDRY INDUSTRY

This component includes establishments primarily engaged in manufacturing iron castings, generally on a job or order basis, for sale to others or for interplant transfer. It includes establishments that cast gray iron, ductile iron, white iron, and malleable iron, and corresponds to SIC 3321 and 3322. Establishments which produce iron castings in a captive relationship to manufacture a specified product are classified in the industry of the specified product, rather than in this component. Some iron castings included in this latter group are to be accounted for in SIC 34, 35, and 37.

Comparing the three foundry components in SIC 33, nonferrous foundries are most numerous, smallest in average number of employees, and have the lowest value of shipments per establishment. Steel foundries are least numerous, largest in average number of employees, and have annual shipments per establishment about three times the value for nonferrous foundries. The iron foundries dominate with about 54% of total employees for the three components and 55% of the total value of shipments.

PROCESS TECHNOLOGY INVOLVED

A representative flowsheet for the production of iron castings is given in Figure 29. The major metallic raw material is ferrous scrap, which is combined with pig iron, ferroalloys, and fluxes and melted in cupolas or electric furnaces. For gray and ductile iron castings, roughly 75% of the metal is melted in cupolas (in which the major fuel is metallurgical coke) and 25% of electric furnaces (usually induction furnaces, but arc furnaces are increasingly being used).

For malleable iron castings (and for alloyed or special irons), electric furnaces account for a higher proportion of melting. The molten iron of controlled composition is poured into molds and cores made in the foundry from aggregates (usually silica sand) bonded with a wide variety of materials that include bentonite, drying oils, sodium silicate, urea-formaldehydes, and furans. The molds and cores may be used baked, unbaked, or chemically hardened, and are used

102

only once, Casting into permanent molds accounts for only a small fraction of the output (except for cast iron pipe where permanent molds are common), and die casting is rare. Castings from the molds are cleaned of gates and risers which are necessary appendages to permit casting, and the metal removed is recycled to the melting operation. In general, roughly 40 to 60% of the metal melted is recycled within the foundry. Heat treatment is applied to only a small portion of gray iron castings, more than half of the ductile iron castings, and to all malleable iron castings.

Of all the ferrous scrap moving in the marketplace, iron foundries consume about 30%. This is about half the tonnage consumed by the steel industry. Because of their high use of scrap, iron foundries are a vital element in the country's efforts to conserve energy through recycling.

FIGURE 29: GENERALIZED FLOWSHEET OF AN IRON FOUNDRY

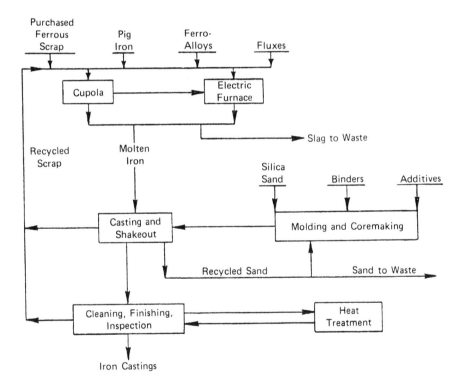

Source: Reference (4)

Another type of flow diagram is shown in Figure 30. The energy values for various unit operations are given in boxes in units that represent millions (10^6) of Btu. Feed materials, products and waste products are underlined with energy values assigned where appropriate. No fuels or electrical energy consumptions are outlined on the flowsheet.

FIGURE 30: PRODUCTION OF UNALLOYED GRAY IRON CASTINGS BY CUPOLA PROCESS

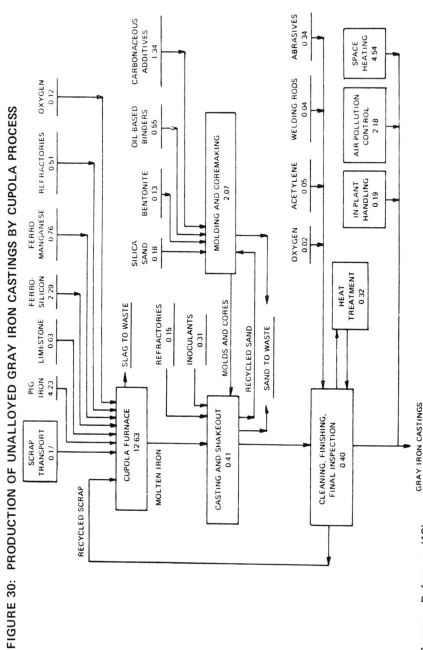

Source: Reference (12)

MAJOR ENERGY CONSERVATION OPTIONS TO 1980

Three major bases of information are available. They are: Bureau of Census data for 1971 from the 1972 Census of Manufactures, a Battelle study for the Bureau of Mines (12), and the foundry industry voluntary reporting system through the American Foundrymen's Society (AFS).

With respect to the Census data, difficulties are encountered, especially in the data for SIC 3321, which accounts for over 90% of the tonnage in this component. Energy-use data for SIC 3321 indicate the consumption of 100.1 x 10^{12} Btu in 1971. This is in essential agreement with Battelle estimates based on other data. However, Battelle was not able to sort out from the Census data the number of tons to which this applies.

They pointed out that the production allocated by Census to SIC 3321 for 1971 seems to range from about 13.5 to 15.5 x 10^6 tons, depending upon the source of data used. Using even the lowest of these tonnages, the average energy use in 1971 calculates as 7.4 x 10^6 Btu per ton, which is clearly low by a factor of 2 or more.

This suggests that the production numbers cited by Census include all production of gray and ductile iron castings, not just the noncaptive portion which in the 1972 Standard Industrial Classification Manual is defined to constitute SIC 332. Using only "shipments for sale" in 1971 (7.6 x 10^6 tons) as the basis for making the per-ton calculation, energy consumption in the gray and ductile iron foundries was 13.2 x 10^6 Btu per ton of castings. This is a more rational number, but it rests on an assumption.

With respect to the Census data for malleable iron foundries, if the above difficulty is present it exists in minor form. Production is indicated at about 0.88 x 10^6 tons in 1971 with an energy use of 17.4 x 10^{12} Btu. This calculates to an average of 19.77 x 10^6 Btu per ton of castings, which is consistent with other data, but which again seems to ignore the fact that some malleable iron production is captive to industries other than SIC 33.

The Bureau of Mines data (12), which are for cupola-melted gray iron only, when converted to the basis being used in this report, indicate an average energy consumption of 18.4 x 10^6 Btu per ton of shipments for cupola-melted gray iron. Adjustment of that value by backing out the coke used in the cupola and adding electricity for melting gives an equivalent value of 15.04 x 10^6 Btu per ton of shipments for electrically melted gray iron. On the basis of 75% melted in cupolas and 25% in electric furnaces, the adjusted average value for gray iron becomes 17.56 x 10^6 Btu per ton.

The foundry industry voluntary reporting system gives 1972 energy use at averages of 12.35 x 10^6 Btu per ton of shipments for gray and ductile iron, and 19.58 x 10^6 Btu per ton for malleable iron. The malleable iron value is consistent with that indicated by the Census data. For malleable iron, the sample of foundries in the voluntary system represents 24 reporting out of a total of 65, and the 24 reporting foundries account for about 76% of total malleable iron shipments. The average output of reporting malleable iron foundries is about 30,000 tons of shipments per year, while the average shipments per foundry for all malleable iron foundries is about 15,000 tons per year. Thus the sample of malleable iron foundries in the voluntary system is reasonably representative of all malleable iron foundries, but is biased toward the larger foundries (which tend to have lower energy use per ton of shipments). The effect of plant size on use of energy is discussed in more detail in the Battelle report to FEA (4).

With respect to gray and ductile iron foundries (which make up about 93% of the shipments in this component), the sample in the voluntary reporting system is heavily skewed toward large foundries. For all gray and ductile iron foundries, about 1,350 accounted for shipments of about 11.85 x 10^6 tons of shipments (excluding ingot molds), for an industry average of about 8,800 tons of shipments per foundry per year.

The 117 foundries in the voluntary reporting system (9% of the number of foundries in the industry) accounted for 3.88 x 10^6 tons of shipments (33% of industry shipments), for an average of 33,000 tons per year for the reporting foundries. Data in the voluntary reporting system show that such bias will understate the actual use of energy by a substantial amount.

Although the foregoing inconsistencies are substantial, information about the use of energy by the component and knowledge of the component permit a rational selection of the probable true use of energy by this component in 1972. The selected value is 16.0 x 10^6 Btu per ton of shipments. The probable error in this base is thought to be less than 6%.

Based on average use of 16.0 x 10^6 Btu per ton of shipments and shipments of 8.77 x 10^6 tons for the component in 1972, total use of energy by the component in 1972 was about 140 x 10^{12} Btu. The distribution of types of energy used by the component in 1972 was about as follows:

| | - - - - - - - - - Percentage of Total Btu- - - - - - - - - | | |
	Census of Manufactures[1]	Bureau of Mines[2]	Component[3]
Coke and coal[4]	36	33	30
Natural gas	33	51	51
Oil and other liquids	15	12	15
Electricity	15	12	15
Other and not specified	13	–	–
Total	100	100	100

[1] For gray and ductile iron castings (SIC 3321).
[2] For gray iron, 75% cupola-melted and 25% electric-melted.
[3] For component, see tabulation below.
[4] Only the Census of Manufactures includes coal (at 11% of the total use of coal and coke).

The above distributions are consistent (if "other" fuels are taken to be natural gas). The AFS industry voluntary survey includes information on this point in the data bank, but to date the system has not been programmed to sort and analyze the data.

The distribution of energy use according to operations for the component was about as shown in Table 33.

TABLE 33: DISTRIBUTION OF ENERGY USE ACCORDING TO OPERATIONS

	10^6 Btu Per Ton Shipped(a)				
	Coke and Coal	Electricity	NG, Oil, Etc.	Total	%
Melting(b)	5.46	1.67	1.60	8.73	48
Molding and Coremaking	--	0.05	1.90	1.95	11
Casting, Shakeout, Cleaning, Handling	--	0.11	0.75	0.86	5
Heat treatment(c)	--	0.30	0.70	1.00	6
Environmental Control	--	0.55	0.50	1.05	6
Makeup Air	--	0.03	4.50	4.53	24
Total	5.46	2.71	9.95	18.12(d)	100
%	30	15	55	100	

(a)This is a computational exercise to synthesize from the best available information the probable distribution of use of energy in this component. It confirms some other data and extends the distribution to operations, a distribution not available from other data. It is based on 82% gray iron, 11% ductile iron, and 7% malleable iron in the component.

(b)70% cupola, 30% electric.

(c)For 10% of the gray iron, 50% of the ductile iron, and all of the malleable iron.

(d)Base for computation of percentages in this exercise only.

Source: Reference (4)

Through the programs of the American Foundrymen's Society (AFS) and the
Iron Castings Society (ICS), iron foundries are being educated on opportunities
for conservation. These programs include special seminars, technical papers,
energy committees, and dissemination of information in other forms.

The reporting system now being administered by AFS has important weaknesses
that are well-known to AFS. These weaknesses do not permit accurate tracking
of conservation attained to date. Since 1972 some of the foundries that have
had intensive energy conservation programs have attained savings in the region
of 15 to 25% per ton of castings, but this is considerably higher than the average for the component.

The following calculations were used to derive the energy efficiency improvement
potential values shown for iron foundries in Table 34. These details should be
considered with awareness that they apply to foundries in SIC 33, and that the
component excludes identifiable foundries allocated to SIC 34, 35, and 37.
These exclusions are mainly large captive foundries.

TABLE 34: ENERGY EFFICIENCY IMPROVEMENT POTENTIAL BY 1980; IRON FOUNDRIES COMPONENT

Technique	Potential Saving, 10^6 Btu/ton of Castings Shipped	
	Technologically Feasible	Technologically Feasible and Economically Practicable
Housekeeping	1.6	1.6
Increase electric melting by additional 25% of total melting	0.8	0.4
Install recuperative hot blast on additional 15% of cupola capacity	0.6	0.3
Install divided blast on additional 40% of cupola capacity	0.4	0.4
Install electronic energy-management systems in foundries making 20% of annual production	0.2	0.2
Convert additional 50% of hot methods for molding and coremaking to cold methods	1.0	0.5
Use insulating sleeves on 30% of risers	0.3	0.1
Preheat charges with waste heat for 40% of electric furnaces	0.2	0.1
Increase yield by 2% by methods in addition to above	0.6	0.6
Total	5.7	4.2

Source: Reference (4)

Housekeeping: Technologically feasible potential was estimated as follows. Two major iron foundries have reported energy conservation improvements in the range of 18 to 23% (per ton of shipments) over the period of 1972 through 1975. Examination of their situations, including on-site visitations and personal discussions, suggest that about half of this is represented by housekeeping improvements.

These are progressive foundries with better than average technical staffs and facilities. It was assumed that the average foundry in this component could conserve 10% of total energy by housekeeping measures over the target period. In the body of the Battelle report to FEA (4) it was estimated that the average use of energy in 1972 by foundries in this component was 16.0×10^6 Btu per ton of shipments. The estimated technologically feasible potential is

$$0.10 \times (16.0 \times 10^6) = 1.6 \times 10^6 \text{ Btu per ton}$$

By definition, housekeeping activities involve little or no investment of capital. Therefore, the entire technologically feasible potential is judged to be economically practicable (1.6×10^6 Btu per net ton of castings shipped).

Increase Electric Melting: Using the conversion factors for this study for Btu equivalent of electricity, electric melting of iron requires less energy than cupola melting. Conversions from cupola melting to electric melting are in progress. It is assumed that over the target period it is technologically feasible for an additional 25% of the total melting capacity to be converted from cupola melting to electric melting. It is thought that the capacity to build new electric furnaces will support such a conversion.

In the Battelle report to FEA (4) it was established that the energy consumption values used for melting iron are 18.4×10^6 Btu per ton of shipments for cupola melting and 15.0×10^6 Btu per ton of shipments for electric melting. The estimated technologically feasible potential is

$$0.25 \times (18.4 - 15.0) \times 10^6 \text{ Btu per ton} = 0.8 \times 10^6 \text{ Btu per ton of shipments}$$

Iron foundries have been comparing the costs of cupola to electric melting conversions for a number of years. The comparisons are greatly affected by what any particular foundry estimates that it must do to existing cupola facilities to bring them into compliance with air-quality regulations.

The economics of conversions are also strongly affected by the tonnage which the foundry plans to melt and the management's view of the preferred scheduling and operating cycle for that particular foundry. Because no specific data are available on the average cost of conversions from cupola melting to electric melting, and because the cost differs widely in different situations, it is estimated subjectively that of the potential which is technologically feasible, 50% will be economically practicable. The estimated economically practicable potential is

$$0.50 \times (0.8 \times 10^6) = 0.4 \times 10^6 \text{ Btu per ton of shipments}$$

Install Recuperative Hot Blast: Heating of the air blown into cupolas for melting iron is used in some foundries today. About 75% of the iron melted in 1972 was melted in cupolas. The usual fuel for heating the air is natural gas.

The practice of recovering heat from the top gas and then using this recovered energy to heat the cupola blast is fairly common in Europe, but rare in the United States. It has been reported that about 5.0×10^6 Btu per ton of iron castings shipped can be recovered from cupola exhaust gases to heat the blast air to about 1000°F.

U.S. foundries have resisted extensive adoption of recuperative hot blast partially on technological grounds. The recuperative recycling system is less flexible in operation than the independently fired systems. Blast heat cannot be obtained until the cupola has been in operation for a period. Control necessities frequently dictate that the recuperative system be backed up with a system based on some other fuel.

Historically, recuperative systems have had problems with dirtying of heat-exchanger surfaces. The technique is most adaptable to long melting campaigns, and many cupolas in this component are operated only a few hours per day. It is estimated subjectively that about 15% of the iron melted in cupolas in this component in 1972 could on reasonable technological grounds be melted with recuperative hot blast by the target period. The estimated technologically feasible potential is

$$0.15 \times 0.75 \times (5.0 \times 10^6) \text{ Btu per ton of castings} = 0.6 \times 10^6$$
$$\text{Btu per ton of castings shipped}$$

An important deterrent to wider adoption of recuperative hot blast in the U.S. is the capital cost and the operating cost of the installations relative to the price of natural gas. From limited information obtained orally from industry sources, it appears that opinions differ considerably whether the capital cost is above or below a level of $10 per million Btu saved annually. Therefore, lacking specific data, it is assumed that 50% of the technologically feasible potential will also be economically practicable. The estimated economically practicable potential is

$$0.50 \times (0.6 \times 10^6) \text{ Btu per ton} = 0.3 \times 10^6 \text{ Btu per ton of castings shipped}$$

Install Divided Blast: Divided blast for blowing air into cupolas for combustion of coke refers to a technique that uses two levels of tuyeres, each with separate control on the amount of air that it admits to the cupola. The technique is not yet widely used in the U.S. At a recent meeting of the American Foundrymen's Society, three iron foundries reported the achievement of a 20% saving in the amount of coke used to melt iron with this technique.

It appears to be technologically feasible to convert about 40% of cupola melting of iron to this technique by the target period. The coke required for cupola melting without divided blast is equivalent to about 5.5×10^6 Btu per ton of iron castings shipped. The estimated technologically feasible potential is

$$0.20 \times 0.40 \times (5.5 \times 10^6 \text{ Btu}) = 0.4 \times 10^6 \text{ Btu per ton of shipments}$$

Reported costs to convert cupolas to divided blast have ranged from $500 to $9,000. Even at $10,000 to convert a small cupola melting 10 tons per hour for 4 hours per day for 200 days per year, the saving in the form of coke amounts to 15,000 million to 20,000 million Btu per year. The investment cost is clearly less than $1 per million Btu saved annually. Thus, all of the technologically feasible potential (0.4×10^6 Btu per ton of castings shipped) is taken to be economically practicable.

Install Electronic Energy Management Systems: Electronic management systems which sense temperatures, operating times, and other factors affecting energy consumption are in use in a number of commercial and office buildings, and increasingly are being used in industrial situations. They are a tool that permits actions to be taken to conserve energy by avoiding waste.

No data are available to establish an average saving potential for such systems for foundries, but a few actual installations suggest that potential savings amount to about 5% of total use of energy. It is assumed that iron foundries that account for about 20% of the total production of castings could find it technologically feasible to install such systems. These are subjective estimates based mainly on oral discussions with foundry personnel familiar with such systems. The estimated technologically feasible potential is

$$0.05 \times 0.20 \times (16.0 \times 10^6) \text{Btu/ton} = 0.2 \times 10^6 \text{ Btu/ton of shipments}$$

Assume that a foundry produces 5,000 tons of castings shipped per month and saves about 5% of its total energy because of the installation of an electronic system that costs $250,000. The annual saving in energy will be about 5,000 $\times$ 12 $\times$ (0.2 $\times 10^6$) = 12,000 million Btu. The investment cost will be about $2 per million Btu saved annually. On this basis, the economically practicable potential is taken to be the same as the technologically feasible potential (0.2×10^6 Btu per ton shipped).

Convert From Hot Methods to Cold Methods for Molds and Cores: Traditional methods for making certain types of molds and cores involve the heating or baking of the molds and cores to achieve bonding between the grains of sand. During the past decade there have been accelerating development and use of binders that harden and set by chemical means which do not require the application of heat.

From familiarity with the application and use of the cold processes relative to the hot processes, it is judged subjectively that technological considerations will allow the conversion of about 50% of such use of hot processes during the target period. The amount of energy used for heating in the manufacture of molds and cores has been estimated at 1.90×10^6 Btu per ton of castings shipped. The estimated technologically feasible potential is

$$0.50 \times (1.90 \times 10^6 \text{ Btu}) = 1.0 \times 10^6 \text{ Btu/ton of shipments}$$

The cost situation with respect to conversions from hot to cold processes is a highly individual matter in each foundry, depending strongly on the type of castings being made and existing facilities in place. Capital cost for the conversion is low, so that decisions as to economic practicability ultimately are made on other considerations of cost, quality, and foundry scheduling. Lacking specific data on an average condition for all iron foundries, it is assumed that about half of the technologically feasible potential will be economically practicable. The estimated economically practicable potential is

$$0.50 \times (1.0 \times 10^6 \text{ Btu}) = 0.5 \times 10^6 \text{ Btu/ton of shipments}$$

Insulating Riser Sleeves: Risers are reservoirs of molten iron attached to castings and necessary for the production of most iron castings. Risers are removed from the castings and recycled to the melting operation. By using insulating sleeves, the risers can be smaller, and thus the risers will constitute less of a recycling load on the energy-intensive melting step.

Risers now account for an estimated 25% of the weight of iron melted to produce iron castings. Based on technical reports in the industry, it is estimated that with the use of insulating sleeves the amount of recycled risers can be lowered to 15%, a lowering of 10% in the amount of metal that needs to be melted. It is assumed that about 30% of the risers can be converted on technological grounds during the target period. The amount of energy used on the average to melt is 8.7×10^6 Btu/ton of castings shipped. The estimated technologically feasible potential is

$$0.10 \times 0.30 \times (8.7 \times 10^6 \text{ Btu}) = 0.3 \times 10^6 \text{ Btu/ton of shipments}$$

Capital costs for the conversion of risers to insulating sleeves are low, but the sleeves are expensive and require extra labor for their stocking and installation in the molds. Cost is a major deterrent to their widespread adoption. Lacking specific data, it is assumed that 30% of the technologically feasible applications will also be economically practicable. The estimated economically practicable potential is

$$0.30 \times (0.3 \times 10^6 \text{ Btu}) = 0.1 \times 10^6 \text{ Btu/ton of shipments}$$

Preheat Scrap for Electric Furnaces: Preheating of ferrous scrap before melting in electric furnaces will save electricity during the melting operation. Preheating is used somewhat in iron foundries, but the usual fuel is oil or natural gas. A significant saving in overall use of energy in the foundry occurs only when otherwise waste heat is used for the preheating of the scrap.

Studies have shown that scrap can be heated, for example, to 1000°F by the exhaust gases from steelmaking furnaces. There are other potentially usable sources of waste heat in many foundries. For this study, it is assumed that scrap can be preheated to 800°F with otherwise waste heat to lower energy consumption in the electric melting furnaces by 20%.

It is further assumed that technologically about 80% of iron foundries in this component will have sources of waste heat to accomplish this preheating. Taking the energy to melt at 1,300 kwh per ton of castings shipped, and electric melting at 30% of all melting, the base of electricity for melting in iron foundries is 1.3×10^6 Btu per ton of castings shipped. The estimated technologically feasible potential is

$$0.20 \times 0.80 \times (1.3 \times 10^6 \text{ Btu}) = 0.2 \times 10^6 \text{ Btu/ton of shipments}$$

The systems for preheating foundry scrap with waste heat are not widely developed and require a substantial amount of equipment. Specific data on capital and operating costs are lacking. It is assumed that half of the technologically feasible installations will also be economically practicable.

The estimated economically practicable potential is

$$0.5 \times (0.2 \times 10^6 \text{ Btu}) = 0.1 \times 10^6 \text{ Btu/ton of shipments}$$

Increase the Yield: The yield of shippable castings in iron foundries averages about 50% of the amount of metal melted. This means that about half of the metal melted recycles back through the melting operation. If the yield is increased, total energy to produce shippable castings declines. Through improved process control and improved design of gating and risering systems, it is judged technologically feasible to increase yield by about 2% of metal melted.

This is a subjective judgement for which supporting data are not available. In the Battelle study (4), the total energy required for the component has been taken at 16.0×10^6 Btu per ton of castings shipped. The estimated technologically feasible potential is

$$0.02/0.50 \times (16.0 \times 10^6 \text{ Btu}) = 0.6 \times 10^6 \text{ Btu/ton of shipments}$$

Because yield would be increased mainly by activities which do not require major capital investments, all of the technologically feasible potential is judged to be economically practicable (0.6×10^6 Btu per ton of shipments).

GOAL YEAR (1980) ENERGY USE TARGET

Based solely on technological feasibility, it is estimated that this component could conserve an average of 5.7×10^6 Btu per ton of castings shipped by 1980. Table 34 lists the various techniques by which this could be accomplished.

The noncaptive iron foundries that make up this component include many that are small and have low profitability. Availability of capital is a major barrier, even when specific investments are calculated to be individually profitable. Because of investments needed to comply with governmental orders (mainly EPA and OSHA), the tendency in the industry is to put available capital into other than energy-conserving investments.

Many investments that are being made are energy-related in that they involve expenditures to convert heating applications to more available sources of energy, but (except for conversions to electricity) often these do not lower Btu requirements. Thus, by computation, these investments are "unprofitable." In spite of this, some are being made. In the case of electricity, so many conversions are being made to electricity that considerable concern is being expressed whether the utilities will in fact be able to supply the increased electrical requirements. In some geographic areas, availability of additional electricity already has limited conversions.

As shown in Table 34, energy savings of 4.2×10^6 Btu/ton of castings shipped are believed to be economically practicable for iron foundries. Of this total, about 38% is accounted for by housekeeping improvements. Applying all of these to the component and projecting them to 1980 gives a total estimated gross potential of 125×10^{12} Btu.

It is estimated that on the average establishments in this component used about 1×10^6 Btu/ton of castings for environmental control in 1972. Confirming data are sparse. The establishments in this component have been singled out for special attention in the National Emphasis Program (NEP) of OSHA. Although primarily a safety program, it can be expected to require more energy use by iron foundries.

Furthermore, additional use of energy will be required to meet government-imposed improvements in environmental control (mostly for air control). Because of these factors, and lacking specific available data, it is estimated that the factors will increase average use of energy by iron foundries by 1×10^6 Btu/ton of castings shipped.

Applying this estimated requirement for environmental control and projecting this to 1980 gives an estimated net potential of 136×10^{12} Btu for this component.

COPPER INDUSTRY

The copper component as considered here consists of three major subcomponents: (a) smelting operations which include the preparation of blister and anode
copper from ore concentrates, (b) electrolytic refining of anode copper, and
(c) the melting and fabrication of copper and brass alloy ingots into such shapes
as plate, sheet, extrusions and bar. The first two subcomponents correspond to
SIC 3331, while the third covers establishments corresponding to SIC 3351.

PROCESS TECHNOLOGY INVOLVED

With few exceptions, domestic copper is mined from very lean sulfide ores which
now contain an average of less than 0.7% copper. In 1972, all 15 primary copper smelting facilities produced copper by either of the two pyrometallurgical
processes. These are (a) roasting, smelting and converting, or (b) smelting and
converting. In 7 of the 15 smelters, concentrate (containing about 25% copper)
is first roasted to convert some of the sulfides to oxides and to eliminate some
sulfur. Most of the energy is derived from the burning of sulfur in the concentrate. The roasted product (calcine) is then charged into the reverberatory furnace with suitable fluxing agents. In 8 other smelters, green feed (unroasted
concentrate) is charged directly into the reverberatory or (in one case) electric
furnaces, also with suitable fluxes. Further detail on processing is given in
Figure 31.

Typically, waste heat from the reverberatory flue gas is recovered in recuperators
and waste-heat boilers. Up to 50% of the fuel energy consumed by the furnace
is recovered and used to generate steam which is either used in the power plant
or consumed in the operation of other plant facilities.

The useful product from the reverberatory furnace is matte, a mixture of copper
and iron sulfides containing from 30 to 60% copper. This is transferred to the
converters (usually of Pierce-Smith design) which are cylindrical and where air
is blown through the matte to oxidize the iron and sulfur from the matte, leaving an impure blister copper containing between 98 and 99% copper. Most of

115

the heat in the converter operation is provided by the oxidation of the iron and sulfur.

FIGURE 31: PRIMARY COPPER SMELTING PROCESSES AND EQUIPMENT

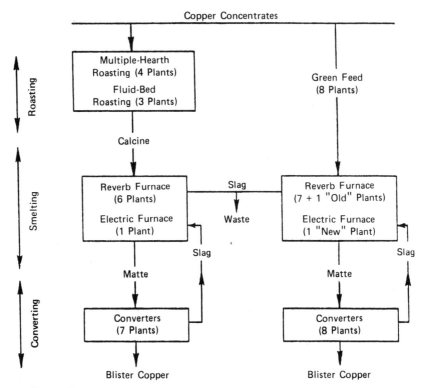

Note: Old and new plants are parallel operations at same site.

Source: Reference (4)

The gases from the converters contain about 4% sulfur dioxide and are passed through various types of duct systems to cool the gas and separate out the dust. In 1972, about half of all of these high-strength sulfur dioxide-bearing gases were also treated for sulfur dioxide removal in acid plants. (The sulfur dioxide content of the reverberatory furnace's flue gas is too low to be suitable for acid manufacture.)

After converting, the blister copper is fire refined to deoxidize and purify it for use as fire-refined copper or for casting into anodes for electrolytic refining. In the fire refining, the molten metal is first treated to complete the oxidation of the residual iron and other minor impurities, then it is reduced by using natural gas to lower the oxygen content to about 0.17%.

The anodes are shipped to the refineries where they are electrolytically refined to cathode copper which is about 99.8+% copper. These are either sold or are cast into refinery shapes, e.g., wire bar and cakes or continuously cast wire rod. Silver, gold, platinum and selenium are recovered as by-products of the electrolysis from the slimes that drop to the bottom of the tanks.

In the manufacture of rod and wire, cast wire bars are purchased or copper is melted and hot-rolled (1300° to 1600°F) to coiled rod. Alternatively, some integrated continuous casting and hot rolling are also being used in some plants. After cleaning, the hot-rolled rod is processed to wire by cold drawing and process annealing.

Brass mill operations include melting and alloying, hot rolling, extrusion, cold rolling and drawing. Melting operations may use electrical energy, but gas-fired reverberatory furnaces are commonly used. Rolling or extruding of some alloys requires heating to 1900°F or even higher. In cold rolling or drawing, softening anneals are also required.

MAJOR ENERGY CONSERVATION OPTIONS TO 1980

Data on energy consumption in various copper ore processing steps appear in reference (13). A flow sheet for copper production is shown in Figure 32, indicating energy values in boxes in units that represent millions (10^6) Btu per net ton of primary product. Feed materials, products and by-products are underlined with energy values assigned where appropriate.

Another flow diagram for copper production showing energy requirements has been published by the U.S. Bureau of Mines (21) and is shown in Figure 33. The data in Figure 33 (21) are in Btu/lb and have to be multiplied by 0.002 to give numbers in 10^6 Btu/ton as are those in Figure 32 (12).

Numerous investigators have reported detailed analyses of energy consumption in the production of copper. A comparison was made of the results obtained from six of these studies including Battelle (12), Bravard (39), Chapman (40), Gordian (11), Kellogg (43) and Snell (44) with values derived from data submitted by the American Mining Congress (AMC) to Battelle in 1976. In order to permit comparison, the data from all sources were reworked slightly to a constant baseline. The energy values reported by the six independent investigators to produce 1 ton of copper ranged from a low of 15.3 x 10^6 Btu (Chapman) to a high of 34.6 x 10^6 Btu (Kellogg).

The 39.3 x 10^6 Btu per ton value derived from the AMC data was significantly higher, because evidently it is the only value of the seven which includes the energy required to operate environmental control devices. This and the fact that these data were specifically derived for the base year of 1972 is why they are preferred as the most representative for the primary copper component. Details on which this comparison was based are given in Table 35.

If one accepts a rounded average of 32.3 x 10^6 Btu per ton (from the Battelle and Snell studies) as representative of the energy required to produce the primary metal only, then the 7.0 x 10^6 Btu per ton difference between this and the AMC value gives an approximation of the energy requirements for the 1972 environmental controls. This amounts to 16% of the total 1972 primary metal energy requirements, a value which is consistent with an estimated 17% value cited by one of the major primary copper producers.

FIGURE 32:   PRODUCTION OF REFINED COPPER BY CONVENTIONAL MINING AND SMELTING PROCESSES

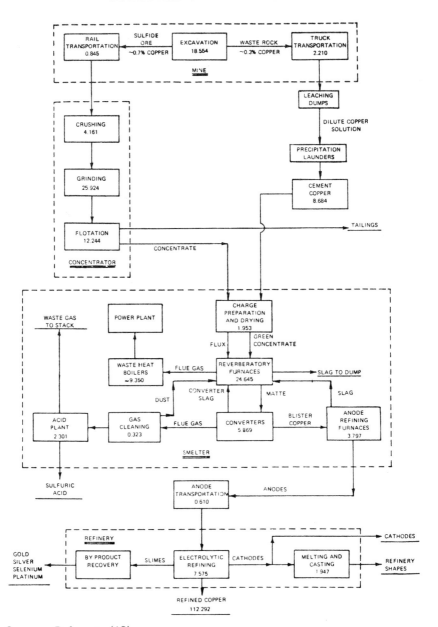

Source: Reference (12)

FIGURE 33: ENERGY CONSUMED AT EACH STAGE OF COPPER PRODUCTION—Btu/lb Copper

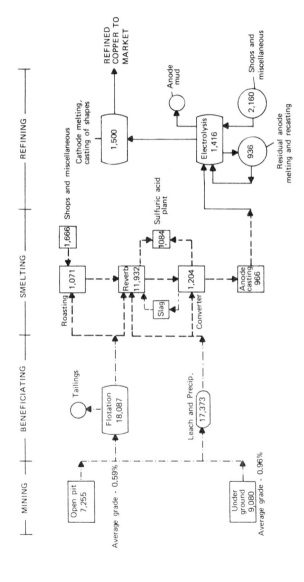

Source: Reference (21)

TABLE 35: ESTIMATES OF ENERGY CONSUMPTION IN THE PRODUCTION OF PRIMARY COPPER

Production Process	American Mining Congress	Battelle	Bravard	Chapman	Gordian	Kellogg	Snell
			Energy Consumption, 10^6 Btu/Ton of Copper				
Smelting							
Reverberatory	-	14.9	12.1	12.1	20.3	-	-
Other	-	7.2	0.2	0.9	1.7	-	-
Subtotal	29.9	22.1	12.3	13.0	22.0	26.2	21.8
Refining							
Furnaces	-	5.6	3.5	1.4	3.5	-	-
Electrolysis	-	5.4	0.7	0.9	0.7	-	-
Subtotal	9.4	11.0	4.2	2.3	4.2	8.4	10.6
Total	39.3	33.1	16.5	15.3	26.2	34.6	32.4

Notes: Electrical energy converted at 3,412 Btu/kwh excepting for Kellogg whose value includes some conversion at 10,580 Btu/kwh.

Bravard and Kellogg data adjusted as discussed in Battelle reference.

All energy data reported in 10^6 Btu/ton.

Values from Battelle, Bravard, Chapman, Gordian and Kellogg do not include energy required to operate environmental control devices or processes.

Source: Reference (4)

In the base year 1972, the AMC reported the smelter and refined copper production as 1.63×10^6 and 1.89×10^6 tons, respectively, which is in good agreement with the corresponding values of 1.65×10^6 and 1.87×10^6 tons, respectively, given by the Bureau of Mines. The total energy required to produce these amounts of smelted and refined copper was calculated as 66.6×10^{12} Btu from the AMC data after applying the standard conversion factors to their data. This is in excellent agreement with the value of 67.9×10^{12} Btu derived for SIC 3331 (primary copper) in the 1972 Census of Manufactures data for 1971 (MC72 SR-6). However, adjusting the Census data for the higher production of 1972 relative to 1971 increases this value to 76×10^{12} Btu.

The Copper and Brass Fabricators Council, Inc. (CBFCI), whose membership accounts for 70% of the total current annual shipments of brass mill products, has reported that a total of 16.4×10^{12} Btu (after applying the standard conversion factors), were required in 1972 to yield $3,011 \times 10^6$ pounds of material produced. Expanding these values to represent the total material produced in 1972 ($3,233 \times 10^6$ pounds) for SIC 3351 gives the total energy requirement of 17.6×10^{12} for this part of the copper component. This is appreciably less than the value of 31.4×10^{12} Btu which was derived from the 1972 Census of Manufactures report for SIC 3351 for 1971 after adjusting for the higher production year of 1972.

Table 36 shows the distribution of types of energy used for the various processes comprising the copper component, using both the trade association and Census of Manufactures data. On the average, the smelting operations lean heavily on natural gas (76.2% of total energy) and electric power is used much less extensively (less than 15% of the totals) in smelting and refining than in fabrication where it accounts for about 23.5% of the total.

TABLE 36: 1972 ENERGY USE DISTRIBUTION FOR COPPER COMPONENT

Process	. . . . .Btu, percent.			
	Electricity	Natural Gas	Other Fuels	Total
Trade association data:				
Smelting	2.2	76.2	21.6	100.0
Refining	12.4	40.0	47.6	100.0
Fabrication	23.5	46.1	30.4	100.0
Total for component	8.8	62.2	29.0	100.0
Census of Manufactures data:				
Smelting and refining	6.8	66.3	26.9	100.0
Fabrication	23.3	53.4	23.3	100.0

Source: Reference (4)

In actuality, the energy-consumption pattern in the real-life smelting operation is not nearly so simple as the line items in Tables 35 and 36 would indicate. As noted earlier, in 1972, there were 15 primary smelters operating in the U.S. with widely differing capacities and using more than six different combinations of roasting, smelting and converting.

The energy requirements for three of the older and more commonly used smelters were compared with those of newer flash-smelting plants. Among these, the Btu required to process a ton of charged concentrate range from a low of 8.57×10^6 for flash smelting to a high of 10.87×10^6 for the older, wet-charge reverberatory process. Details are given in Table 37.

TABLE 37: ENERGY REQUIREMENTS FOR ALTERNATIVE PRIMARY COPPER RECOVERY PROCESSES*

	Calcine-Charged Reverb	Green-Charged Reverb	Flash Smelting	Electric Furnace
Roasting	0.52	–	–	–
Drying	–	–	0.52	0.52
Smelting	3.20	5.70	1.80	5.52
Waste heat (credit)	(0.96)	(1.45)	(0.81)	(0.29)
Converting	1.40	1.62	1.33	1.62
Gas cleaning and SO_2 recovery	1.86	1.62	2.35	2.32
Leaching	–	–	–	–
Anode casting	0.78	0.78	0.78	0.78
Electrorefining	2.60	2.60	2.60	2.60
Electrowinning	–	–	–	–
Electrolyte purge	–	–	–	–
Total	9.40	10.87	8.57	13.07

*Expressed as 10^6 Btu/ton of concentrate and cement copper. Electric power converted using 10,500 Btu/kwh. See text for other qualifications.

Source: Reference (4)

Developments in copper smelting during the last two decades do not indicate a high concern for energy savings. Concern in the component has focused on the need for the most effective and least expensive air-pollution control systems and for overall cost reductions through process simplifications. As a result, two general trends have emerged.

In existing smelters, roasting operations have been minimized or phased out in favor of wet charging of reverberatory furnaces. This has lowered operating and maintenance costs at the expense of higher fuel consumption. In other cases, flash smelting furnaces are being installed, again eliminating roasters, but also replacing the reverberatory furnace with a more efficient unit. Predictably, smelters of this design presently show the lowest energy consumption in existing copper smelters.

In recent years, much emphasis has been placed on development of a continuous smelting process to conserve energy and to minimize air pollution. The new facility in Utah is the only example of this type in the United States.

Most of the major companies comprising this component are well aware of the pressing need for energy conservation and have implemented energy-saving programs.

Virtually the entire smelting and refining subcomponents of the primary copper industry have representation on the Subcommittee on Energy Conservation of the American Mining Congress, which submitted its first voluntary report on energy consumption on April 5, 1976. In addition to smelting and refining, this report also includes energy-consumption data for mining and concentrating of copper ores, which are outside the consideration of this present SIC 33 study. The principal conclusion of the report, with the reporting system devised, was that for 1975 the total energy consumption for this primary copper industry had increased by 21.1% over the base year 1972. The following factors were listed as contributing to this increased consumption of energy: (a) decreasing quality (i.e., lower copper content) in the ore, (b) lower production in 1975, and (c) regulatory changes concerning both the environment and safety.

The Copper and Brass Fabricators Council, Inc., representing 27 domestic brass mill companies accounting for about 85% of the total U.S. production, has also recently agreed to participate in the DOC and FEA voluntary industrial energy conservation program, and hopes to achieve an aggregate industry-wide energy savings of 8% per unit of production by the year 1980.

Initially, the data given in Table 38 were prepared on the basis of raw data from three sources in an effort to compare the 1972 energy efficiencies of the 15 primary copper smelters. The estimated 1972 production breakdown was derived by prorating the 1972 total figure of 1.63×10^6 tons among these smelters according to the capacities reported for them. An initial estimate of the energy requirements of each smelter was then assumed (Columns 2 and 3 in Table 38) on the basis of the different smelter characteristics using the energy data of Table 39, which were normalized from the study of Kellogg and Henderson. Column 4 was then derived by multiplication of the values in Columns 1 and 3.

The selected value of the 1972 total energy consumed by the smelting subcomponent, exclusive of the energy required for environmental control, is 37.4×10^{12} Btu. This was determined as follows from the American Mining Congress (AMC) data submitted in the first voluntary energy conservation progress report:

$$\begin{array}{l} 48,810 \times 10^9 \text{ total Btu for smelting} \\ \underline{11,410 \times 10^9} \text{ Btu for environmental controls} \\ 37,400 \times 10^9 \end{array}$$

The $11,410 \times 10^9$ value was obtained by first substracting the average of the Battelle and Snell values (32.3×10^6 Btu per ton) from the AMC value of 39.3×10^6 (see Table 35), then multiplying the difference by the 1972 smelter production of 1.63×10^6 tons. This selected value of $37,400 \times 10^9$ Btu was then prorated among the 15 smelters using the data from Column 4 in Table 38 as the basis. Column 6 was then derived by dividing the values of Column 5 by those of Column 1.

Estimates of the relative energy required by the various smelters in 1972 were computed similarly. Thus, the literature values in Column 8 of Table 38 were first multiplied by the 1972 tons from Column 1 to obtain the Column 9 totals. These were then used to prorate the total of $1,369 \times 10^9$ Btu against the actual $11,410 \times 10^9$ Btu required for the 15 smelters (see above). Thus, the values of Column 9 were multiplied by 11,410/1,369 or 8.3, and the Btu total so obtained for each smelter was divided by its 1972 production to give the values shown in Column 10.

TABLE 38:   ESTIMATED AND CALCULATED VALUES OF ENERGY REQUIRED IN THE 1972 SMELTING OF COPPER

Company/ Location	(1) Estimated 1972 Production* 10³ tons	(2) Process**	(3) Initial Estimate 10⁶ Btu/ ton Cu**	(4) Total 10⁹ Btu	(5) Adjusted Estimate Total 10⁹ Btu	(6) Adjusted Estimate 10⁶ Btu/ ton Cu	(7) Type Equipment***	(8) Environmental Controls First Estimate 10⁶ Btu/ ton†	(9) 10⁹ Btu	(10) Adjusted Estimate 10⁶ Btu/ton Cu
Asarco, El Paso	87.7	B	15.321††	1,344	1,828	20.84	G	0.330	28.9	2.7
Asarco, Hayden	157.9	B	15.321††	2,419	3,290	20.84	G,A	1.700**	268.4	14.2
Asarco, Tacoma	87.7	B	15.321††	1,344	1,828	26.84	G,A	1.700**	149	14.2
Phelps Dodge, Douglas	128.1	B	15.321††	1,963	2,670	20.84	-	-	0	0
Phelps Dodge, Morenci	159.7	B	14.997	2,395	3,257	20.39	G,A	1.208	193.2	10.1
Phelps Dodge, Ajo	13.2	A	19.512	258	351	26.59	-	-	0	0
Kennecott, Hayden	70.2	B	14.997	1,053	1,432	20.40	G,A	1.208	84.7	10.1
Kennecott, Hurley	80.9	A	19.512	1,579	2,147	26.54	G	0.580	47	4.9
Kennecott, McGill	37.7	A	19.512	736	1,000	26.52	G	0.580	21.9	4.9
Kennecott, Garfield	228	A	19.512	4,449	6,051	26.54	G,A	1.319	301	11.0
Cities Service, Copperhill	13.2	C	11.170	147	200	15.15	G,A	1.326	17	10.8
Anaconda, Anaconda	177.2	B	14.997	2,657	3,614	20.39	G	0.330	58	2.7
Magma, San Manuel	160.5	A	19.512	3,132	4,260	26.54	G	0.580	92.8	4.8
White Pine, White Pine	124.5	A	19.512	2,429	3,303	26.53	G	0.580	72.2	4.8
Inspiration, Miami	103.5	B	15.321	1,586	2,157	20.84	G	0.330	34.5	2.8
Total	1,630			27,491	37,388				1,369	

*Values prorated from 1.63 x 10⁶ total using capacities listed in Reference (4).
**Process identified in Table 39. Energy cited excludes values for environmental controls.
***G designates gas cleaning equipment and A designates single-contact acid plants, in operation in 1972.
†Estimated energy cost of implementing environmental controls, derived from Reference (2) and shown in Table 39.
††Values include 0.324 x 10⁶ Btu/ton extra, to compensate for multiple hearth roasters.

Source: Reference (4)

TABLE 39: NORMALIZED ENERGY REQUIRED TO PRODUCE ONE TON OF ANODE COPPER

...... Energy Requirements, 10^6 Btu/ton* ....

Item	(A) Wet Charge Reverberatory	(B) Hot Calcine Charge Reverberatory	(C) Electric Smelting Dry Concentrate
Smelting	15.176	11.402	6.834
Converting	2.826	2.085	2.826
Anode production	1.346	1.346	1.346
Miscellaneous	0.164	0.164	0.164
Subtotal	19.512	14.997	11.170
Gas cleaning	0.580	0.330	0.274
Acid manufacture**	0.739	0.878	1.052
Subtotal	1.319	1.208	1.326
Grand total	20.831	16.205	12.496

*Electrical energy converted using 3,412 Btu/kwh.
**Single-contact process.

Source: Reference (4)

The following paragraphs outline some of the specific energy conservation options in copper production.

Conversion to Fluidized Bed Roasters: In 1972, the five smelters listed below were assumed to be the only ones which employed multiple hearth roasting. Kellogg and Henderson (45) have shown that fluidized bed roasting is less energy intensive, and furnished data indicating that a saving of 0.32 x 10^6 Btu per ton of copper is possible compared to multiple hearth roasting. On this assumption, the initial Btu per ton values for these five smelters (from Column 3 in Table 38) were reduced by 0.32 x 10^6 Btu to give the corrected initial estimates shown in Table 40.

TABLE 40: ENERGY ESTIMATES FOR FLUIDIZED BED ROASTING

Company Location	Corrected	Prorated	Original	Difference	Potential Energy Savings, 10^9 Btu
Ascarco,					
El Paso	15.0	20.40	20.84	0.44	38.6
Hayden	15.0	20.40	20.84	0.44	69.4
Tacoma	15.0	20.40	20.84	0.44	38.6
Phelps Dodge,					
Douglas	15.0	20.40	20.84	0.44	56.4
Anaconda,					
Anaconda	14.68	40.00	20.39	0.39	69.1
Total					272.1

....Energy Required, 10^6 Btu/Ton of Cu ....

Source: Reference (4)

These were then prorated using the ratio of 37,388/27,491 derived from Table 38. The difference between these and the original Btu per ton values represents the potential savings. These were converted to 10^9 Btu values, by multiplication with the 1972 production values, and added to give the total of 272.1×10^9 Btu for all five smelters. This value, in turn, was then multiplied by 8.3 to give an adjusted potential saving of 2.26×10^{12} Btu for this technique, assuming that all five smelters were able to completely convert to the fluidized bed roasters by 1980.

Technologically, because of the design and construction lead times involved between now and 1980, it is estimated that only about 10% of the 1971 multiple hearth roaster capacity can be converted to fluidized bed roasters. This leads to a technologically feasible potential savings of $0.1 \times 2.26 \times 10^{12} = 0.2 \times 10^{12}$ Btu. Dividing this by the 1972 total of 1.62×10^6 tons of fabricated products (derived from data provided by the Copper and Brass Fabricators Council) gives the technologically feasible potential saving of 0.12×10^6 Btu per ton.

Sufficient data of acceptable quality are not available to justify an estimate of the capital cost of installing fluidized bed roasters to replace multiple hearths on a component-wide basis. Therefore, the required judgment of economic practicability was made on the assumption that conversions would not save sufficient energy to justify such installations based solely on the saving in the cost of energy, and the estimation that other demands for capital within the copper plants would give such installations a low priority for implementation. On these assumptions and estimates, economic practicability during the target period was taken as zero.

Installation of Noranda Process Smelter: The Noranda Process smelter at Garfield, Utah is expected to require 10.4×10^6 Btu to process 1 ton of concentrates versus a requirement of 11.5×10^6 Btu for the reverberating-converter smelter located on this site. This represents a potential savings of 1.1×10^6 Btu per ton of concentrates processed. Because the new smelter is intended to completely replace the old one, an equivalent annual capacity of 1×10^6 tons of concentrate has been assumed. Multiplication of the $(1.1 \times 10^6) \times (1 \times 10^6)$ values gives a potential savings of 1.1×10^{12} Btu. Dividing this by 1.62×10^6 tons of 1972 fabricated products yields the technologically feasible potential of 0.68×10^6 Btu per ton. The technologically feasible potential is taken to be economically practicable because the installation is being implemented.

Installation of Outokumpu Flash Smelter: According to an industry spokesperson, the smelter at Hidalgo, New Mexico, based on the Outokumpu flash smelting process, is expected to require 6.3×10^6 Btu per ton of new metal-bearing materials. This value, which was derived using a factor of 10,500 Btu per kwh for electrical energy conversion, includes an estimated 2.1×10^6 Btu per ton for environmental controls, leaving a balance of 4.2×10^6 Btu to smelt 1 ton of new metal-bearing material. The estimated 1980 capacity of the smelter is 642,000 tons of new metal-bearing material, which is expected to yield 155,000 tons of copper. From these figures, the quantity of energy required annually would be $(4.2 \times 10^6) \times (642 \times 10^3) = 2.7 \times 10^{12}$ Btu. Dividing this by 155,000 tons of copper gives a preliminary calculated energy consumption of 17.42×10^6 Btu per ton of copper.

In an effort to adjust for the 10,500 Btu per kwh factor used in these calculations, it was assumed that about 10% (1.74 x 10^6 Btu per ton) might represent electrical energy, leaving a balance of 15.68 x 10^6 nonelectric Btu per ton. Multiplying the 1.74 x 10^6 by 3,412/10,500 gives the recalculated electrical energy of 0.57 x 10^6 Btu per ton, which, added to 15.68 x 10^6 value, gives an adjusted value of 16.25 x 10^6 Btu per ton of copper from the new smelter.

In 1972, the AMC reported a total of 48.8 x 10^{12} Btu were consumed in smelting 1.63 x 10^6 tons of copper for an average of 29.938 x 10^6 Btu per ton. In 1980, the new smelter will require (155 x 10^3) x (16.25 x 10^6) = 2.5 x 10^{12} Btu. This gives the following 1980 total values:

	Copper Produced, tons	Energy, Btu
1972	1.63 x 10^6	48.8 x 10^{12}
New capacity	0.16 x 10^6	2.5 x 10^{12}
Total	1.79 x 10^6	51.3 x 10^{12}

Dividing 51.3 x 10^{12} Btu by 1.79 x 10^6 tons gives the value of 28.659 x 10^6 Btu per ton at 1980 efficiency. Subtracting this from the 1972 value of 29.938 x 10^6 gives an efficiency improvement value of 1.279 x 10^6 Btu per ton. Multiplying this by the 1972 smelter production of 1.63 x 10^6 tons gives a potential savings of 2.1 x 10^{12} Btu. Dividing this by 1.62 x 10^6 tons gives 1.30 x 10^6 Btu per ton of fabricated products, which is both technologically feasible and economically practicable, because the change is being implemented commercially.

Substituting Fuel Oil for Gas: According to Kellogg and Henderson (45), the substitution of fuel oil for natural gas in the hypothetical operation of a green or wet-charge reverberatory furnace can decrease the fuel rate from 5.68 x 10^6 Btu per ton of concentrate to 4.81 x 10^6 Btu per ton. In the Kellogg and Henderson study, a value of 16.54 x 10^6 Btu was assigned to the fuel oil consumed in the production of 1 ton of copper.

In 1972, the use of fuel oil for this purpose was probably nil, because of the low price and widespread availability of natural gas. However, this technique is judged to be technically feasible, on the assumption that sufficient fuel oil will be available in 1980.

For 1972, it is assumed that five of the six smelters employing green feed to a reverberatory furnace (i.e., using a smelting process equivalent to "A" in Table 39) were technologically suitable for using this technique by 1980, if sufficient fuel oil is available. The 1972 capacity of these five smelters was taken from Table 38 as 416.8 x 10^3 tons. (The Kennecott-Garfield smelter, which also employed Process A in 1972, was excluded from these calculations because this will be replaced by the new Noranda smelter by 1980.)

From Kellogg and Henderson's data (45), the estimated requirement to produce 1 ton of copper using natural gas is: 5.68/4.81 x (16.54 x 10^6) = 19.52 x 10^6 Btu. Subtracting the 16.54 x 10^6 value from this gives 2.98 x 10^6 (3 x 10^6 Btu per ton) as the potential saving for effecting the fuel oil substitution. Multiplying 3 x 10^6 Btu per ton by the potentially affected 1980 capacity of 416.8 x 10^3 tons leads to the estimated component potential of 1.3 x 10^{12} Btu. Dividing this by 1.62 x 10^6 tons of 1972 fabricated products gives the technologically feasible potential of 0.80 x 10^6 Btu per ton of fabricated products.

By 1980, it is deemed economically practicable that those smelters which are presently using natural gas to process green concentrate will be able to use this technique to maintain production on a standby basis for an average of 1 month per year; e.g., when the supply of gas becomes seasonally critical. On this basis, the technique has the economically practicable potential of saving $\frac{1}{12}$ x (1.3×10^{12}) = 0.1 x 10^{12} Btu. Dividing by 1.62 x 10^6 tons gives an economically practicable potential of 0.06 x 10^6 Btu per ton of fabricated products.

Oxygen-Enrichment of Air in Converters: As discussed by Kellogg and Henderson (45), the use of oxygen-enriched air in converters is an effective way of decreasing the consumption of natural gas, providing that suitable measures are involved to counter the higher flame temperatures which are encountered, and which tend to shorten the furnace refractory life. According to an industry spokesman, techniques have now been developed which are expected to render this technique technologically feasible by all of the smelters by 1980, assuming sufficient oxygen is available. The energy cost of the oxygen required was ignored in these calculations because the oxygen will most likely be purchased from outside vendors.

To quantify this technique, Battelle data from a recent Bureau of Mines study were used. These indicated that 5.6 x 10^6 Btu of a total of 22 x 10^6 Btu required to smelt 1 ton of copper are consumed in the converter. Of the 5.6 x 10^6 Btu, 0.85 x 10^6 Btu were derived from natural gas, which, therefore, represents 3.8% of the total energy consumed in smelting 1 ton of copper. It is assumed that up to half of this amount of natural gas (1.9%) might be saved by using oxygen-enriched air. Technologically, it is more likely to expect that 90%, rather than 100%, of all converters might utilize this technique by 1980. On this basis, the total technologically feasible potential is calculated as follows: $(0.019)(29.9 \times 10^6)(0.9)(1.63 \times 10^6)$ = 0.8 x 10^{12} Btu, where 29.9 x 10^6 represents the total Btu per ton of copper consumed in smelting 1.63 x 10^6 tons of copper produced in 1972. Dividing 0.8 x 10^{12} Btu by 1.62 x 10^6 tons of fabricated products gives the technologically feasible potential value of 0.49 x 10^6 Btu per ton.

By 1980, it is estimated that it may be economically practicable for about one-third of the smelter capacity to effect the equipment and process modifications needed to use oxygen-enriched air in the converters. The potential savings of this technique are: $(0.019)(29.9 \times 10^6)(0.33)(1.63 \times 10^6)$ = 0.30 x 10^{12} Btu. On the basis of 1.62 x 10^6 tons of products, this is 0.19 x 10^6 Btu per ton of fabricated products.

Conversion of Reverberatory Furnaces for Cathode Smelting to ASARCO Shaft Furnace: As pointed out by Kellogg (46), the ASARCO shaft furnace has a considerably higher energy efficiency than the reverberatory furnaces which are used in the melting of cathode copper. Specifically, values computed from this reference (by applying the conversion factor of 3,412 Btu per kwh) give: 1,042 Btu per pound of copper melted with a reverberatory furnace equipped with a waste heat boiler and 678 Btu per pound of copper melted in an ASARCO shaft furnace plus a holding furnace. Thus, the potential saving of shifting from the reverberatory to the ASARCO shaft furnace system is 364 Btu per pound of copper melted.

According to information provided by a company spokesman, it is estimated that about 30% of the 1.89×10^6 tons of refined copper produced in 1972 was melted in the reverberatory-type furnaces. It is also estimated that, by 1980, it might be technologically feasible for all of these reverberatory furnaces to be replaced by the ASARCO shaft furnaces. On this basis, the technologically feasible potential savings are computed as follows: $364 \times (1.89 \times 10^6) \times (2 \times 10^3) \times 0.3 =$ 0.4×10^{12} Btu. Dividing this by 1.62×10^6 tons of fabricated products gives a technologically feasible potential of 0.25×10^6 Btu per ton.

On the basis of economic considerations, it is estimated that up to one-half of the reverberatory furnaces which were being utilized for cathode smelting in 1972 might be replaced with the ASARCO shaft furnace by 1980. On this basis, the economically practicable potential savings are $0.5 \times (0.25 \times 10^6) =$ 0.12×10^6 Btu per ton of fabricated products.

Housekeeping: The potential savings from improved housekeeping practices were derived by considering the energy utilization in each of the three subcomponent operations; i.e., smelting, refining and fabrication.

In 1972, the smelting component consumed 37.19×10^{12} Btu as natural gas and 1.06×10^{12} Btu as electricity. It is estimated that savings of 10 and 5%, respectively, (totaling 3.8×10^{12} Btu) of these energy sources could be gained by 1980 by instituting practices which include: (a) improved combustion controls, (b) improved heat recuperation, (c) stopping leaks of steam, compressed air, hot water, etc., (d) turning off lights, appliances and motors when not in use, and (e) installation and repair of insulation.

In 1972, the refining component consumed 7.1×10^{12} Btu as natural gas and 2.20×10^{12} Btu as electricity. It is estimated that savings equivalent to 10% of the natural gas energy (0.7×10^{12} Btu) could be conserved from these sources by adopting practices which include: (a) improved combustion controls, (b) stopping leaks of steam, compressed air, etc., and (c) scheduling use of electrical equipment to minimize peak demand.

In 1972, the fabrication component consumed about 8.1×10^{12} Btu in natural gas and 9.5×10^{12} Btu in other fuels. It is estimated that savings of 20 and 10%, respectively, (totaling 2.6×10^{12} Btu) could be effected from these sources by instituting practices which include: (a) installation and repair of insulation, (b) improved combustion controls, (c) more efficient use of electric motors, and (d) scheduling use of electric equipment to minimize peak demand.

Collectively, the aggregated potential housekeeping savings are 7.1×10^{12} Btu. These are regarded as both technologically feasible and economically practicable. Dividing this value by 1.62×10^6 tons gives the component potential of 4.38×10^6 Btu per ton of fabricated products.

GOAL YEAR (1980) ENERGY USE TARGET

Projected 1980 production figures for the elements of this component are shown on the following page.

Item	1980 Production, 10^6 tons
Smelter production	1.8
Refined production	2.0
Mill products	1.8

Solely on the basis of technological feasibility, it is estimated that this component could conserve an average of 8.0 x 10^6 Btu per ton of copper products by 1980 (see Table 41).

TABLE 41: ENERGY EFFICIENCY IMPROVEMENT POTENTIAL FOR 1980: COPPER COMPONENT

Technique	Potential Saving, 10^6 Btu/Ton of . . . . Fabricated Products	
	Technologically Feasible	Technologically Feasible and Economically Practicable
Conversion of multiple hearth to fluidized bed roasting	0.12	0.0
Installation of Noranda process smelter at Garfield, Utah	0.68	0.68
Installation of Outokumpu flash smelter at Hidalgo, N.M.	1.30	1.30
Substitution of fuel oil for natural gas in smelting	0.80	0.06
Oxygen-enrichment of air in converters	0.49	0.19
Conversion of reverberatory furnaces for cathode smelting to ASARCO shaft furnace	0.25	0.12
Housekeeping	4.38	4.38
Total	8.02	6.73

Source: Reference (4)

To develop these energy-saving projections, the following factors have been presumed to apply: (a) continued availability of projected energy needs and (b) ability to operate at assumed production rate, e.g., assuming stability in labor and the absence of a major strike.

It should also be noted regarding the Noranda process smelter and the Outokumpu flash smelter that appreciable on-line experience will be required to verify the efficiencies that were used in estimations here and to prove their validity. It should also be noted that no energy costs were assessed against the component for the production of the oxygen required, either in the air-enrichment of converters in general, or for the Noranda process in particular. In most cases, the oxygen required is or will be supplied by outside vendors, not by SIC 33 establishments.

The technologically feasible potential energy savings for seven practices or techniques were listed in Table 41, which also listed the economically practicable savings that were estimated for these same practices. These estimated savings amount to a total of 6.7 x 10^6 Btu per ton of fabricated products.

Applying all of the economically practicable savings to the aggregated component and projecting them to 1980 gives a total estimated gross potential of 80×10^{12} Btu.

The passage by Congress of the Clean Air Act of 1967 and the Amendments of 1970, followed in April, 1971, by the establishment of federal ambient air standards for sulfur dioxide have had tremendous implications for the copper-smelting establishments. In addition, in August, 1973, the Environmental Protection Agency (EPA) published a preliminary draft of background information gathered in support of a proposed New Source Performance Standard (NSPS). These standards, which would apply to new grass-roots smelters and to modified existing smelters propose that gases from any affected facility (i.e., a facility subject to NSPS) could not be discharged to the atmosphere if:

(a) They contain SO_2 concentrations in excess of 650 ppm by volume.

(b) They exhibit 10% opacity or greater (later changed to 20% opacity or greater) if a sulfuric acid plant was utilized to control SO_2 emissions.

The background information document and the proposed standard raised several issues between the EPA, other governmental agencies and the industry. These issues are listed below without comment:

(a) NSPS imply that the conventional reverberatory smelting cannot be used in the future for new grass-roots smelters. Because there are no substitutes for this technology in treating certain raw materials, the standard would affect custom smelters and result in nonutilization of certain domestic resources.

(b) Alternative technology such as electric smelting would not be cost effective in some instances.

(c) Other alternatives such as flash smelting produce high-grade mattes which have a lower scrap melting ability. This technology could result in decreased utilization of scrap and other nonsulfide, copper-containing materials.

(d) The application of NSPS to modification of existing smelters is unclear. Most interpretations of modification imply constraints on the flexibility of existing smelters to expand.

(e) The numerical value of the proposed SO_2 standard is based on and achievable by only one technology: double absorption acid plants.

(f) This numerical value rules out other potential technologies (e.g., the Outokumpu elemental sulfur process) and could inhibit future process development.

(g) The insistence on acid-making technology might not be cost-effective since smelters are distant from traditional acid markets. Acid disposal near the smelter via leaching and neutralization might have deleterious environmental impact and might be prohibited in the future by water pollution regulations.

Because of the many uncertainties posed by NSPS, it was assumed that for the target period only the more moderate emission controls, comparable to the post-1972 ambient air standards would be required, with no efforts toward implementation of NSPS by that date.

Accordingly, the following energy requirements were selected for the EPA and OSHA effects on this component over the period from 1972 to 1980:

Subcomponent	Additional Energy Requirement
Smelter production	12×10^6 Btu/ton of copper
Refined production	0.06×10^6 Btu/ton of copper
Mill products fabricated	0.11×10^6 Btu/ton of products

Applying these requirements to the aggregated subcomponents and projecting these to 1980 gives a total estimated net potential of 102×10^{12} Btu for this component.

To assist in the SIC 33 energy target definition, the potential energy savings for each of the three subcomponents were itemized as shown in Table 42. These values were then used along with the values of 1972 energy output and projected 1980 production for the three subcomponents to compute the projected energy efficiency improvement potential for the entire copper component. The results are given in Table 43. For the same reasons as discussed in the aluminum section, the totals line for the copper component in Table 43 contains four artificially derived numbers as indicated by the footnotes.

TABLE 42: TECHNOLOGICALLY FEASIBLE AND ECONOMICALLY PRACTICABLE 1980 POTENTIAL ENERGY SAVINGS FOR COPPER COMPONENT

Technique	Potential Savings, 10^{12} Btu
Smelting (1.63×10^6 tons in 1972):	
Installation of Noranda Process Smelter	1.1
Installation of Outokumpu Flash Smelter	2.1
Substituting fuel oil for natural gas	0.1
Oxygen enrichment of air in converters	0.3
Housekeeping	3.8
Subtotal	7.4
Refining (1.89×10^6 tons in 1972):	
Conversion to ASARCO shaft furnaces	0.2
Housekeeping	0.7
Subtotal	0.9
Fabrication (1.62×10^6 tons in 1972):	
Housekeeping	2.6
Subtotal	2.6
Total	10.9

Source: Reference (4)

TABLE 43: COMPUTATIONS TO DETERMINE PROJECTED 1980 ENERGY EFFICIENCY IMPROVEMENT POTENTIAL FOR COPPER COMPONENT

Subcomponent	(1) 1972 Energy/ Output (10^6 Btu/ton)	(2) Projected 1980 Production (10^6 tons)	(3) 1980 Energy Consumption Using 1972 Efficiency (10^{12} Btu)	(4) Projected 1980 Energy/Output (10^6 Btu/ton) Gross	(5) Net	(6) Projected 1980 Energy Consumption (10^{12} Btu) Gross	(7) Net
Smelting	29.9	1.8	53.82	25.36	37.36	45.65	67.25
Refining	9.4	2.0	18.80	8.92	8.98	17.84	17.96
Fabrication	10.9	1.8	19.62	9.30	9.41	16.74	16.94
Total	51.2**	1.8*	92.24	44.57***	56.75†	80.23	102.15

*Value selected to represent total 1980 component output.
**Determined by dividing total of Column (3) by total of Column (2).
***Determined by dividing total of Column (6) by total of Column (2).
†Determined by dividing total of Column (7) by total of Column (2).

Source: Reference (4)

FERROALLOY INDUSTRY

This component (SIC 3313) includes establishments primarily engaged in the manufacture of ferrous and nonferrous additive alloys by electrometallurgical or metallothermic processes. Practically all of these additive alloys are ferroalloys sold to and consumed by the basic steel industry for refining and alloying a variety of grades of steel.

PROCESS TECHNOLOGY INVOLVED

Practically all of the ferroalloys in this component are made in submerged-arc electric furnaces. A very small tonnage of specialty ferroalloys is made in small open-top electric furnaces with aluminum as the reducing agent. As can be noted in Figure 34, the production of ferroalloys is very much a single unit operation without many steps.

FIGURE 34: FLOWSHEET—SUBMERGED-ARC FERROALLOY OPERATIONS

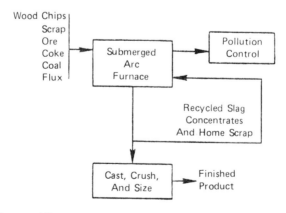

Source: Reference (4)

The furnace is essentially a fixed steel crucible which is lined either with refractory or a carbonaceous material. Three carbon electrodes are submerged in the charge which consists of ore, coke and/or coal, and fluxes (if necessary). Heat is generated by the resistance of the charge to the flow of electricity. The coke and/or coal reacts with the ore to produce the particular metallic element (e.g., manganese) and carbon monoxide gas. The gas flows upward through the charge and into a hood suspended over the top of the furnace. The gas and indrawn air are exhausted by fans to a pollution-control facility to remove particulates. The alloy is tapped from the furnace at intervals of about 3 to 4 hours, and is cast into molds to solidify. It is then crushed and sized and packed for shipment.

MAJOR ENERGY CONSERVATION OPTIONS TO 1980

Very little information is available on total energy consumption in the ferroalloy industry (4). For the base year 1972, no statistical data are available for the fuels used. However, data on fuels and energy consumption for 1971 and for total ferroalloy production for 1972 were available from the Census of Manufactures. Despite the inconsistency inherent in using data for different years, calculations were made which indicated an average energy consumption of 31.9×10^6 Btu per net ton of product. However, it appeared that in the Census data the amount of coke used was much too low and the amount of coal much too high.

Therefore, based on information on the amount of fuels and electrical energy consumed in making each type or grade of ferroalloy, an energy analysis was made based on 1972 data on the amount of each of the alloys produced in that year. The energy consumption for each type or grade was then aggregated and divided by the total production for 1972 to give an average energy consumption of 31.1×10^6 Btu per net ton of product. This is the value used for the 1972 base year. Table 44 shows the distribution of types of energy consumed in 1972 to produce ferroalloys in the United States.

The ferroalloy industry has not, up to date, been reporting any energy data. However, the Ferroalloy Association has been discussing this subject with the Department of Commerce and expects to participate in a voluntary reporting program as soon as guidelines and criteria have been established.

In submerged-arc electric furnace production of ferroalloys, the two major items of energy consumption are solid fuels (e.g., coal) and electrical energy. The carbon from the solid fuels serves strictly as the chemical reductant to produce metal from oxidic ores. The electrical energy provides the heat so that high temperatures can be maintained and also provides the thermodynamic energy requirement to reduce the oxidic ores to metal.

There is little opportunity to lower the amount of the solid fuels (assuming good furnace control) because the amount required is fixed by the particular chemical reaction for producing the metallic element from a particular ore or combination of ores. For example, in the production of ferrosilicon, the basic chemical reaction is: $2C + SiO_2$ (quartzite) $= Si + 2CO$ (gas). This chemical equation says that to produce 1 net ton of silicon requires 0.86 net ton of carbon, and there is no way to decrease this requirement. The carbon monoxide gaseous product of the reaction has a potential heating value, and this is discussed later.

TABLE 44: SUMMARY OF 1972 TOTAL ENERGY CONSUMPTION BY TYPE IN THE FERROALLOY COMPONENT

Type	1972 Energy Consumption, 10^{12} Btu	Percent of Total
Electrical energy[a]	36.2	49
Coke[b]	18.6	25
Coal[c]	12.9	17
Wood chips[d]	4.8	6
Natural gas[e]	2.4	3
Total	74.9	100

(a) Electrical energy at 3,412 Btu per kwh.
(b) Energy value of coke at 24.8×10^6 Btu per net ton.
(c) Energy value of coal at 23.65×10^6 Btu per net ton.
(d) Wood chip heating value at 7.3×10^6 Btu per net ton.
(e) Natural gas energy value at 1,032 Btu per cubic foot.

Source: Reference (4)

In recent years, there has been a definite trend to the replacement of small furnaces with larger furnaces with better controls. Some decrease in the electrical energy requirement per ton of product is achieved with the larger furnaces, particularly when they replace small furnaces in the 10 to 20 Mw range. However, no quantitative documentation is available as to electrical energy efficiency as a function of furnace size for the major ferroalloys. It is considered technically feasible to install a sufficient number of large, well-controlled furnaces by 1980 to achieve an energy saving averaging 2.8×10^6 Btu/ton of ferroalloys made by this component.

The following paragraphs summarize some of the energy conservation options in the ferroalloy industry (4).

Increased Use of Larger Electric Furnaces: The use of larger electric furnaces during the 1972 to 1980 period is assumed to increase on an economically practicable basis with an estimated annual energy saving of 1%. It is considered technologically feasible (within the constraints of the capacity of builders of electric furnaces) to double the installation of new large furnace capacity during the last two years of the 1972 to 1980 period. The energy requirement in 1972, averaged for the various types of ferroalloys, was 31.1×10^6 Btu per net ton of product. Therefore, it is considered economically practicable that for the seven years the energy saving potential would be $(7 \times 0.01) (31.1 \times 10^6) = 2.2 \times 10^6$ Btu per net ton of product. Thus, the energy saving potential estimated to be technologically feasible would be $2.2 \times 10^6 + (2 \times 0.01 \times 31.1 \times 10^6) = 2.8 \times 10^6$ Btu per net ton of product.

Conversion of Some Open-Hooded Furnaces to "Sealed" Mode of Operation: From a technological point of view, the biggest opportunity for energy conservation in the ferroalloy industry is in the recovery and utilization of the furnace off-gas which contains about 80% carbon monoxide and 10% hydrogen.

Some data have been reported on heat recovery from ferroalloy-furnace gases. In Sweden, "steam hoods" have been used on "semisealed" furnaces which produce 75% ferrosilicon. The energy recovery is equivalent to 15 to 20% of the furnace electrical load. Other information concerning a Japanese sealed furnace producing 75% ferrosilicon indicates that use of the gas as a fuel would provide a saving of electrical energy of about 15% after adjusting for an electrical conversion efficiency of 33%. Another method of energy recovery from the off-gas from a sealed furnace in Japan is the preheating of manganese ore by using the gas as a fuel in a rotary kiln. By this method, the electric energy requirement to produce a ton of ferromanganese is lowered by 20%.

In the United States, however, there is very little movement toward the use of sealed furnaces. Indirect information indicates that one existing facility is in the process of being converted to a sealed condition, and the use of the gas as a fuel to produce electricity will lower the consumption of purchased electricity by about 15%.

The reason that U.S. ferroalloy producers do not want to use sealed furnaces is that such furnaces are less flexible than comparable open-hooded furnaces. Flexibility of a ferroalloy furnace means the capability to change the furnace from the production of one type or family of ferroalloys to another type which requires different smelting conditions. A number of different types of ferroalloy products can be produced in a furnace that has high flexibility. Sealed furnaces are considered less flexible than open-hooded furnaces, however.

The retrofit of existing furnaces or the installation of new furnaces for operation in the sealed mode is not considered economically practicable by 1980. This is because the economic impact of certain factors (aside from the capital cost for the gas-collecting, handling, and conversion facilities) is at present uncertain. These factors are:

> A sealed furnace will in most cases be used to produce one specific type of ferroalloy continuously. What is the cost if the furnace has to be shut down because market demand for the alloy is down and inventories are high?

A recent study on the future potential for sealed furnaces in the United States concluded that if a sealed furnace was changed from one product to another the change-over time could exceed that of an open furnace by several weeks. What would the cost impact be for this situation?

What is the additional operating cost to take care of the sealed-furnace requirements for improved burden preparation (compared to an open furnace) and greater attention to operating conditions?

What are the dollar savings with sealed furnaces because considerably less gas needs to be cleaned and, therefore, a smaller air-pollution control facility is required?

However, it appears likely that after the target period some of the large ferro-alloy producers will install sealed furnaces dedicated to the manufacture of one type of ferroalloy.

The economic practicability of sealed furnaces has not been analyzed for ferro-alloy furnaces, and domestic ferroalloy producers show no significant move in that direction. This is discussed above. However, in the Battelle report to FEA (4), it was assumed that it is technologically feasible to convert 10% of 1972 ferroalloy production (2.2×10^6 net tons, excluding ferrophosphorus) to sealed mode of operation. This would be equal to 0.22×10^6 tons. The saving of electrical energy which could be achieved by using the off-gas as a fuel to produce electricity is estimated at 15%. In 1972, 10.599×10^9 kwh of electricity were consumed in the ferroalloy industry. Thus the potential energy saving is estimated as (0.15) $(10.599 \times 10^9$ kwh$)$ $(3{,}412$ btu/kwh$)$ = 5.42×10^{12} Btu. The energy saving per ton of product is:

$$\frac{5.42 \times 10^{12}}{2.2 \times 10^6} = 2.5 \times 10^6 \text{ Btu/ton}$$

Because installation and capital costs would be very high relative to the amount of energy conserved, no energy saving is considered economically practicable by 1980 for this type of innovation.

Housekeeping: The potential energy saving in the ferroalloy component from housekeeping is considered to be small and is taken at 2%. The average energy consumed in 1972 was 31.1×10^6 Btu/ton of product. Therefore, the potential energy saving, which is considered both technologically feasible and economically practicable is (0.02) (31.1×10^6) = 0.6×10^6 Btu/ton of product.

GOAL YEAR (1980) ENERGY USE TARGET

The economically practicable energy-conservation potential which is related to the use of larger and better-controlled furnaces for the 1972 to 1980 period is estimated at 2.2×10^6 Btu per net ton of ferroalloy (Table 45). Housekeeping improvements are expected to add an additional saving of 0.6×10^6 Btu/ton.

Applying all of the economically practicable savings to the aggregated component and projecting them to 1980 gives a total estimated gross potential of 82×10^{12} Btu.

Applying the estimated requirement for environmental control and projecting this to 1980 gives a total estimated net potential of 86×10^{12} Btu for this component.

TABLE 45: ENERGY EFFICIENCY IMPROVEMENT POTENTIAL BY 1980

Technique	Potential Saving, 10^6 Btu Per Ton of Ferroalloys	
	Technologically Feasible	Technologically Feasible and Economically Practicable
Increase in use of larger submerged-arc furnaces with improved controls	+2.8	+2.2
Conversion of some open-hooded furnaces to "sealed" mode of operation	+2.5	0.0
Housekeeping	+0.6	+0.6
Total	+5.9	+2.8

Source: Reference (4)

New air pollution control facilities have been and continue to be installed widely throughout the ferroalloy industry. Water pollution control facilities have been much less important. Most of the air pollution facilities consist of baghouses with a few electrostatic precipitators. With open-top furnaces (which are the usual type), a large volume of air is drawn up into an open hood by high-powered fans to combust and cool the furnace gases. Relatively large amounts of energy are required for the operation of baghouses. For instance, in a recent installation the electrical requirement for a baghouse was 2,250 hp to handle gases from a 25 Mw furnace.

If the furnace load during operation is 20 Mw, the energy requirement for the cleaning facility is over 8% of the furnace energy. This high energy consumption for air pollution control on ferroalloy furnaces is believed typical. Because electrical energy is 50% of total energy (see Table 44), the energy requirement for pollution control during the period 1972 to 1980 is taken at 4% of total energy, which is equivalent to 1.2×10^6 Btu per net ton of ferroalloy. Table 46 gives a summary of the energy-conservation potential and projected 1980 energy per unit, including the requirements for environmental control, for the ferroalloy component.

TABLE 46: ENERGY CONSERVATION POTENTIAL AND PROJECTED 1980 ENERGY PER UNIT FOR FERROALLOY COMPONENT

1972 Energy/Unit, 10^6 Btu/net ton	Projected 1980 Production, 10^6 net tons	1980 Energy Consumption Using 1972 Efficiency, 10^{12} Btu	Projected 1980 Energy/Unit 10^6 Btu/net ton Gross	Net
31.1	2.9	90.2	28.3	29.5

Source: Reference (4)

NONFERROUS FOUNDRY INDUSTRY

This component includes establishments engaged in manufacturing nonferrous castings, generally on a job or order basis, for sale to others or for interplant transfer. It includes establishments that manufacture aluminum alloy, copper alloy, magnesium alloy, nickel alloy, titanium alloy, and zinc alloy castings, and corresponds to SIC 3361, 3362, and 3369. Establishments that produce nonferrous castings in a captive relationship to manufacture a specified product are classified in the industry of the specified product, rather than in this component.

PROCESS TECHNOLOGY INVOLVED

A representative flowsheet for the production of nonferrous castings is given in Figure 35. The major metallic raw material is nonferrous ingot, which is combined with recycled scrap, alloy additions, and fluxes and melted in reverberatory, crucible, or electric furnaces. About 46% of the aluminum is melted in reverberatory furnaces, 34% in crucible furnaces, and 20% in electric furnaces.

About 55% of copper alloys is melted in crucible furnaces, 28% in electric furnaces, and 17% in reverberatory furnaces. Within the die-casting segment of this component, about 80% of the metals is melted in crucible and reverberatory furnaces and the remaining 20% in electric furnaces. The nonferrous alloy of controlled composition is poured into molds, which are different for the three major processes used in manufacturing nonferrous castings. These are sand casting, permanent-mold casting, and die casting.

Sand molds and the necessary cores are made from aggregates, usually silica sand, bonded with bentonite clays, drying oils, sodium silicate, urea-formaldehydes, and furans. The molds and cores are used in a baked, unbaked, or chemically hardened condition, and are used only once. Permanent molds are made from gray iron, steel, or graphite and are used repeatedly. Depending on the number of castings required, permanent molding may be hand operated or mechanically operated. In the die-casting process, a steel mold is mounted in a machine and the molten metal forced into the mold cavity under pressure.

FIGURE 35: GENERALIZED FLOWSHEET FOR NONFERROUS CASTINGS

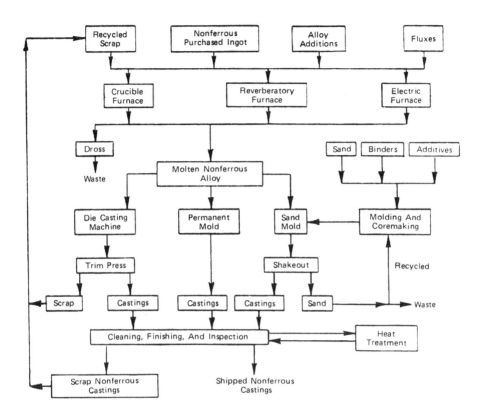

Source: Reference (4)

Die casting is used for parts where the number of castings needed is large, such as for automobile or domestic appliance use. Castings from the molds are cleaned of gates and risers, which are the required channels and reservoirs required in making castings. The metal removed is recycled to the melting operation. Depending on the process, the recycled metal can vary between 40 and 60%. Aluminum, magnesium, and copper alloy castings are heat treated depending on the alloy used and on the properties required in the finished casting.

Nonferrous foundries do not, as a general practice, purchase scrap on the open market, because of the many compositions of scrap and the difficulties of separating the various alloys. Scrap that is not recycled within a nonferrous foundry is usually sold to a secondary smelter, where the scrap is melted, refined, and cast into ingots of controlled composition for sale to nonferrous foundries.

MAJOR ENERGY CONSERVATION OPTIONS TO 1980

Through the programs of the American Foundrymen's Society (AFS) and the American Die Casting Institute (ADCI), the nonferrous foundries are being educated on the opportunities for conservation. The ADCI has established an Energy Committee which has a program underway directed toward conservation in die-casting plants.

Two major bases of information are available. They are [1] Bureau of Census data for 1971 from the 1972 Census of Manufactures, and [2] the foundry industry voluntary reporting system through the American Foundrymen's Society. A third source was also used for die castings. This was information supplied by the American Die Casting Institute to Battelle (4).

The 1971 Census energy data adjusted to 1972 shipments show an energy consumption of 47.8×10^6 Btu per ton of aluminum castings shipped, AFS data indicates an energy consumption of 48.6×10^6 Btu, while Battelle (4) calculations show energy consumption of 36.5×10^6 Btu per ton which was the selected value for this report.

A similar comparison for SIC 3362 shows the Census energy data at 20.3×10^6 Btu per ton of castings shipped, compared to AFS data of 24.0×10^6 Btu, and Battelle (4) calculations of 26.5×10^6 Btu per ton which was the selected value for this report. Census data for SIC 3369, for castings not elsewhere classified, show an energy consumption of 31.0×10^6 Btu per ton of castings shipped compared to the Battelle (4) value of 32.4×10^6 Btu per ton. The latter value was selected.

The energy distributions for SIC 3361, 3362 and 3369 and the component total were developed from Census data and are shown in Table 47.

TABLE 47: ENERGY DISTRIBUTION FOR NONFERROUS FOUNDRY INDUSTRY—BASED ON TYPE OF FUEL USED

Type of Energy	. . . . Percent of Total Btu			
	SIC 3361	SIC 3362	SIC 3369	Component
Natural gas	73	55	53	67
Electricity	16	19	23	18
Distillate oil	1	7	7	3
Residual oil	1	4	4	2
Coal	1	2	0*	1
Other and not specified	8	13	13	9
Total	100	100	100	100

*Less than 0.1%.

Source: Reference (4)

On the basis of information developed for this report, an estimated distribution of energy among the various operations is shown in Table 48.

TABLE 48: ENERGY DISTRIBUTION FOR NONFERROUS FOUNDRY INDUSTRY—BASED ON TYPE OF OPERATION CONDUCTED

Operation	. . . . Percent of Total Btu			
	SIC 3361	SIC 3362	SIC 3369	Component
Melting	29	33	12	25
Molding	2	5	0	2
Casting	10	2	6	8
Cleaning	2	2	1	2
Heat treating and inspection	2	11	0	2
In-plant handling	7	9	4	6
Environmental control	7	4	8	7
Makeup air	41	34	69	48
Total	100	¹100	100	100

Source: Reference (4)

Energy requirements for air pollution control in this component undoubtedly will be increasing as the EPA and OSHA turn their attention from the heavy industries and increase their emphasis on companies in this component. By 1980, it is estimated that the additional controls will require about 0.4×10^6 Btu per ton of castings. Applying this estimate and projecting it to 1980 gives a total estimated net potential of 41×10^{12} Btu for this component.

GOAL YEAR (1980) ENERGY USE TARGET

On the basis of technological feasibility, it is estimated that by 1980, this component could conserve an average of 5.2×10^6 Btu per ton of castings shipped. A listing of the various techniques by which this could be accomplished is given in Table 49.

TABLE 49: ENERGY IMPROVEMENT POTENTIAL BY 1980; NONFERROUS FOUNDRIES COMPONENT

Technique	Potential Saving, 10^6 Btu/Ton of . . . . Castings Shipped	
	Technologically Feasible	Technologically Feasible and Economically Practicable
Housekeeping	1.5	1.5
Bulk transport of molten metal	2.0	0.2
Improvement of furnace efficiency	1.5	1.3
Preheating of furnace charges	0.2	0.2
Total	5.2	3.2

Source: Reference (4)

With respect to the availability of capital, nonferrous foundries face the same constraints as described previously for iron foundries.

As shown in Table 49, four techniques having an aggregated energy savings of 3.2×10^6 Btu per ton of casting shipped were deemed to be economically practicable for the nonferrous foundry component by 1980. About 47% of this total is accounted for by housekeeping improvements and 41% by improvements in melting furnaces.

Applying all of the economically practicable savings to the component and projecting these to 1980 gives a total estimated gross potential of 40×10^{12} Btu.

STEEL FOUNDRY INDUSTRY

This component includes establishments primarily engaged in manufacturing steel castings (carbon and alloy), generally on a job or order basis, for sale to others or for interplant transfer. Establishments that produce steel castings in captive relationship to manufacture a specified product are classified in the industry of the specified product, rather than in this component. The amounts of steel castings taken to be included in this latter group are accounted for in SIC 34, 35, and 37. This component corresponds to SIC 3323 (prior to 1972) and to SIC 3324 (steel investment castings) and SIC 3325 (other steel castings) since 1972. The investment castings account for only about 1.4% of the tonnage in the component.

PROCESS TECHNOLOGY INVOLVED

The major metallic raw material is ferrous scrap, which is combined with pig iron, ferroalloys, and fluxes and melted in electric furnaces. By far most of the tonnage is melted in direct-arc furnaces. The steel melting and refining practice is similar to the practice in the steel industry, involving refining with oxygen and deoxidation with aluminum and other materials, but the scale of melting is smaller in foundries.

The molten steel is transported in ladles and poured into molds and cores made in the foundry from aggregates (usually silica sand, but sometimes a more refractory material) bonded with a wide variety of materials that include bentonite, other clays, drying oils, sodium silicate, Portland cement, formaldehydes, and furans. Before use the molds are dried, baked, or chemically hardened, and are used only once. Casting into permanent molds accounts for only a tiny fraction of the output. For investment castings, the mold-making procedure is quite complex and specialized.

Castings removed from the molds are cleaned of gates and risers which are necessary appendages to permit casting, and the metal removed is recycled to the melting operation. In general, roughly 50 to 70% of the metal melted is recycled

FIGURE 36: PRODUCTION OF CARBON STEEL CASTINGS

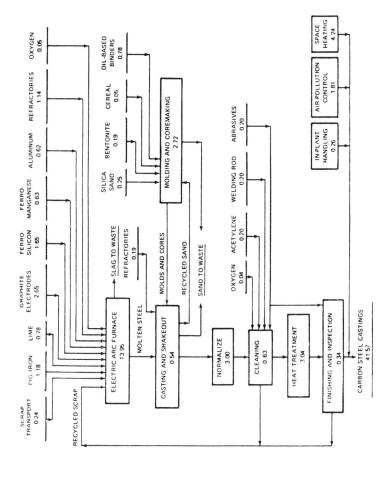

Source: Reference (12)

within the foundry. Some type of heat treatment is applied to almost all castings, and many are given several heat treatments, of which the different types are annealing, stress relieving, normalizing, and quench and temper. Cleaning and repair of defects, usually by welding, is a major activity in steel foundries; so is inspection, which sometimes involves radiography and other nondestructive testing to meet high quality standards.

Figure 36 is a block flow diagram of the steel foundry industry showing the energy values for various unit operations in the boxes in units that represent millions (10^6) of Btu per net ton of primary product. Feed materials, products and by-products are underlined with energy values assigned where appropriate. No fuels or electrical energy consumption are itemized on the flowsheet.

MAJOR ENERGY CONSERVATION OPTIONS TO 1980

Two major bases of information are available for the steel foundry industry (4). They are [1] Bureau of Census data for 1971 from the 1972 Census of Manufactures, and [2] the foundry industry voluntary reporting system through the American Foundrymen's Society (AFS).

The Census data show an energy consumption of about 36.5×10^{12} Btu for SIC 3323 in 1971. Casting tonnages for 1971 as reported by the Bureau vary slightly, but are taken to be 1.58×10^6 tons of steel castings. On this basis, the average use of energy was 23.1×10^6 Btu/ton of castings. However, this computation runs into the same complication as discussed in connection with the Census data under the heading "Iron Foundries."

The tonnage figure of 1.58×10^6 refers to all castings, whereas SIC 3323 is by definition in the 1972 Standard Industrial Classification Manual supposed to exclude captive foundries which serve other industries. On the latter basis using Census data the amount of steel castings "for sale" was 1.32×10^6 tons, and the resulting average is 27.6×10^6 Btu/ton of castings.

The AFS foundry industry voluntary reporting system gives 1972 energy use at an average of 26.69×10^6 Btu/ton of steel castings. This is obtained from a sample of 30 steel foundries out of about 390 that make up the industry. Thus the sample represents 8% of the number of foundries. The sample accounts for 19% of the tonnage. Thus the sample is biased toward larger foundries. The average production of foundries in the sample is about 10,300 tpy, while the industry average is about 4,000 tpy. Because of the effect of plant size upon energy consumption, the average use of energy in steel foundries nationally is probably understated in the AFS survey.

The two best sources on energy consumption by steel foundries in 1972 as noted above are taken to be: [1] the 1971 Census data based on castings for sale (27.6×10^6 Btu/ton), and [2] the AFS survey data for 1972, which are believed to be slightly understated at 26.69×10^6 Btu/ton. The base used here is 28.0×10^6 Btu/ton, which with component production of 1.55×10^6 tons, gives total use of energy by the component in 1972 of about 43×10^{12} Btu.

The Census energy data allow the following estimate to be made of the distribution of sources of energy for this component.

	Percentage of Total Btu
Natural gas	51
Electricity	22
Fuel oil[a]	6
Coal and coke[b]	6
Other and not specified	15
Total	100

(a) About 55 % is residuals, balance distillates
(b) About 93 % is coal

The AFS survey includes energy source information in the data bank, but this has not yet been programmed for retrieval and analysis. No data are available on the distribution of energy among operations in the manufacture of steel castings. However, estimates based on a limited sample (4) provide the following:

	Percentage of Total Btu for the Component
Melting	24
Molding and coremaking	13
Casting, shakeout, cleaning, handling, inspection	7
Heat treatment	30
Environmental control	5
Makeup air	21
Total	100

Steel foundries participate in the voluntary reporting system of the American Foundrymen's Society and in the energy conservation seminars, educational activities, and committee activities of AFS, the Steel Founders' Society of America, and the Cast Metals Federation. The voluntary reporting system indicates a decrease in energy consumed from 26.69×10^6 Btu/ton of steel castings shipped in 1972 to 22.73×10^6 Btu/ton in 1975; a decrease of about 15%. Uncertainties and other factors that mask energy conservation attainments in the AFS reporting system must be taken into consideration. From familiarity with steel foundries and their conservation efforts, it is judged that the amount of conservation actually attained probably is about half the reported value.

On the average, establishments in this component used about 1×10^6 Btu/ton of castings for environmental control in 1972. This is an estimate because confirming data are meager. As described for iron foundries, steel foundries also are subject to special emphasis in the OSHA "National Emphasis Program" (NEP).

Also, their needs for energy to meet air-pollution control regulations will increase. Lacking specific available data, it is estimated that average use of energy by this component will increase about 1 x 10^6 Btu/ton of castings in order to meet such special circumstances. Applying this estimate and projecting it to 1980 gives a total estimated net potential of 45 x 10^{12} Btu for this component.

GOAL YEAR (1980) ENERGY USE TARGET

Based solely on technological feasibility, it is estimated that this component could conserve an average of 6.2 x 10^6 Btu/ton of castings shipped by 1980. Table 50 lists the various techniques by which this could be accomplished.

TABLE 50: ENERGY EFFICIENCY IMPROVEMENT POTENTIAL BY 1980

Technique	Potential Saving, 10^6 Btu per ton of castings shipped	
	Technologically feasible	Technologically feasible and economically practicable
Housekkeeping	2.8	2.8
Install electronic energy-management systems in foundries making 20% of annual production	0.3	0.2
Convert additional 40% of "hot" methods for molding and coremaking to cold methods	0.8	0.6
Use insulating sleeves on 30% of risers	0.3	0.1
Preheat charges with waste heat for 40% of electric furnaces	0.3	0.1
Increase yield by 2% by methods in addition to above	1.2	1.2
Other	0.5	0.3
Total	6.2	5.3

Source: Reference (4)

Steel foundries face the same constraints with respect to the availability of capital as were described previously for iron foundries. Table 50 listed seven techniques which showed a 1980 estimated economically practicable savings of 5.3 x 10^6 Btu/ton of castings shipped. About 53% of this total was accounted for by housekeeping improvements. Applying all of these economically practicable savings to the component and projecting these to 1980 gives a total estimated gross potential of 43 x 10^{12} Btu.

OTHER PRIMARY
NONFERROUS METALS INDUSTRY

This component covers establishments primarily engaged in smelting and refining nonferrous metals other than copper, lead, zinc, and aluminum. It does not include establishments engaged primarily in rolling, drawing, or extruding these primary metals, nor does it include the production of bullion at the site of the mine. It corresponds to SIC 3339.

PROCESS TECHNOLOGY INVOLVED

Magnesium

In 1972, most of the domestic production of magnesium was conducted using Dow's electrolytic process which is shown schematically in Figure 37. Some magnesium was made by an electrolytic process, but one which includes solar evaporation to concentrate brine from the Great Salt Lake in Utah. By 1980, one company is expected to produce magnesium by the Magnetherm process.

Silicon

As shown in Figure 38, a balanced mixture of high-quality quartzite and various carbon-bearing materials is heated by an electrode submerged in the charge to reduce the quartzite to silicon rather than preferentially to silicon carbide. A large quantity of silicon is lost as a volatile silicon monoxide. Periodically, the molten silicon is drained from the arc furnace and cast.

Silver

Silver is produced mostly as a by-product in the production of copper, lead, and zinc. A variety of procedures is used, depending on the major source material being processed. Ultimately, however, the pure silver product is obtained via an electrolytic process. Figure 39, which illustrates the flow chart for recovering silver from copper slimes, is representative of the procedures used.

FIGURE 37: PRODUCTION OF MAGNESIUM METAL FROM SEAWATER

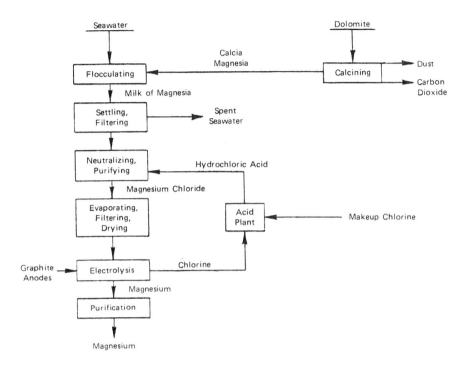

FIGURE 38: PRODUCTION OF SILICON METAL

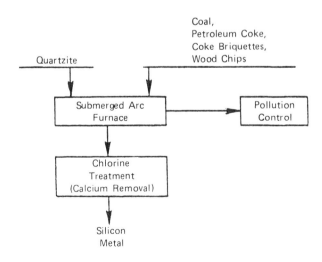

Source: Reference (4)

FIGURE 39: PRODUCTION OF SILVER BARS FROM COPPER ANODE
SLIMES

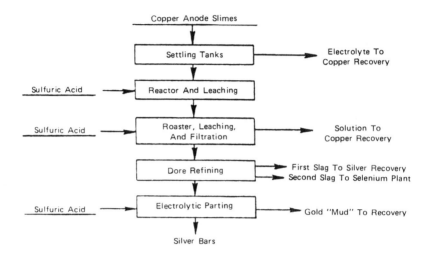

Source: Reference (4)

Titanium

Rutile (95% TiO_2) is chlorinated to produce titanium tetrachloride which is
reduced with magnesium (Kroll process) or sodium (two-stage reduction) to
titanium sponge. The reactor is cooled and the massive titanium sponge is removed, granulated to less than $\frac{3}{8}$ inch by chipping or crushing, leached, dried,
and sealed in metal containers.

The magnesium chloride or sodium chloride from the reactors is electrolyzed
to recover and recycle the chlorine and the magnesium or sodium. The flowsheets for the two processes for producing titanium sponge are shown in Figures 40 and 41.

MAJOR ENERGY CONSERVATION OPTIONS TO 1980

Magnesium

The Bureau of Mines' study gave a value of 199.9 x 10^6 Btu per ton of mag-
nesium. The input data were from actual plant practice. A study by Oak Ridge
National Laboratory gave a value of 233.2 x 10^6 Btu per ton of magnesium.
Much of their input data was indirect.

Therefore a value of 200 x 10^6 Btu was selected as the most representative of
the amount of energy to produce a ton of magnesium. About twice as much
energy in the form of natural gas is required as for electrical energy. About
90% of the total electrical energy was used in the electrolytic cells. About
51% of the total energy from the use of natural gas was used in evaporating,

filtering, and drying the magnesium chloride prior to feeding it into the electrolytic cells; 18% was used in the electrolytic cells; 11% in calcining the dolomite; 9% in neutralizing and purifying the magnesium hydroxide; 9% in the acid plant; and 1% in purification before casting the magnesium. (Details in Table 51.)

FIGURE 40: PRODUCTION OF TITANIUM SPONGE BY THE KROLL PROCESS

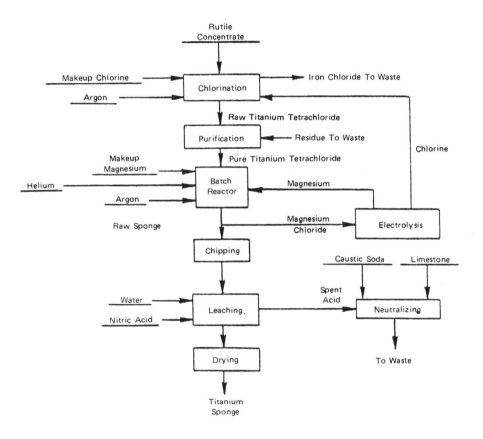

Source: Reference (4)

FIGURE 41: PRODUCTION OF TITANIUM SPONGE BY THE TWO-STAGE SODIUM-REDUCTION PROCESS

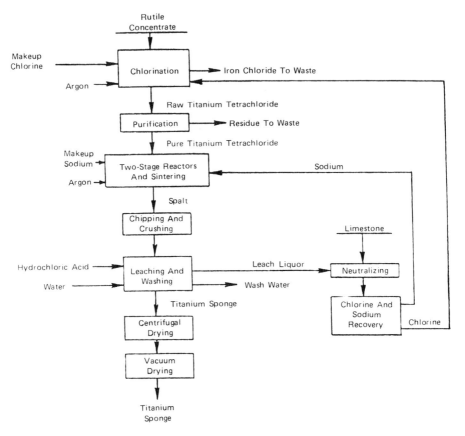

Source: Reference (4)

TABLE 51: ENERGY FORMS AND AMOUNTS OF ENERGY CONSUMED TO PRODUCE PRIMARY MAGNESIUM

	Unit	...Bureau of Mines Study...		Oak Ridge National .... Laboratory Study	
		Unit/Ton of Magnesium	10^6 Btu/Ton of Magnesium	Unit/Ton of Magnesium	10^6 Btu/Ton of Magnesium
Purchased electricity	kwh	19,063	65.1	14,547	49.6
Natural gas	ft^3	130,599	134.8	177,880	183.6
Total			199.9		233.2

Source: Reference (4)

Older sections of the major producer's plants for producing magnesium from seawater are being continuously upgraded with more efficient equipment and process controls. The current value of 200 x 10^6 Btu per ton of magnesium is expected to drop to 186 x 10^6 Btu as a result of a 10% savings in thermal energy by 1980. New processes having the potential for markedly lowering energy requirements are being developed, but are not expected to be in production until after 1980. Production by 1980 is projected to be 146,200 tons.

The Rowley, Utah, plant operating with the brine process is expected to reach an annual production of 25,000 tons by 1977 with an estimated energy requirement of 182 x 10^6 Btu per ton of magnesium. (See Table 52.) The new Magnetherm process plant is expected to be operating at 24,000-ton capacity by 1980 with an estimated energy requirement of 141 x 10^6 Btu per ton of magnesium. (See Table 52.)

From the above, the production in 1980 is projected to be 195,000 tons and the weighted average for energy consumption per ton is 180 x 10^6 Btu. Housekeeping is estimated to be able to save 7 x 10^6 Btu per ton in addition.

TABLE 52: ENERGY FORMS AND AMOUNTS OF ENERGY CONSUMED TO PRODUCE PRIMARY MAGNESIUM FROM BRINE AND BY THE MAGNETHERM PROCESS

| | Unit | ....Brine-Electrolysis .... | | ... Magnetherm Process. ... | |
		Unit/Ton of Magnesium	10^6 Btu/Ton of Magnesium	Unit/Ton of Magnesium	10^6 Btu/Ton of Magnesium
Purchased electricity	kwh	12,775	43.6	17,688	60.4
Coal and natural gas	–	–	138.2	–	80.3
Total			181.8		140.7

Source: Reference (4)

Silicon

Electricity accounted for 51% of the total energy of 105 x 10^6 Btu required to produce a ton of silicon. About 89% of the total electrical energy was consumed in operating the submerged-arc furnaces. Various carbon-bearing materials charged into the arc furnace accounted for 49% of the total energy of 105 x 10^6 Btu per ton of silicon. (Details in Table 53.)

The only energy conservation anticipated by 1980 is about 5 x 10^6 Btu per ton via housekeeping. Silicon will also be obtained as a by-product in the production of magnesium by the Magnetherm process, but no information is available on how the by-product silicon is to be produced.

Silver

About 965 x 10^6 Btu were required to produce a ton of silver. Electricity accounted for 13% of the total energy, natural gas contributed 60%, and 26%

TABLE 53: ENERGY FORMS AND AMOUNTS OF ENERGY CONSUMED
TO PRODUCE PRIMARY SILICON

	Unit	Unit/Ton of Silicon	10^6 Btu/Ton of Silicon
Purchased electricity	kwh	15,617	53.3
Metallurgical coal	ton	0.66	17.2
Petroleum coke	ton	0.66	17.2
Coke briquettes	ton	0.17	4.4
Wood chips	ton	1.77	12.9
Total			105.0

Source: Reference (4)

from steam. Of the total energy from natural gas, 54% was used in the refining
furnaces; 20% in roasting, leaching, and filtering; and 14% in the leaching reactors. Of the total electrical energy, 43% was used in the refining furnaces; 34%
in roasting, leaching, and filtering; and 21% in the leaching reactors. Steam was
used in all of the processing steps except the refining furnaces. Eighty percent
of the total steam energy was used in the leaching reactor and in the roasting,
leaching, and filtering step. (Details in Table 54.)

TABLE 54: ENERGY FORMS AND AMOUNTS OF ENERGY TO PRODUCE
PRIMARY SILVER

	Unit	Unit/Ton of Silver	10^6 Btu/Ton of Silver
Purchased electricity	kwh	36,802	125.6
Natural gas	ft^3	559,320	557.2
Steam	lb	187,030	261.8
Total			944.6

Source: Reference (4)

It is estimated that 97 x 10^6 Btu per ton can be saved by preheating furnace
combustion air with stack gas and 48 x 10^6 Btu per ton by housekeeping programs.

Titanium

Bravard, et al (39) gave a value of 172.7 x 10^6 to produce a ton of titanium
sponge by the Kroll process whereas Battelle study (47) gave a value of 90.6 x 10^6
Btu. Much of the input data for Bravard was indirect, while the Battelle data
were from plant practice. Therefore, the value of 90.6 x 10^6 Btu per ton of
titanium was selected for production by the Kroll process. For the process
using two-stage reduction with sodium, Reference (47) gave a value of 97.4 x 10^6

Btu per ton of titanium. About 65% of the production was by the Kroll process. Thus, the weighted average value for a ton of titanium is taken as 93 x 10^6 Btu.

Table 55 summarizes the amounts and forms of energy used to produce titanium sponge by the various processes.

TABLE 55: ENERGY FORMS AND AMOUNTS OF ENERGY TO PRODUCE TITANIUM SPONGE

	Unit	. Two-Stage Reduction . .		. . . . . Kroll (47)		. . . . . Kroll (39)	
		Unit/Ton of Titanium	10^6 Btu/Ton of Titanium	Unit/Ton of Titanium	10^6 Btu/Ton of Titanium	Unit/Ton of Titanium	10^6 Btu/Ton of Titanium
Purchased electricity	kwh	25,373	86.6	22,505	76.8	40,720	139.0
Natural gas	ft^3	10,470	10.8	13,340	13.8	-	-
Propane	gal	-	-	-	-	353	33.7
Total			97.4		90.6		172.7

Source: Reference (4)

The two processes are expected to continue to be used to 1980. A marked expansion in demand would be required to initiate construction of a more economical, energy-conserving, electrolytic process. A marked increase in demand is not foreseen by 1980. By 1980 a savings of about 5 x 10^6 Btu per ton can be realized by housekeeping measures and 3 x 10^6 Btu per ton from improvements in current processing technology.

Combined Primary Nonferrous Metals

The data available on energy consumption per ton and total energy consumed in 1972 for the primary nonferrous metals in SIC 3339 are sparse. It appears that the annual total energy consumption for this component in 1972 was about 43.2 x 10^{12} Btu and that the annual production was about 289,666 tons. This gives an average energy consumption of 149.1 Btu per ton.

In summary, the selected values for the segments of this component are listed below, and indicate that the total energy consumed by the component in 1972 was about 43.2 x 10^{12} Btu.

Subcomponent	1972 Production, tons	1972 Energy Output, 10^6 Btu/ton	1972 Energy Consumed, 10^{12} Btu
Magnesium	120,823	200	24.2
Silicon	116,376	105	12.2
Silver	2,658	965	2.6
Titanium	20,267	93	1.9
Others	29,542	77	2.3
SIC 3339	289,666	149.1*	43.2

*Weighted average.

The projected production and potential for energy use by this component by 1980 are listed below. The projected growth in the production of magnesium by 1980 is expected to be at a greater rate than for other metals.

Subcomponent	Projected 1980 Production, tons	Projected 1980 Energy, 10^6 Btu per ton
Magnesium	195,000	173
Silicon	141,000	100
Silver	3,200	820
Titanium	24,500	85
Others	35,700	73
SIC 3339	399,400	138*

*Weighted average.

GOAL YEAR (1980) ENERGY USE TARGET

The data for other primary nonferrous metals are given in Table 56.

TABLE 56: COMPUTATIONS TO DETERMINE PROJECTED 1980 ENERGY EFFICIENCY IMPROVEMENT POTENTIAL FOR OTHER PRIMARY NONFERROUS METALS

Metal	(1) 1972 Energy Output (10^6 Btu/ton)	(2) Projected 1980 Production (10^6 tons)	(3) 1980 Energy Consumption Using 1972 Efficiency (10^{12} Btu)	(4) Projected 1980 Energy Output (10^6 Btu/ton) Gross	(5) Net	(6) Projected 1980 Energy Consumption (10^{12} Btu) Gross	(7) Net
Magnesium	200	0.195	39	173	175	33.7	34.1
Silicon	105	0.141	14.8	100	101	14.1	14.2
Silver	965	0.0032	3.1	820	825	2.6	2.6
Titanium	93	0.0245	2.3	85	85.5	2.1	2.1
Others	76	0.0357	2.7	72	73	2.6	2.6
Total	155*	0.3994	61.9	138.0**	139.2***	55.1	55.6

*Total of Column (3) divided by total of Column (2).
**Total of Column (6) divided by total of Column (2).
***Total of Column (7) divided by total of Column (2).

Source: Reference (4)

NONFERROUS PROCESSING INDUSTRY

This component consists of two subcomponents. The first includes establishments primarily engaged in the rolling, drawing, and extruding of nonferrous metals other than copper and aluminum (SIC 3356). The second includes establishments primarily engaged in the drawing and insulating of wire and cable of nonferrous metals from purchased wire bars, rods, or wire (SIC 3357).

PROCESS TECHNOLOGY INVOLVED

The SIC 3356 subcomponent is a mixture of establishments that fabricate plate, sheet, strip, bar and tubing from a variety of metals. In 1972, the total amounts of metal consumed by these establishments were about as follows:

Item	Metal Consumed, 10^3 tons
Nickel and nickel alloys	49.8
Zinc and zinc alloys	38.9
Magnesium and magnesium alloys	22.7
Lead and lead alloys	21.9
Copper and copper alloys	17.2
Tin and tin alloys	8.8
Tungsten and tungsten alloys	6.1
Aluminum and aluminum alloys	6.1
Molybdenum and molybdenum alloys	3.3
Other nonferrous	131.1
Total	305.9

The SIC 3357 subcomponent is largely dominated by electrical wire and cable manufacturing plants whose products range from base wire through stranded, insulated conductors, insulated residential and commercial wiring, and complex power, transmission, and communications cables.

In 1972, the total materials consumed by these establishments were about as follows:

Item	Metal Consumed, 10^3 tons
Copper	1,333
Aluminum	565
Others	14
Total	1,912

It is projected that the consumption of these materials will increase about at the rate of 2.8% compounded annually over the period 1972 to 1980.

MAJOR ENERGY CONSERVATION OPTIONS TO 1980

The metals processed by the SIC 3356 subcomponent have widely diverse properties ranging from the low-melting and very soft metals, such as lead and zinc, which require very little energy for melting, casting, and the manufacture of wrought products, to the high-melting metals, such as tungsten, that can only be worked at elevated temperatures.

In 1972, the nonferrous rolling and drawing establishments consumed a total of 12.7×10^{12} Btu, according to the Census of Manufactures. The largest fuel consumption (65.8% of the total) was from natural gas, which is used mainly in annealing and heat-treating furnaces. About 26% of the energy was electricity which is used to power rolling mills, extrusion presses, some melting furnaces (especially for nickel, titanium, molybdenum, and tungsten), as well as to operate some special vacuum annealing furnaces.

In 1972, the drawing and insulating subcomponent consumed about 22.8×10^{12} Btu. The energy consumption by this segment is dominated by natural gas and electricity which account for 42.9 and 33.2% of the total, respectively. Most of the gas goes to either space heating or annealing furnaces. The electricity is used chiefly to power fabrication machinery. Details are given in Table 57.

Over the period 1971 to 1973, several producers of nickel and tungsten alloys (reporting under SIC 3356) responded to a survey. The results indicated that for a few selected companies energy savings of 15 to 36% per unit of product could be expected via the institution of good housekeeping practices. Although many establishments in this component are known to be conducting energy conservation programs, data to measure progress are not available.

There are no new processing changes applicable before 1980 which are expected to result in any significant energy savings in the nonferrous processing component. Therefore, the only energy conservation anticipated is that which will result from the institution of better housekeeping practices. It is expected that the aggregated savings which result by this means will be about 1.1×10^6 Btu per ton.

TABLE 57: ENERGY USE DISTRIBUTION FOR NONFERROUS PROCESSING COMPONENT*

Element	. . . . Btu, percent			
	Electricity	Natural Gas	Other Fuels	Total
SIC 3356	25.6	65.8	8.6	100.0
SIC 3357	33.9	42.9	23.2	100.0
Totals	31.0	51.0	17.9	100.0

*Based on data contained in *Nonferrous Metal Mills and Miscellaneous Primary Metal Products,* Census of Manufacturers, MC 72(2)-33D, March 1975.

Source: Reference (4)

GOAL YEAR (1980) ENERGY USE TARGET

Target data for the nonferrous processing industry are shown in Table 58.

TABLE 58: COMPUTATIONS TO DETERMINE PROJECTED 1980 ENERGY EFFICIENCY IMPROVEMENT POTENTIAL FOR NONFERROUS PROCESSING INDUSTRY

Subcomponent	(1) 1972 Energy Output (10^6 Btu/ton)	(2) Projected 1980 Production (tons)	(3) 1980 Energy Consumption Using 1972 Efficiency (10^{12} Btu)	(4) Projected 1980 Energy Output (10^6 Btu/ton) Gross	(5) Net	(6) Projected 1980 Energy Consumption (10^{12} Btu) Gross	(7) Net
SIC 3356	41.5	0.38	15.77	37.2	37.5	14.14	14.25
SIC 3357	12.0	2.31	27.72	11.4	11.5	26.33	26.57
Total	16.1*	2.70	43.49	15.0**	15.0***	40.47	40.82

*1972 total of 35.5×10^{12} Btu divided by 1972 total production of 2.2×10^6 tons.
**Total of Column (6) divided by total of Column (2).
***Total of Column (7) divided by total of Column (2).

Source: Reference (4)

MISCELLANEOUS
METAL PRODUCTS INDUSTRY

This component includes establishments engaged in the heat treating of metal for the trade and establishments engaged in manufacturing primary metal products not elsewhere classified. Establishments engaged in heat treating correspond to SIC 3398, and those engaged in primary metal products not elsewhere classified correspond to SIC 3399. Manufacturers of metal powders are the only establishments considered under SIC 3399, because in the value of product shipped they account for 64% of the production.

PROCESS TECHNOLOGY INVOLVED

A generalized flowsheet for heat treating of metals is given in Figure 42. Aluminum, copper, magnesium, nickel, titanium, and steel alloys are heat treated. The nature of the treatment depends on the alloy chemistry and the properties required in the finished parts. Almost without exception, heat-treating establishments do not own the metals or parts that are heat treated within their plants.

The flowsheet shown in Figure 42 represents the more simple methods for heat treating. Many heat-treating procedures are complex and time consuming. In addition to heat treatment, the procedures may include carburizing, nitriding, or cyaniding to produce selective surface conditions for steels. Special heat-treating procedures such as flame and induction hardening may be used to selectively harden only portions of the parts.

The manufacture of powders is accomplished by two methods. Iron powders are generally manufactured by melting the iron and atomizing it with high-pressure water, drying, and screening the powder. Other powders are usually manufactured by chemical methods which reduce the oxides to metallic powders or which precipitate powders from chemical solutions. These latter methods are technically sophisticated and require a high degree of control in the manufacturing processes.

FIGURE 42: GENERALIZED FLOWSHEET FOR HEAT TREATING

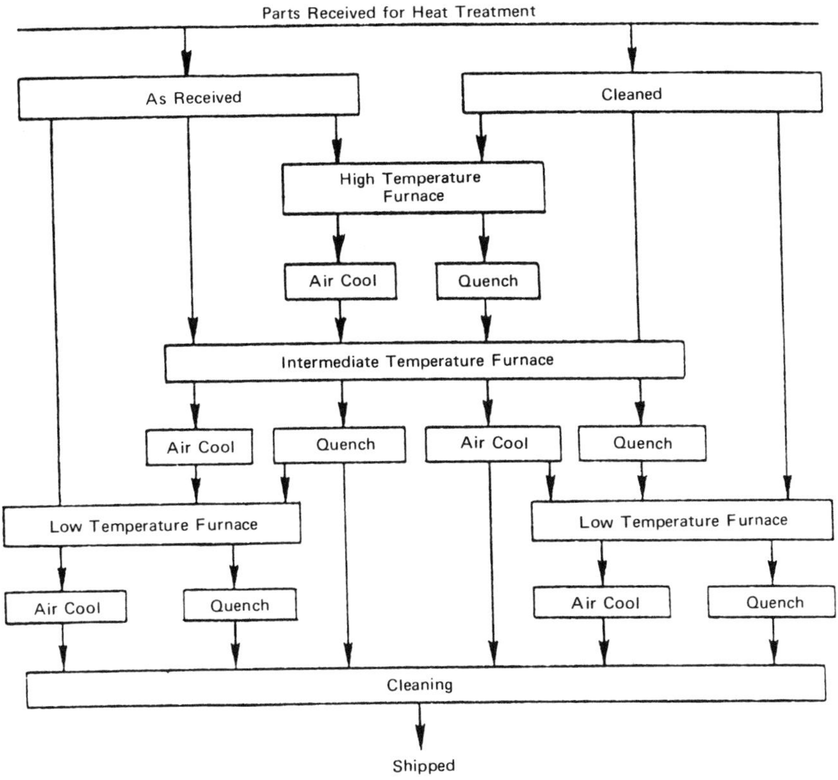

Source: Reference (4)

MAJOR ENERGY CONSERVATION OPTIONS TO 1980

There are no sources of information within Census reports that will permit an assessment to be made of energy consumption per ton of product shipped for SIC 3398 (heat treating for the trade) and SIC 3399 (primary products not elsewhere classified). Data pertaining to total energy for the combined subcomponents are listed, however, and were used to provide an estimate for the total 1972 energy consumption for this component of 34.5×10^{12} Btu.

The Metal Treating Institute has indicated that in a sampling over an 8-month period from 70 heat-treating establishments (out of about 950), the average energy reported consumed was 9.7×10^6 Btu per ton of product shipped. The energy requirement varied from a low of 3.3×10^6 Btu to a high of 45.8×10^6 Btu per ton of product shipped. An average value of 25×10^6 Btu per ton has been assumed.

Energy distribution for SIC 3398 and 3399, based on Census data, is as follows:

Type of Energy	Percent of Total Btu for SIC 3398 and 3399
Natural gas	70
Electricity	14
Distillate oil	2
Residual oil	5
Coal	1
Other and not specified	8
Total	100

Information concerning energy consumption for SIC 3398 (heat treating) suggests that there is a 450% difference between the average and the highest value. This indicates a possible high technological potential for energy conservation. However, the sample for data is small and information is not available as to the number of establishments in the high and low groupings. Total energy consumed for this component in 1972 was estimated to be 34.5×10^{12} Btu. Weight of metal heat treated was estimated on a basis of heat-treating costs of $0.25 per pound of product shipped. Product shipped was valued at $381.1 million which leads to an estimated total of 0.76×10^6 tons of product shipped.

The total Btus assigned to SIC 3398 is 19.1×10^{12}. The difference between 34.5×10^{12} and 19.1×10^{12} Btu was assigned to SIC 3399. Total product shipped for SIC 3399 was assumed to be twice that for powdered metals which leads to an estimated 15.4×10^6 Btu per ton. Because heat treating accounts for a large amount of energy consumed by this component, and because there is an indicated large potential for improvement in SIC 3398, it is estimated that an overall component lowering of energy use can be achieved by 1980 amounting to about 12%. Heat-treating furnace improvements are assumed to account for 8% of the conservation and 3% for housekeeping improvements. The remaining 1% is assumed for SIC 3399, because most of the products manufactured by establishments in this code have rather sophisticated energy-conserving technologies. Taken together, these potential savings are equivalent to 1.7×10^6 Btu per ton of aggregated products for this component.

The Metal Treating Institute has initiated an energy-consumption reporting program and is reporting data to the Commerce Department. The American Society for Metals, a technical metallurgical society, has also initiated seminars pertaining to the conservation of energy in the heat treatment of metals.

GOAL YEAR (1980) ENERGY USE TARGET

The economically practicable energy savings potential has been estimated to be 17.0×10^6 Btu per ton of product (based on a weighted average).

The 1980 net energy savings potential for this industrial segment is thus estimated at 55×10^{12} Btu.

SECONDARY NONFERROUS SMELTING AND REFINING INDUSTRY

This component includes establishments engaged in recovering nonferrous metals and alloys from new and used scrap and dross. Also included are establishments engaged in the recovery and alloying of precious metals, and plants engaged in recovering of tin by smelting and refining or by chemical processes. The component corresponds to SIC 3341.

PROCESS TECHNOLOGY INVOLVED

Figures 43, 44, 45 and 46 show representative procedures for the processing of secondary zinc, copper, lead and aluminum, respectively. In general, each of these processes is initiated with various methods of segregating and separating the scrap and lowering its bulk by some comminution process.

FIGURE 43: FLOWSHEET FOR PROCESSING OF SECONDARY ZINC

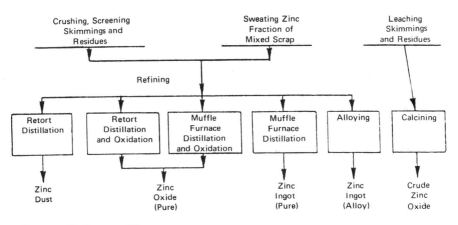

Source: Reference (4)

FIGURE 44: FLOWSHEET FOR PROCESSING SECONDARY COPPER

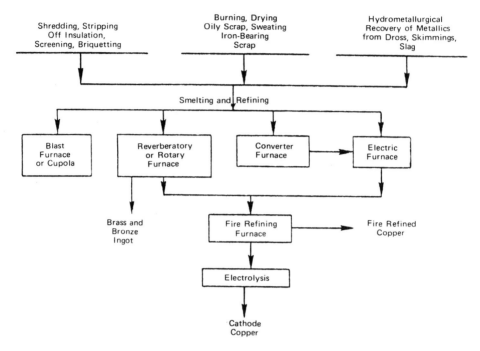

FIGURE 45: FLOWSHEET FOR PROCESSING OF SECONDARY LEAD

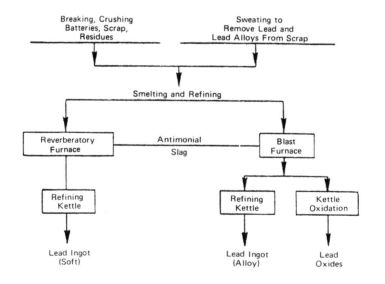

Source: Reference (4)

FIGURE 46: FLOWSHEET FOR PROCESSING SECONDARY ALUMINUM

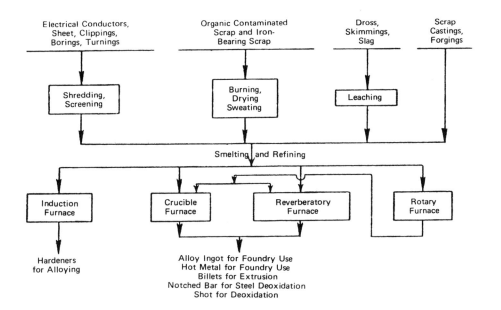

Source: Reference (4)

The scrap is then subjected to a pyrometallurgical smelting and refining operation to yield a metallic product which is of direct utility. In the case of copper, some of the fire-refined material is also electrolytically purified to make it suitable for electrical applications.

MAJOR ENERGY CONSERVATION OPTIONS TO 1980

Based on an analysis of the best available data, selected energy use values for 1972 in terms of 10^6 Btu per ton are 27 for secondary aluminum, 15 for secondary copper, 11 for secondary lead, and 14 for secondary zinc. These values were interpolated from data available for 1971 and for 1973. Details of the data, which are not consistent nor extensive, are given in Tables 59, 60, 61 and 62.

For the component, the total energy use in 1972 was about 32.6×10^{12} Btu. This value and the annual total production of 1,838,605 tons result in an average energy consumption of 17.7×10^6 Btu per ton of product. Details of the calculation are given in Table 63.

An attempt to use the data from the Census of Manufactures gave an unrealistic value of 5.8×10^6 Btu per ton of product, which corresponds to an annual consumption of 20.8×10^{12} Btu (4).

TABLE 59: ENERGY FORMS AND AMOUNTS OF ENERGY CONSUMED TO PROCESS SECONDARY ALUMINUM

| | Unit | . . . . . . 1971* . . . . . . | | . . . . . . 1973* . . . . . . | |
		Unit/Ton of Product	10^6 Btu/Ton of Product	Unit/Ton of Product	10^6 Btu/Ton of Product
Purchased electricity	kwh	93	0.32	104	0.35
Natural gas	ft^3	22,754	23.48	24,370	25.15
Propane, butane	gal	4.67	0.45	3.91	0.37
Middle distillates	gal	3.51	1.15	7.33	1.11
Residual fuel oil	gal	5.62	0.84	5.35	0.80
Gasoline	gal	0.73	0.09	0.67	0.08
Lubricants	gal	0.02	0.003	0.01	0.002
Btu/ton of product			26.33		27.86

*Based on a survey of eight plants.

TABLE 60: ENERGY FORMS AND AMOUNTS OF ENERGY CONSUMED TO PROCESS SECONDARY COPPER

| | Unit | . . . . . . 1971* . . . . . . | | . . . . . . 1973* . . . . . . | |
		Unit/Ton of Product	10^6 Btu/Ton of Product	Unit/Ton of Product	10^6 Btu/Ton of Product
Purchased electricity	kwh	261	0.89	400	1.36
Natural gas	ft^3	8,805	9.09	10,408	10.74
Propane, butane	gal	3.32	0.32	3.38	0.32
Middle distillates	gal	8.02	1.11	9.68	1.34
Residual fuel oil	gal	12.22	1.83	9.54	1.43
Gasoline	gal	0.62	0.08	0.62	0.08
Lubricants	gal	0.01	0.002	0.01	0.002
Coke	ton	0.019	0.49	0.031	0.82
Btu/ton of product			13.81		16.09

*Based on a survey of thirteen plants.

TABLE 61: ENERGY FORMS AND AMOUNTS OF ENERGY CONSUMED TO PROCESS SECONDARY LEAD

| | Unit | . . . . . . 1971* . . . . . . | | . . . . . . 1973* . . . . . . | |
		Unit/Ton of Product	10^6 Btu/Ton of Product	Unit/Ton of Product	10^6 Btu/Ton of Product
Purchased electricity	kwh	68	0.23	75	0.26
Natural gas	ft^3	1,931	1.99	2,155	2.22
Propane, butane	gal	2.39	0.23	0.66	0.06
Middle distillates	gal	0.14	0.02	0.11	0.02
Residual fuel oil	gal	5.39	0.81	4.62	0.69
Gasoline	gal	0.57	0.07	0.38	0.05
Lubricants	gal	0.04	0.005	0.02	0.003
Coke	ton	0.443	11.52	0.265	6.89
Btu/ton of product			14.88		10.19

*Based on a survey of eleven plants.

Source: Reference (4)

TABLE 62: ENERGY FORMS AND AMOUNTS OF ENERGY CONSUMED
TO PROCESS SECONDARY ZINC

| | | . . . . . . 1971* . . . . . . | | . . . . . . 1973* . . . . . . | |
	Unit	Unit/Ton of Product	10^6 Btu/Ton of Product	Unit/Ton of Product	10^6 Btu/Ton of Product
Purchased electricity	kwh	114	0.39	116	0.40
Natural gas	ft^3	4,021	4.15	12,349	12.74
Propane, butane	gal	0.09	0.01	0.09	0.01
Middle distillates	gal	5.34	0.74	4.11	0.57
Residual fuel oil	gal	6.51	0.97	6.37	0.95
Gasoline	gal	1.26	0.16	1.77	0.22
Lubricants	gal	0.03	0.004	0.00	0.00
Btu/ton of product			6.42		14.89

*Based on a survey of fifteen plants.

TABLE 63: 1972 PRODUCTION AND ENERGY CONSUMPTION FOR
SECONDARY NONFERROUS SMELTING AND REFINING

Alloyed Products	Production (tons)	Energy Consumption per Ton (10^6 Btu)	Annual Energy Consumption (10^{12} Btu)
Aluminum	680,064	27.0	18.4
Copper	344,908	15.0	5.2
Germanium*	13**	1.4***	0.00002
Gold	72**	1.0†	0.00007
Iridium	No data	–	–
Lead	627,152	11.0	6.9
Magnesium	15,662**	3.0***	0.05
Nickel	27,863**	4.7†	0.01
Platinum	8.8**	19.8†	0.0002
Selenium*	15**	0.2***	0.000003
Silver	2,157**	1.6†	0.003
Tin	2,199**	0.2***	0.0004
Zinc	138,491	14.0	1.9
Total	1,838,605	17.7††	32.5

*The production figure is estimated.
**Metal content rather than product weight.
***Calculated on the basis of melting the scrap to a moderate superheat temperature.
†Calculated as in (***) even though recovery is also by chemical or electrochemical
separation techniques.
††Weighted average.

Source: Reference (4)

A breakdown according to major sources of energy for the four major subcomponents is as follows:

	. . . . . Percentage of Total Btu Used in 1972				
	Aluminum	Copper	Lead	Zinc	Total
Coke	–	1	11.0	–	12.0
Natural gas, propane, butane	36.0	16	4.0	18.0	74.0
Purchased electricity	0.5	2	0.5	0.5	3.5
Oil, gasoline, lubricants	3.0	4	1.0	2.5	10.5
Total	39.5	23	16.5	21.0	100.0

In summary, the selected 1972 production and energy-efficiency values for the elements of this component are listed below.

Subcomponent	1972 Production, tons	1972, 10^6 Btu/ton	Annual Energy Consumed, 10^{12} Btu
Aluminum	680,064	27.0	18.4
Copper	344,908	15.0	5.2
Lead	627,152	11.0	6.9
Zinc	138,491	14.0	1.9
Other	47,990	4.2	0.2
Total	1,838,605	17.7	32.6
		(Weighted Avg.)	

Some housekeeping measures were started in 1973 by some elements of this component. One secondary zinc plant reported a lowering of about 14% in fuel consumption and an increase in production by 10% by installing meters on each furnace for fuel monitoring and by reducing thermal losses from the furnaces.

GOAL YEAR (1980) ENERGY USE TARGET

It is assumed that since 1972 the average plant has already achieved about a 5% decrease in energy consumption per ton of product through housekeeping measures, and that the improvement will continue by an additional 6% by 1980. Also by 1980, widespread adoption of improvements in the operation of blast furnaces and improvements in melting, smelting, and refining furnaces have assumed potential to add another 10% saving in energy consumption per ton of product.

Collectively, the potential savings from all of these sources is equivalent to 2.8×10^6 Btu per ton of aggregated product. By 1980, the projected combined production for this component is expected to be 2.2×10^6 tons (4).

PRIMARY ZINC INDUSTRY

This component includes establishments engaged in smelting zinc from the ore, and in refining zinc by any other process. It corresponds to SIC 3333.

PROCESS TECHNOLOGY INVOLVED

Flowsheets for electrothermic, electrolytic, and retort processes are presented in Figures 47, 48 and 49. The energy values for various unit operations are given in the boxes in units that represent millions (10^6) of Btu. Products, major material inputs, by-products, and waste products are underlined with energy values assigned where appropriate. A summation of all of the values shown on the flowsheet gives the total energy requirement per net ton of primary product and is shown at the bottom of the flowsheet. No fuels or electrical energy consumptions are itemized on the flowsheet.

In the electrothermic process, a charge of sintered concentrate, coke, and sand is heated in a vertical shaft by passing electric current through the charge. Zinc vapor is withdrawn from the top and collected as a liquid in a condenser and cast into slabs.

In the electrolytic process, the zinc in roasted ore is leached with sulfuric acid. The zinc-bearing solution is filtered and purified and the zinc is recovered by electrolysis. The zinc cathodes are melted and cast into slabs.

In the vertical-retort process, concentrate and roaster dust are pelletized with a clay binder and roasted. The resulting calcine is mixed with coal and sintered. The sinter is mixed with coal and a clay binder and pressed into briquettes, which are heated at a low temperature to improve their strength while in the retort.

The retort is heated externally with natural gas and recycled CO. The reduction reaction in the retort produces zinc vapor and CO, which flow out of the top of the retort where the zinc vapor is condensed to a liquid and cast into slabs.

171

FIGURE 47: PRODUCTION OF ZINC BY THE ELECTROTHERMIC PROCESS

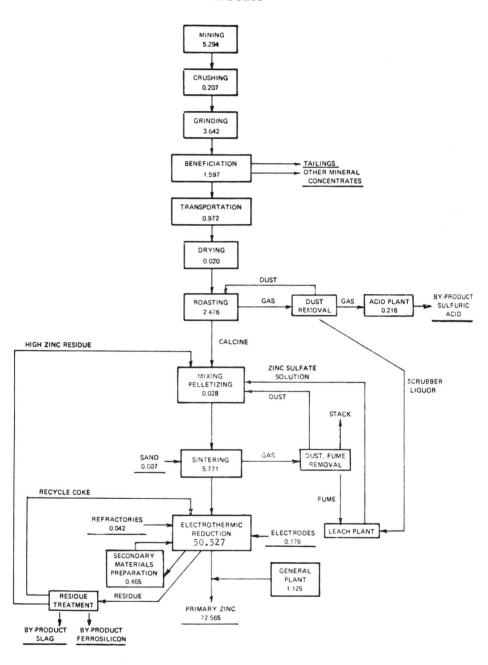

Source: Reference (4)

FIGURE 48: PRODUCTION OF ZINC BY THE VERTICAL RETORT
PROCESS

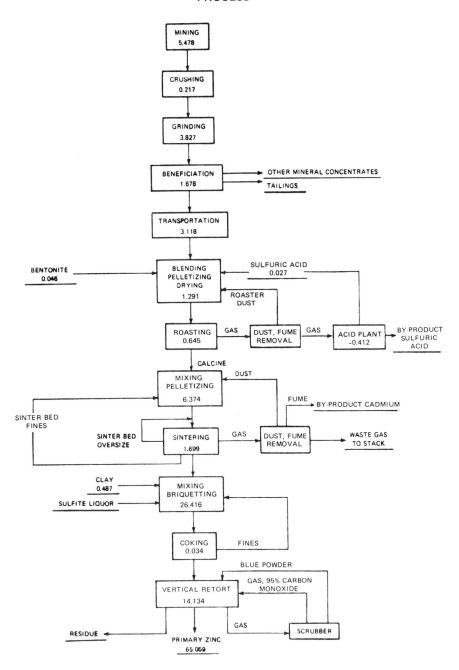

FIGURE 49: PRODUCTION OF ZINC BY THE ELECTROLYTIC PROCESS

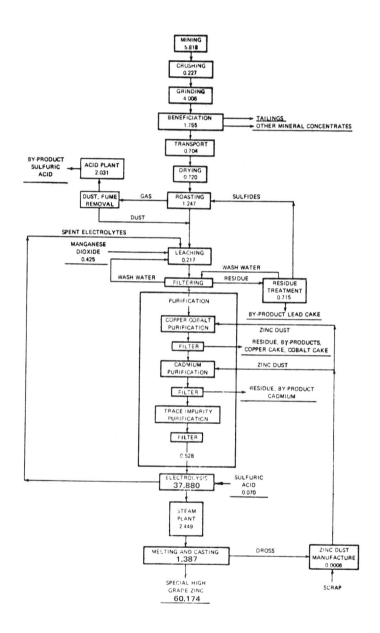

Source: Reference (4)

TABLE 64: ENERGY FORMS AND AMOUNTS OF ENERGY CONSUMED IN THE PRODUCTION OF PRIMARY ZINC

	Unit	Electrothermic Unit/Ton of Zinc	10^6 Btu/Ton of Zinc	Electrolytic Unit/Ton of Zinc	10^6 Btu/Ton of Zinc	Retort Unit/Ton of Zinc	10^6 Btu/Ton of Zinc	1972 Census Unit/Ton of Zinc	10^6 Btu/Ton of Zinc
Purchased electricity	kwh	3,247	11.1	4,496	15.3	525	1.8	1,225	4.2
Natural gas	ft³	2,327	2.4	5,150	5.3	14,205	14.6	15,991	16.5
Coke	ton	0.19	3.9	-	-	-	-	0.17	4.1
Coke breeze	ton	0.19	3.9	-	-	-	-	-	-
Residual fuel oil	gal	-	-	-	-	8.37	1.2	-	-
Distillate fuel oil	gal	-	-	-	-	-	-	0.31	0.04
Anthracite coal	ton	-	-	-	-	0.4	10.2	-	-
Bituminous coal	ton	-	-	-	-	0.79	18.7	0.47	11.1
Propane	gal	-	-	8.14	0.8	-	-	-	-
Total			34.3		21.5		46.5		35.9

Note: Production in 1972 was 19% electrothermic, 42% electrolytic, 39% retort; weighted average was 33.6×10^6 Btu/ton of zinc.

Source: Reference (4)

MAJOR ENERGY CONSERVATION OPTIONS TO 1980

The energy required to produce zinc by the electrothermic, electrolytic, and retort processes is 34.3 x 10^6 Btu per ton, 21.4 x 10^6 Btu per ton, and 46.5 x 10^6 Btu per ton respectively. (Details in Table 64.)

The weighted average for the production by the three processes in 1972 was 33.6 x 10^6 Btu per ton and the annual consumption was 21.3 x 10^{12} Btu. Data from the Census of Manufactures gave a value of 35.9 x 10^6 Btu per ton of product (slab, remelt slab, dust, residue, and other zinc products) in 1971. Because the energy per net ton can vary depending upon the production by process for a given year, the value of 33.6 x 10^6 Btu per ton was selected for 1972.

By 1980, the projected production is 285,000 tons of zinc by the electrothermic process, 669,000 tons by the electrolytic process, and 110,000 tons by the vertical retort.

GOAL YEAR (1980) ENERGY USE TARGET

For the electrothermic, electrolytic, and vertical-retort processes, it is estimated that the energy use per ton will be decreased by housekeeping measures by 2.1 x 10^6 Btu, 0.9 x 10^6 Btu, and 2.8 x 10^6 Btu, respectively. Process improvements are assumed to decrease energy use by 1.4 x 10^6 Btu, 2.1 x 10^6 Btu, and 1.9 x 10^6 Btu per ton, respectively. Table 65 summarizes the target data for the zinc industry.

TABLE 65:   COMPUTATIONS TO DETERMINE PROJECTED 1980 ENERGY
EFFICIENCY IMPROVEMENT POTENTIAL
FOR PRIMARY ZINC INDUSTRY

	(1)	(2)	(3)	(4)	(5)	(6)	(7)
Subcomponent	1972 Energy/ Output (10^6 Btu/ton)	Projected 1980 Production (tons)	1980 Energy Consumption Using 1972 Efficiency (10^{12} Btu)	Projected 1980 Energy/Output (10^6 Btu/ton) Gross	Net	Projected 1980 Energy Consumption (10^{12} Btu) Gross	Net
Electrothermic	34.3	0.285	9.8	30.8	32.5	8.8	9.3
Electrolytic	21.4	0.669	14.3	18.5	20.3	12.3	13.6
Vertical retort	46.5	0.110	5.1	41.8	43.8	4.6	4.8
	27.4*	1.064	29.2	24.2**	26.0***	25.7	27.7

*Total of Column (3) divided by total of Column (2).
**Total of Column (6) divided by total of Column (2).
***Total of Column (7) divided by total of Column (2).

Source: Reference (4)

PRIMARY LEAD INDUSTRY

This component includes establishments primarily engaged in smelting lead from the ore, and in refining lead by any process. It corresponds to SIC 3332.

PROCESS TECHNOLOGY INVOLVED

At the smelter, lead concentrates are mixed with fluxes and with recycle products such as dust from collection systems and slag. The mixture is pelletized and sintered to remove sulfur as SO_2 and to remove other readily oxidized impurities. The sinter is charged to a blast furnace with coke and fluxes to yield bullion (impure lead).

A series of refining steps removes copper as sulfide in the dross and removes antimony, arsenic, and tin by oxidizing the molten bullion. The softened lead is treated with zinc dust to remove gold and silver. Remaining zinc is removed by a vacuum process. Bismuth is removed as a dross by adding calcium and magnesium. Residual calcium and magnesium as well as traces of zinc, antimony, and arsenic are removed by treatment with caustic soda and sodium nitrate. The refined lead is cast into pigs. For flowsheet details, see Figure 50.

MAJOR ENERGY CONSERVATION OPTIONS TO 1980

The energy use pattern in the smelter section and refinery section respectively of a plant for the production of refined lead are shown in Figures 51 and 52.

The energy values for various unit operations are given in the boxes in units that represent millions (10^6) of Btu. Products, major material inputs, by-products, and waste products are underlined with energy values assigned where appropriate. A summation of all the values shown on the flowsheet gives the total energy requirement per net ton of primary product and is shown at the bottom of the flowsheet. No fuels or electrical energy consumption are itemized on the flowsheet.

177

FIGURE 50: GENERALIZED FLOWSHEET OF A LEAD SMELTER AND
REFINERY

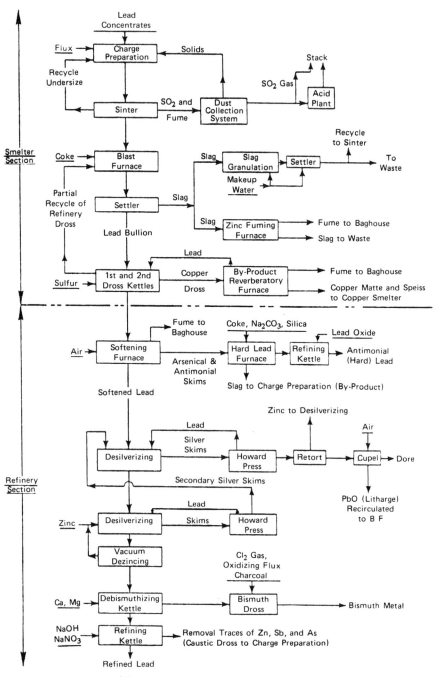

Source: Reference (4)

FIGURE 51: PRODUCTION OF REFINED LEAD–SMELTER SECTION

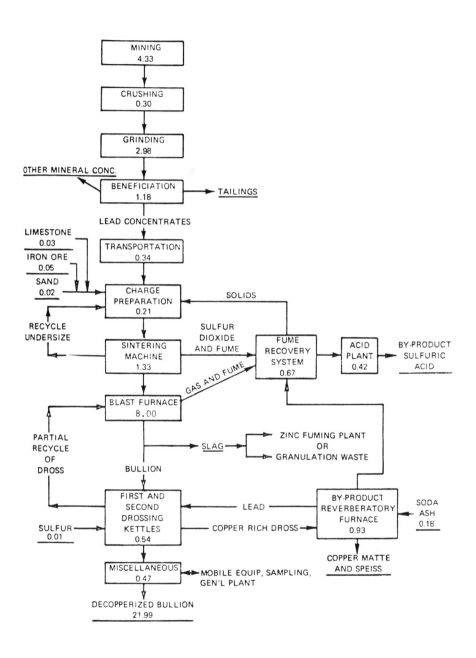

Source: Reference (12)

FIGURE 52: PRODUCTION OF REFINED LEAD–REFINERY SECTION

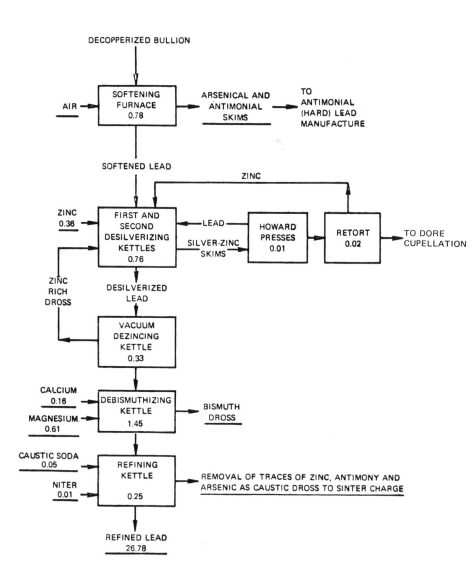

Source: Reference (12)

Because of inconsistencies in the best available data, it is necessary to make an informed judgment as to the probable total use of energy by this component in 1972. The selected value (based on an average use of 13×10^6 Btu per ton of lead and a production of 695,659 tons) is 9.0×10^{12} Btu.

The selected breakdown according to major sources of energy is as follows:

	Percentage of Total Btu Used in 1972		
	Smelting	Refining	Total
Coke, Breeze, and Coal	47	--	47
Natural Gas, Propane, Butane	21	25	46
Purchased Electricity	5	1	6
Oil, Gasoline, Lubricants	1	--	1
Total	74	26	100

The available data on the energy forms and the amounts consumed to produce a ton of lead are meager and somewhat inconsistent. Values of 11.97×10^6 Btu per ton in 1971 and 11.19×10^6 Btu per ton in 1973 are based on information supplied by three smelters and refiners to Battelle (4). A value of 13.77×10^6 Btu per ton in 1973 was obtained from data supplied by five smelters and refiners to Battelle (4). A value of 16.17×10^6 Btu per ton was calculated from Census of Manufactures statistics for 1971.

A value of 13.03×10^6 Btu per ton is a result of an adjustment of the data to account for the difference in energy consumption for the two types of ore that are used by the industry. Details on these various sources of data are given in Table 66.

The data from the three-plant survey indicate a small decrease in total energy usage as a result of a slight reduction in the consumption of natural gas in 1973 as compared with 1971. However, the five-plant survey shows consumption of natural gas in 1973 was twice the amount reported in the three-plant survey. The higher use of natural gas also was indicated by the Census data.

In the five-plant survey, about 1% of the total energy was consumed for space heating and 5% for converting the collected SO_2 to sulfuric acid. These energy usages were not included in the three-plant survey.

Specific data have not been identified. However, it is estimated that since 1972 this component has achieved modest savings of 7% in its consumption of energy per ton of lead, mainly through improved housekeeping measures. This amounts to about 0.9×10^6 Btu per ton.

GOAL YEAR (1980) ENERGY USE TARGET

Based solely on technological feasibility, it is estimated that this component could conserve an average of 2.5×10^6 Btu per ton of lead by 1980 using the techniques listed in Table 67.

TABLE 66: ENERGY FORMS AND AMOUNTS OF ENERGY CONSUMED TO PRODUCE REFINED PRIMARY LEAD

	Unit	...... 1971 ...... Unit/Ton of Lead	10^6 Btu/Ton of Lead	...... 1973 ...... Unit/Ton of Lead	10^6 Btu/Ton of Lead	...... 1971 ...... Unit/Ton of Lead	10^6 Btu/Ton of Lead	...... 1973 ...... Unit/Ton of Lead	10^6 Btu/Ton of Lead	...... 1973 ...... Unit/Ton of Lead	10^6 Btu/Ton of Lead
Purchased electricity	kwh	239	0.82	244	0.83	255	0.87	257	0.88	253	0.86
Natural gas	ft³	3,509	3.62	2,727	2.81	5,286	5.46	6,147	6.34	4,869	5.02
Propane, butane	gal	10	0.96	9	0.86	–	–	–	–	–	–
Middle distillates	gal	0.31	0.04	3	0.42	1.20	0.17	0.40	0.06	0.40	0.06
Residual fuel oil	gal	–	–	–	–	0.17	0.03	–	–	–	–
Gasoline	gal	0.13	0.02	0.17	0.02	–	–	–	–	–	–
Lubricants	gal	0.06	0.01	0.06	0.01	–	–	–	–	–	–
Coke	ton	0.25	6.50	0.24	6.24	0.28	7.28	0.23	5.98	0.25	6.45
Coke breeze	ton	–	–	–	–	–	–	0.025	0.51	0.029	0.59
Coal	ton	–	–	–	–	0.09	2.36	–	–	–	–
Btu/ton Lead			11.97		11.19		16.17		13.77		12.98

Source: Reference (4)

TABLE 67: ENERGY EFFICIENCY IMPROVEMENT POTENTIAL BY 1980—
           PRIMARY LEAD COMPONENT

| Technique | Potential Saving, 10^6 Btu per Ton of Lead | |
	Technologically Feasible	Technologically Feasible and Economically Practicable
Housekeeping		
Smelting	0.6	0.6
Refining	0.3	0.3
Equipment and Processes[9]		
Smelting	1.0	0.5
Refining	0.6	0.2
Total	2.5	1.6

Source: Reference (4)

CHEMICALS
AND ALLIED PRODUCTS INDUSTRY

A major information source for this portion of this volume is the energy conservation target study carried out in 1976 by Battelle Columbus Laboratories (5) for the Federal Energy Administration.

The chemical and allied products industries manufacture thousands of products, many of which are manufactured with totally different technologies. The chemical and allied products industries as defined in the most recent (1972) Standard Industrial Classification manual include:

281	Industrial inorganic chemicals
282	Plastic materials, synthetic resins, synthetic rubber, synthetic and other man-made fibers, except glass
283	Drugs
284	Soaps, detergents, and cleaning preparations, perfumes, cosmetics, and other toilet preparations
285	Paints, varnishes, lacquers, enamels and allied products
286	Industrial organic chemicals
287	Agricultural chemicals
289	Miscellaneous chemical products

Some detail as to the specific chemical products involved in these various industrial classifications is given below.

Product Classifications of 4-Digit Level with Selected Products/Processes Indicating Scope (5)

```
281    Industrial inorganic chemicals
       2812    Alkalies and chlorine
               Chlorine
               Caustic soda
               Soda ash
       2813    Industrial gases
```

(continued)

184

```
            Oxygen, nitrogen, argon
            Acetylene
      2816  Inorganic pigments
            Titanium dioxide
            Zinc oxide
      2819  Industrial inorganic (not elsewhere classified)
            Sulfuric acid
            Alumina
            Sodium tripolyphosphate
            White phosphorus
            Potassium chloride
            Aluminum sulfates
            Sodium sulfate
            Sodium silicate
            Hydrofluoric acid
            Enriched uranium

282   Plastic materials, synthetic resins, synthetic rubber,
         synthetic and other man-made fibers, except glass
      2821  Plastic materials, synthetic resins, and
               nonvulcanizable elastomers
            Polyethylene
            Polypropylene
            Acrylonitrile-butadiene-styrene resins
            Polyvinyl chloride
            Polystyrene
            Phenolics
      2822  Synthetic rubber
            Styrene-butadiene rubber
      2823  Cellulosic man-made fibers
            Rayon
      2824  Synthetic organic fibers, except cellulosic
            Polyesters
            Nylon
            Acrylics

283   Drugs
      2831  Biological products
      2833  Medicinal chemicals and botanical products
      2834  Pharmaceutical preparations

284   Soaps, detergents and cleaning preparations, perfumes,
         cosmetics, and other toilet preparations
      2841  Soap and other detergents, except
               specialty cleaners
      2842  Specialty cleaning, polishing, and
               sanitation preparations
      2843  Surface active agents, finishing agents, sulfonated
               oils and assistants
      2844  Perfumes, cosmetics, and other toilet preparations

285   Paints, varnishes, lacquers, enamels, and allied products
      2851  Paints, varnishes, lacquers, enamels, and allied
               products

286   Industrial organic chemicals
      2861  Gum and wood chemicals
      2865  Cyclic (coal tar) crudes, and cyclic intermediates,
               dyes and organic pigments (lakes and toners)
```

(continued)

                              Aniline
                              Dyes and pigments
                              Coal tar products
                              Cyclohexane
                              Ethylbenzene
                              Phenol
                              Phthalic anhydride
                              Styrene
                              Terephthalic acid
              2869   Industrial organic chemicals, n.e.c.
                              Acetaldehyde
                              Acetic acid and anhydride
                              Acetone
                              Acrylonitrile
                              Adipic acid
                              Methanol
                              Olefins
                              Carbon disulfide
                              Carbon tetrachloride
                              Fluorinated hydrocarbons
                              Ethanol
                              Ethylene oxide and glycol
                              Formaldehyde
                              Hexamethylenediamine
                              Isopropyl alcohol
                              Vinyl acetate
                              Plasticizers
                              Tetraethyl lead and TML
                              Chlorinated ethylenes
                              Propylene oxide

      287   Agricultural chemicals
              2873   Nitrogenous fertilizers
                              Ammonia
                              Nitric acid
                              Urea
              2874   Phosphatic fertilizers
              2875   Fertilizers, mixing only
              2879   Pesticides and agricultural chemicals, n.e.c.

      289   Miscellaneous chemical products
              2891   Adhesives and sealants
              2892   Explosives
              2893   Printing ink
              2895   Carbon black
              2899   Chemicals and chemical preparations, n.e.c.

PROCESS TECHNOLOGY INVOLVED

The process technology involved in each of the divisions of SIC 28 will be discussed under the specific division in question.

MAJOR ENERGY CONSERVATION OPTIONS TO 1980

Tables 68 and 69 show more detailed information of fuel utilization by unit operation and process in the chemical industry.

TABLE 68: FUEL UTILIZATION BY PROCESS AND OPERATION IN THE CHEMICAL INDUSTRY

Process and Operation	. . . . Fuel Usage (10^{12} kcal/year)[a]					
	Purchased Electricity[b]	Coal	Petroleum Products	Natural Gas	Feedstock	Total
Chlorine/caustic soda						
Electrolysis	33	. . . . .30			x	63±10
Compression	4	. . . . 3			x	7±2
Evaporation		. . . . .26			x	26±5
Other		. . . . 4			x	4
	37	10	3	50		100±10
Ethylene/propylene[c]						
Direct heating[e]		. . . . .49			x	49±10
Compression	3	. . . . .38			x	41±8
Distillation		. . . . 6			x	6±2
Feedstock	x	x	x	x	272[d]	272±30
	3	4[f]	1[f]	88[f]	272[d]	368±40
Ammonia						
Compression	7	. . . . .20			x	27±5
Direct heating[e]		. . . . .41			x	41±8
Feedstock	x	x	x	x	87[g]	87±10
	7	–	5	56	87[g]	155±16
Ethylbenzene/styrene						
Direct heating		. . . . 5			x	5.5±1
Distillation	1	. . . . 7			x	7.5±2
Feedstock	x	x	x	x	22[h]	22±4
	1	1	0.3	11	22	35±4
Carbon black						
Direct heating				9		9±2
Drying				1		1
Feedstock	x	x	x	x	25[i]	25±5
				10	25[i]	35±5
Sodium carbonate[j]						
Compression	–	. . . . 3			x	3±1
Drying	–	. . . . 3			x	3±1
Distillation	–	. . . . 3			x	3±1
Direct heating	–	. . . . .2.5			x	2.5±1
	3	1	7.5		x	11.5±2
Oxygen/nitrogen						
Compression	19.5	0.3	0.3	0.4	x	20.5±2
Cumene						
Process		0.3	0.3	1.5	x	2.1±0.5
Feedstock	x	x	x	x	8.5[h]	8.5±2
	x	0.3	0.3	1.5	8.5	10.6±1
Phenol/acetone[k]						
Distillation		. . . . 3.7			x	3.7±1
Other	0.4	. . . . 1			x	1.4±0.5
	0.4	1	0.3	3.4	x	5.1±0.5
Total	68	20	11	227	414	

(continued)

TABLE 68: (continued)

aFor the year 1973.

bFuel value of purchased electricity using a conversion factor of 2,500 kcal/kwh.

cApproximately 55% of the propylene produced in 1973 was a by-product of ethylene production.

d89 x 10^{12} kcal as ethane, 95 x 10^{12} kcal as propane, 87 x 10^{12} kcal as naphtha.

eDirect heating includes heating of steam which enters into the process stream.

fConsiderable gaseous by-products are produced in the ethylene process which can be used as fuel. They are not credited to the ethylene process in this analysis. Their 1973 fuel value was 55 x 10^{12} kcal.

gNatural gas feedstock.

hBenzene feedstock.

i22.5 x 10^{12} kcal as natural gas.

jSynthetic sodium carbonate only. Approximately 50% of the U.S. production in 1973 was synthetic.

kThe cumene oxidation process only. This process accounted for approximately 87% of the U.S. production of cumene in 1973.

Source: Reference (1)

Processes accounting for approximately 48% of the total chemical process (non-feedstock) energy usage are analyzed in Table 68. The total chemical industry energy consumption by operation (Table 69) was estimated using the analyzed process information plus published information on total energy usage in the chemical industry. Feedstock coverage in analyzed processes was much more complete. Approximately 77% of published total feedstock consumption was accounted for in the chemical processes which were analyzed in Table 68.

TABLE 69: FUEL UTILIZATION BY OPERATION IN THE CHEMICAL INDUSTRY

Operation	Energy Consumption Processes Analyzed, 10^{12} kcal/yr*	All Chemical Processes, 10^{12} kcal/yr*
Direct heating	106 ± 15	140 ± 40
Compression	99 ± 25	190 ± 50
Distillation	20 ± 5	100 ± 50
Electrolysis	63 ± 10	90 ± 20
Evaporation	27 ± 4	65 ± 20
Drying	4 ± 1	10 ± 5
Feedstock	413 ± 50	490 ± 50
Other	–	75
Total	732 ± 75	1,160 ± 120

*For the year 1973

Source: Reference (1)

Energy conservation approaches outlined in a Dow Chemical Company report (3) include:

(1)   Operation modification
(2)   Research and development

(continued)

    (3)   Design modification
    (4)   Insulation
    (5)   Maintenance
    (6)   Process integration
    (7)   Process modification
    (8)   Market modification
    (9)   Waste utilization

Table 70 shows where these energy conservation approaches can be applied. All processes analyzed appear to have operations where energy losses could be decreased. A more detailed analysis of the processes and the approaches would be necessary to determine the economic feasibility of implementing the approaches.

TABLE 70: CHEMICAL INDUSTRY ENERGY CONSERVATION STUDY SUMMARY

Process	Energy Intensive Operations	OM	R&D	DM	I	M	PI	PM	MM	WU
Chlorine	Electrolysis	x	x	x						x
Caustic soda	Evaporation			x	x					
	Overall process		x					x		
Ethylene	Furnace combustion			x	x	x				
	Compression			x		x	x			
	Refrigeration			x			x			
	Overall process		x	x						
Ethylbenzene	Distillation			x						
	Overall process		x	x						x
Styrene	Furnace combustion			x	x	x				
	Distillation			x						
	Overall process		x	x						x
Phenol/acetone	Compression			x		x				
	Distillation			x						
	Furnace combustion			x	x	x				
	Overall process		x	x						x
Cumene	Distillation			x						
	Overall process		x							
Sodium carbonate	Kiln calcining	x		x	x					
	Compression			x		x	x	x		
	Calcining (drying)			x	x					
	Overall process							x		
Carbon black	Furnace combustion			x	x					
	Drying			x	x					
	Overall process							x		
Oxygen/nitrogen	Compression			x	x					
	Distillation									

OM – Operation modification   PI – Process integration
R&D – Research and development   PM – Process modification
DM – Design modification   MM – Market modification
I   – Insulation   WU – Waste utilization
M   – Maintenance

Source: Reference (3)

The analyses of the 10 chemical processes were made under the assumption that the plants were well designed and operated. In actual practice several operational and design problems commonly occur. Table 71 shows problems associated with three large energy consumers, furnaces, compressors, and distillation columns.

TABLE 71: OPERATIONAL AND DESIGN PROBLEMS IN ENERGY INTENSIVE EQUIPMENT

Common Operational and Design Problems in High Energy Consumption Equipment	Measures to Overcome Problems
(1) Furnace combustion	
(a) Improper air/fuel ratio	Provide instrumentation to measure oxygen content in flue gas (automatic controls)
(b) Leaks in furnace stacks	Maintenance
(2) Compression	
(a) Leaky compressor by-pass valves	Maintenance
(b) Overdesign of motor or turbine	Do not over-design
(c) Improper suction pressure	Do not over-design
(d) Increasing clearance to lower output	Reduce compressor speed to lower output
(e) Use of less expensive and less efficient turbines and compressors	Realize the value of high efficiency when selecting equipment
(3) Distillation	
(a) Erratic control of columns	Automatic control
(b) Excessive reflux resulting in excessive component separation	Produce minimum quality material
(c) Improper feed tray	Any change in process operation could result in a change in the optimum feed tray
(d) Nonoptimum distillation scheme	Consider energy saving possibilities such as multifeeds, side product draw, or cascade distillation schemes

Source: Reference (3)

In addition to energy intensive operations a common energy wastage problem area is heat transfer equipment.

Table 72 lists equipment where problems commonly occur along with some possible measures to overcome the problems.

TABLE 72: OPERATIONAL PROBLEMS WITH HEAT TRANSFER EQUIPMENT

Common Problems with Heat Transfer Equipment	Measures to Overcome Heat Transfer Problems
(1) Steam traps	
(a) Faulty operation	Monitoring required
(b) Leaking traps	Maintenance
(c) Misdesign	Need proper application and sizing
(2) Steam tracing	
(a) Leaks	Maintenance
(b) Unnecessarily high steam temperature	Substitute another fluid such as Dow SR-1 for steam
(3) Heat exchangers	
(a) Fouling	Maintenance
(b) Higher than necessary temperature separation between fluid streams	Design for low temperature differences by increasing heat transfer surface area
(c) Complete reliance on water cooling	Air cooling requires less

Source: Reference (3)

The energy-conservation options broadly classified by Battelle (5) include:

Housekeeping and maintenance
Process control and energy balancing
Equipment modification or replacement in present processes
Building of new plants with present or new technology

Ideally the economic feasibility of each option and each process would be evaluated using such criteria applicable to the chemical industry as return on investment, alternative investment opportunities, capital availability, capacity utilization, cost differentials. Wherever possible the data-collection efforts will be directed at this level of detail. In some cases simpler criteria such as simple payback period may have to be used. It is apparent already that the practices for making technical and economic evaluations differ from one 4-digit class to another and from one company to another within a subindustry. Judgment will have to be applied to make generalizations.

Following is a listing in the estimated order of importance for conserving energy in the chemicals industry of the major conservation techniques that are being used, or are intended to be used more aggressively in the immediate future.

Reduce excess air in firing steam boilers and process heaters. There is usually rapid payout requiring only better instrumentation for measurement and control.

Increase maintenance (traps and leaks) and insulation on steam systems. Payout is quick due to 10 to 20% reduction in steam needed.

Increase recovery of waste heat from process streams and stack gases. Careful analysis of costs, retrofit problems, and unit downtime is required, but experience indicates many excellent payouts.

Recover wastes for fuel values. This may become complicated, but usually pays out compared to other disposal techniques.

Optimize process controls with respect to energy saving at higher energy costs. This applies to both designs and existing operations.

One study using a similar approach to techniques lists the above techniques in slightly different rank order plus others of lesser significance. It concluded that 28% of energy used by the chemical industry could be saved within five years (1).

Housekeeping: There is constant reference in published information and seminars on energy conservation to savings obtained by better housekeeping. It is also used in the Battelle study (5) despite the lack of clear definition of what is meant by the term. As used here, it refers to the myriad of small things which can be done with little effort, little expense, and little change in basic policies or operation guidelines. Usually they can also be done quite quickly. It includes attention to lighting levels, HVAC systems (thermostats, ventilation rates), minor maintenance (steam leaks, steam traps, repairing insulation), adherence to old or new operating standards (excess air in combustion, reflux rates, temperature levels), and avoiding obvious losses (closing warehouse doors, steam condensate return instead of to sewer). Each plant must make its own specific list to follow.

Much of the early reporting on energy savings gains was largely from housekeeping. Some early specific examples were spectacular, but when an overall program running for several years is analyzed it appears that new practices compared to old yield about 20% reduction in energy used. Note carefully that this does not mean a 20% reduction in specific energy consumption (SEC). The amount of total energy subject to savings from better housekeeping may range from as little as 5% in some plants to 80% in others.

Product and Intermediate Product Specifications: This is an area for energy savings which has received relatively little attention to date. It refers to higher purity or quality than is needed for intended use, and the often high increment of energy used to produce this extra quality.

Almost every energy conservation manager knows of some changes which could save energy for his company without apparent detriment. Any change in product specifications might be detrimental to marketing, and therefore must be carefully evaluated. For intermediate products to be further processed or combined with other products, the entire system must be carefully checked out. The cumulative effects of small changes on product yields or production rates may well offset any energy savings.

Except for a few companies with advanced energy management programs and techniques, the Battelle study (5) has not considered product specification changes as a significant factor in reducing energy use by January 1, 1980. For the longer term, it could provide a significant further improvement in many industries.

Methods for Computing and Tracking Energy Savings Programs: It is not surprising in view of the many interrelated factors effecting conservation that there is no universally agreeable system for reporting energy use improvements, nor monitoring it. Each industry, and companies within the industry, have adapted to the information available to it. A fairly common objective of industry trade associations it to find a method, suitable to its members, for reporting to the Federal Energy Administration/Department of Commerce programs. Typically, even a single company needs to use several methods, and a good discussion with examples is that by R.E. Doerr of Monsanto (48). We have not found any single practical method that presents a true picture under all circumstances.

Efficiency of Power Plants for Large Chemical Complexes: It is well demonstrated that in a chemical plant complex that can utilize large amounts of relatively low pressure steam, it is possible to erect a steam boiler-turbo-generator combination which exceeds 70% thermal efficiency. By contrast, modern utility power plants run in the range of 33% to perhaps 40% thermal efficiency. The difference in efficiency is the inability of the utility power plant to utilize the low level heat.

Despite the efficiency difference which means less energy use, building a central power plant is often not considered to be justified by a chemical company compared to a simpler steam boiler plant and purchased electricity. The major parameters that influence the economics of decisions are:

> Relative investment due to economics of large size. An industrial plant would seldom be larger than that equivalent to 200 to 300 Mw. Utility plants will be several-fold larger.

> Project life: An industrial complex is seldom analyzed for longer than 15 to 20 years, whereas a utility will look at 30 to 40 years.

> Acceptable rate of return on investment: The utility accepts a lower, but relatively guaranteed rate.

Equally important to many companies is avoiding some investment. The ideal situation for most chemical companies would be for a utility to make the investment and sell both steam and electric power to the chemical plants complex.

All large companies are considering coal as an alternate to fuel oil for either steam boiler plant or steam-electric power installations of the future. These will not be built by January 1, 1980, although a few could be built by 1982 to 1985. In addition to the economic criteria and use of capital mentioned before, use of coal presents another set of impediments to chemical companies:

> The need for long term coal contract commitments.

> The unknowns of cost of alternate means of transportation.

> Long term effects of the Clean Air Act.

> Long lead time to open new coal mines, and requirements of the Coal Leasing Act.

The significance of the foregoing condensed and over-simplified outline with respect to energy conservation action is two-fold. First, decisions to build more energy efficient steam-power units may be deferred, or go the less efficient way. Second, a chemical company because of its different business orientation will require expectation of a substantially greater spread between the value of oil and coal ($/million Btu) than a utility expects before it can justify the investment in coal-fired facilities.

Problem Areas in Energy Conservation

(1) Manpower: Manpower assigned to developing energy savings projects appears to be a very real limitation in many large companies, and is almost invariably so in small companies. In the few large companies with advanced long range plans adequate manpower to meet program goals has been assigned. More frequently Battelle (5) estimated by indirect evidence that less than half the manpower thought to be needed is assigned now. No analysis was attempted of the reasons which cause this low level assignment of people, but considering project lead time, Battelle (5) in many cases discounted what would be achieved by January 1, 1980 because of this factor.

(2) Decision Delays: It appears that company decisions to proceed with larger energy-saving projects are coming much slower today than was anticipated by planners during 1973 to 1974 according to Battelle (5). The two major reasons for this slowdown seem to be that the 1975 low operation period changed priorities and caused rethinking of projects, and that national policy decisions which had been anticipated have not occurred. Without doubt, lower earnings during 1975 also contributed to this change, but availability of funds for investment in energy savings projects does not appear to have been the principal reason.

GOAL YEAR (1980) ENERGY USE TARGET

The methodology in the Battelle study for FEA (5), in brief, was that estimates of energy efficiency targets would be developed by 4-digit establishment class. The steps can be defined as follows:

Step 1   Develop a framework for projections of the growth for segments of the chemical industry and for aggregation of energy efficiency targets from the 4-digit level to the 2-digit level.

Step 2   Identify energy usage and conservation options.

Step 3   Evaluate technical and economic feasibility.

Step 4   Estimate an energy efficiency target for each of the 4-digit level industries.

Step 5   Calculate an energy efficiency target for SIC 28.

The growth estimates used are summarized in Table 73.

TABLE 73: PRELIMINARY ESTIMATES OF THE GROWTH OF 4-DIGIT (PRODUCT) CLASS OF CHEMICAL AND ALLIED PRODUCTS INDUSTRIES

Standard Industrial Classification	Ratio: Value of Shipments in 1979.5 / Value of Shipments in 1972 (both expressed in constant 1972 dollars)
28 Chemicals and Allied Products	1.341
2812 Alkalies and Chlorine	1.332
2813 Industrial Gases	1.260
2816 Inorganic Pigments	1.384

(continued)

TABLE 73: (continued)

Standard Industrial Classification	Ratio: $\dfrac{\text{Value of Shipments in 1979.5}}{\text{Value of Shipments in 1972}}$ (both expressed in constant 1972 dollars)
2819 Industrial Inorganic Chemicals n.e.c.	1.211
2821 Plastics, Synthetic Resins, etc.	1.458
2822 Synthetic Rubber	1.271
2823 Cellulosic Man-Made Fibers	1.125
2824 Synthetic Organic Fibers ex Cellulosic	1.558
2831 Biological Products	1.891
2833 Medicinal Chemicals and Botanicals	1.165
2834 Pharmaceutical Preparations	1.234
2841 Soap, Detergents, etc.	1.209
2842 Specialty, Cleaning, Polishing, Sanitation	1.442
2843 Surface Active Agents, etc.	1.325
2844 Perfumes, Cosmetics, etc.	1.286
2851 Paints, Varnishes, Lacquers, etc.	1.278
2861 Gum and Wood Cemicals	1.390
2865 Cyclic (Coal Tar) Crudes, etc.	1.441
2869 Industrial Organic Chemicals, n.e.c.	1.414
2873 Nitrogenous Fertilizers	1.056*
2874 Phosphatic Fertilizers	1.194
2875 Fertilizers, Mixing only	1.403
2879 Pesticides and Ag Chemicals, n.e.c.	1.350
2891 Adhesives and Sealants	1.632
2892 Explosives	.995
2893 Printing Ink	1.380
2895 Carbon Black	1.297
2899 Chemicals and Chemical Prepn. n.e.c.	1.359

*Preliminary estimate is unreasonably low which may be caused by anomaly in Census data. Independent methods of estimating the 1979.5/1972 ratio for nitrogeneous fertilizers is being developed.

Source: Reference (5)

It is apparent that the methodology developed for SIC 28 must also take into account the differences in data compiled by establishment class compared with data compiled by product class as pointed out by Battelle (5). Certain data are collected and summarized in accordance with the SIC classification of the establishment in which products are manufactured. Other data are collected and summarized by 4-digit product classifications which include data on products regardless of the manufacturing establishment classification in which they are produced. It is necessary to be explicit at all times regarding the basis of the data being considered, i.e., whether on an establishment basis or a product basis.

As an example Figure 53 shows the value of shipments for SIC 2812 (Establishment) class and the value of shipments for SIC 2812 (Product) class. In simple terms this shows that establishments in SIC 2812 (Establishment) class ship a substantial amount of products which are not classified in SIC 2812. It also shows that of the total amount of products classified as alkalis and chlorine about a third is produced in establishments not classified as chloralkali producers.

FIGURE 53:   DATA ON SHIPMENTS BY ESTABLISHMENTS CLASSIFIED
AS SIC 2812 AND SHIPMENTS OF PRODUCTS CLASSIFIED AS 2812
REGARDLESS OF WHERE MANUFACTURED

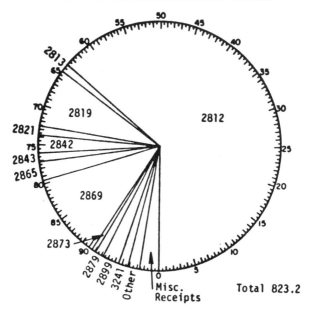

SIC	Shipments	% of Total
2812	525.7	64
2813	10*	1.2
2819	62.6	7.6
2821	10*	1.2
2842	20*	2.4
2843	10*	1.2
2865	20*	2.4
2869	91	11
2873	6.1	0.7
2879	10*	1.2
2899	20*	2.4
3241	10*	1.2
Other	2.1	1.3
Misc.	16.6	2.0

*
Exact figures not cited in
data, estimation required.

(continued)

FIGURE 53: (continued)

Shipments of Chloralkali Products

Total 805.7

SIC	Shipments	% of Total
2812	525.7	65
2819	30.7	3.8
2611	8*	∿1
2865	8*	∿1
2869	225.2	27.9
2873	8*	∿1

Source: Reference (5)

Data on an establishment basis are necessary for economists to describe the manufacturing sector of the economy in relation to the economy as a whole and to other manufacturing, resource, and service sectors. Technologists, industry specialists, and marketing researchers deal with products and processes which can be analyzed and projected only from data summarized on a product basis. The 1971 and 1974 surveys of energy consumption conducted by the Bureau of Census were collected on an establishment basis. There is no way of directly checking the reasonableness of those composite data. Any technical evaluation of energy usage necessarily must deal with specific products and processes and data must be summarized on a product basis.

Since some data are on an establishment basis and others are on a product basis, it is necessary to develop some kind of method for converting data summarized by SIC product class to an SIC establishment basis and from an SIC establishment basis to an SIC product basis at the 4-digit level. Such a method has been developed and is an essential part of the methodology described in the Battelle report (5) to FEA.

Unfortunately few, if any, sizeable manufacturing establishments produce products falling into only one 4-digit code; they usually cover several. Furthermore, multiestablishment chemical companies are often in many loosely related lines of business with products covering a broad spectrum of 4-digit codes. These companies rarely manage energy on a product basis, but rather on an establishment basis.

They attack the problem of reducing energy use within each establishment on whatever division into parts is rational for that establishment (areas; delineated plants; overall establishment systems like power or steam generation, buildings, warehouses, etc.). It is rare that such breakdown produces information applicable in its entirety to a single product or 4-digit SIC code of products. Yet this system is entirely rational for the purpose of understanding, managing and achieving more efficient energy use.

If improvement goals for products, or 4-digit SIC industries, rationally amalgamated into a goal for an establishment, or a company of many establishments, does not roughly check with what a company or its establishments can expect to achieve by its energy management methods, then the energy savings goals obtainable by January 1, 1980 would be suspect. Battelle (5), therefore, considered it prudent to examine, as far as time would permit, what success has been obtained and what is expected by larger companies with well-organized energy management programs. The energy efficiency goals for SIC 28 are summarized in Table 74.

TABLE 74: CHEMICAL INDUSTRY ENERGY EFFICIENCY GOALS BY ESTABLISHMENT CLASS

Class	SIC (Establishment) Name	. . 1979.5 Energy Use . . With 1972 Technology (10^{12} Btu)	With 1979.5 Technology (10^{12} Btu)	Goal* (percent)
2812	Alkalies and Chlorine	265.419	242.045	8.8
2813	Industrial Gases	110.442	101.551	8.1
2816	Inorganic Pigments	95.630	87.593	8.4
2819	Industrial Inorganic Chemicals, n.e.c.	552.198	529.645	4.1
Total 281		1,023.689	960.834	6.2
2821	Plastics, Synthetic Resins, etc.	323.542	265.728	17.9
2822	Synthetic Rubber	86.438	65.026	24.8
2823	Cellulosic Man-Made Fibers	71.113	63.464	10.8
2824	Synthetic Organic Fibers, ex Cellulosic	303.456	234.602	22.7
Total 282		784.549	628.820	19.8
2831	Biological Products	n.a.	n.a.	n.a.
2833	Medicinal Chemicals and Botanicals	n.a.	n.a.	n.a.
2834	Pharmaceutical Preparations	n.a.	n.a.	n.a.
Total 283		147.340	118.024	19.9
2841	Soap, Detergents, etc.	48.946	41.446	15.3
2842	Specialty, Cleaning, Polishing, Sanitation	14.650	12.639	13.7
2843	Surface Active Agents, etc.	17.236	14.945	13.3

(continued)

TABLE 74: (continued)

Class	SIC (Establishment) Name	1979.5 Energy Use With 1972 Technology (10^{12} Btu)	With 1979.5 Technology (10^{12} Btu)	Goal* (percent)
2844	Perfumes, Cosmetics, etc.	12.353	10.556	14.5
Total 284		93.185	79.586	14.6
2851	Paints, Varnishes, Lacquers, etc.	30.476	26.934	11.6
2861	Gum and Wood Chemicals	11.223	10.084	10.1
2865	Cyclic (Coal Tar) Crudes, etc.	259.467	210.784	18.9
2869	Industrial Organic Chemicals, n.e.c.	1,544.642	1,281.494	17.0
Total 286		1,815.332	1,502.362	17.2
2873	Nitrogenous Fertilizers	395.955	336.281	15.1
2874	Phosphatic Fertilizers	167.754	137.241	18.2
2875	Fertilizers, Mixing only	15.703	12.191	22.4
2979	Pesticides and Agricultural Chemicals, n.e.c.	50.345	42.682	15.2
Total 287		629.757	528.395	16.1
2891	Adhesives and Sealants	18.251	16.417	10.0
2892	Explosives	69.662	48.608	30.2
2893	Printing Ink	3.274	2.963	9.5
2895	Carbon Black	·64.478	39.069	39.4
2899	Chemicals and Chemical Preparation, n.e.c.	71.508	58.228	18.6
Total 289		227.173	165.285	27.2
Total 28		4,751,501	4,010,240	15.6

*The percentages, although shown to a tenth of a percent, are expected to be rounded to integers.

Source: Reference (5)

Upper Limits on Reducing Specific Energy Consumption (SEC) by January 1, 1980

For the large-scale, continuous chemical process in which chemical conversion or separation of chemical components is the major objective, theoretical thermodynamic considerations define a minimum energy requirement for ideal processes. Estimates on this theoretical basis suggest that one need use only 25 to 35% of the energy used today for many chemical processes. There are no ideal processes, however. Technology forecasters envisioning the best that might be attainable within the next 25 to 50 years, without any restraints on capital funds, ultimate costs, manpower or political, sociological, or societal restrictions, guess that a 40 to 50% reduction is within reality (5).

The time span of the Battelle study (5) is the much nearer future, 4 or 5 years from present practice. Because of the lead time of 2 to 5 years to study, engineer, and construct improvements, this means that upper limits must only consider well-proven or near-proven technology changes, or operation practices. One example of a published analysis (49) using such criteria concluded that the chemical industry could reduce specific energy consumption 31%, and would have 1 to 3 year payouts on its investments based on estimated fuel costs during the next five years. This applies to energy used in the process operations, not to energy for support services or human comfort. Battelle's own experience and industry comments support considering 25 to 30% reduction in SEC as an entirely practical goal for a representative group of chemical industry processes. There are, of course, notable exceptions. For example, no way is currently known to achieve any quick, large reduction in SEC for air separation, or chlorine-soda ash from present practice. Equally, there are examples where very large reductions in SEC are possible as in manufacturing carbon black, and in new versus old ammonia plants using natural gas feedstocks.

Another good example to support this general level of achievement in reducing SEC comes from historical data from the petroleum refining industry which uses many processes identical in concept and size range to large portions of the chemical industry. Using Nelson factors, well respected in the industry, it can be shown that new high efficiency refineries compared to typical refineries built 10 to 15 years ago are on average 28% more efficient in SEC regardless of overall complexity.

Some companies whose mixture of manufacturing processes should be fairly typical of the foregoing consider an SEC reduction of 30% a bit too high either because the analytical basis or economic criteria were not considered sufficiently stringent. Few seem to object to 25% as a technically and economically feasible goal for a representative mixture of chemical process unit operations that make up many of the large-energy consuming chemical manufacturing centers.

Practical Limits on Reducing SEC by January 1, 1980

There are some very real reasons why the upper limits on SEC reduction that were discussed above cannot realistically be expected to be achieved. The most important of these reasons are listed below with an accompanying brief discussion of each.

Energy for Support Services: This energy is not included in most SEC studies, which consider process energy uses only. The energy used for space conditioning, human comfort, safety of personnel and equipment, warehousing, offices, testing laboratories, etc. ranges from 5 to 80% of the total energy used in a variety of chemical manufacturing establishments. While there are examples of high reductions in use of this energy, experience supported by numerous company observations is that a 20% sustained saving from 1972 practice is a good record. Even this requires a constant search for improvements, and a high degree of motivation for personnel because it is often difficult to control without sophisticated, tamper-proof systems.

On the other hand, where volume of production increases without expanding the energy requirements for support services, SEC for the plant is automatically reduced without there being any real improvement in the efficiency of using energy.

Downtime: A good continuous process unit, well maintained, operates from 85 to 95% of the time, a typical number being 93%. When the unit is out of service it still consumes energy for cooling down, warming up and coming to specification operation, and for maintenance, or standby protection of equipment. This can easily consume 2 to 5% of the annual energy use.

Off-Peak Performance: This refers to operation at less than the clean, new, design, or performance test condition where operating criteria are established. Most plants run this way for large parts of their operating time. The more sophisticated energy management programs are aiming at significant savings by less off-peak performance.

Most SEC target data are obtained from plant performance tests when clean and new, or from design criteria. In many instances these data do not allow for the fact that most operation is poorer than these optimum conditions, due to such factors as: partially fouled heat exchangers, or furnace tubes, worn pumps and compressors, or process control settings set for process stability rather than optimum energy use. Well-developed plant programs can generate significant energy cost savings by proper management of cleaning and maintenance schedules, and changes in process control practices.

Operating Rates (% Capacity Utilization): Operating rates change with product demand, and are often far from the optimum point for minimum specific energy consumption. Effects on some industries can be very large as during 1975 when the business slowdown cut some operating rates to 50% to 75% of normal. For some establishments, SEC rose enough to eliminate the savings of all energy improvements in the prior two years.

SEC is almost never equal at all operating rates, and is often very significantly poorer for operating rates as little as 10 to 15% up or down from the optimum point.

Many people with limited knowledge of the chemical process industries have unwittingly assumed that energy use was more or less directly proportional to the amount of product produced up to the maximum capacity of production. In Figure 54, diagonal line 1 represents this thinking graphically.

Plant operators and engineers have all known that this was never true, but few realized the real impact of operating at nonoptimum rates until the energy shortages forced them to measure it. A few plants had data in the 50% up operating range (large plants seldom operate at lower levels) which when plotted show lines like 2 and 3. When extrapolated back to 0% operation, these still showed high energy use. In general terms, this low production energy use is that needed for inefficient use of pumps and compressors, minimum rates of fuel combustion, minimum heat levels, inefficient heat recovery, and the like.

If the worst cases were like lines 2 and 3, the effect would not be overwhelming, but some cases have been analyzed that are like lines 4 and 5 on Figure 54. In these cases, reducing production causes very little reduction in total energy used in the plant. SEC rises very rapidly.

When one also considers an establishment with a high percentage of energy use for support services—almost unchanging with the establishment's production rate, and combines this with processes such as illustrated in lines 4 and 5, even a small drop in production rate can lead to a large increase in SEC for that establishment.

FIGURE 54: ENERGY CONSUMPTION VERSUS OPERATING RATE

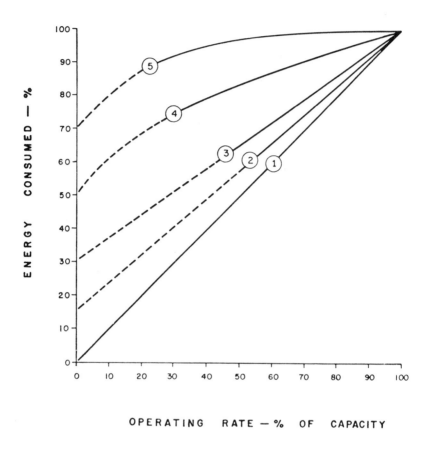

OPERATING   RATE — %   OF   CAPACITY

Source: Reference (5)

In fact, various combinations of these factors happened to a broad segment of the chemical industry during 1975 and have continued to some extent in 1976. Many companies had shown good progress in SEC compared to 1972 during 1973 and 1974; improvements of 5 to 10% each year were recorded. Suddenly, 1975 showed less improvement, no improvement, or negative results. The answer is mainly that due to a business recession, these plants were running at production rates from 50 to 75% of that run in 1972.

The Battelle study (5) has drawn particular attention to both capacity utilization and energy for support services, for industries that seem most sensitive to these factors, in making estimates of energy improvement goals. Except where specifically noted, however, the goals estimated for January 1, 1980 assume an operating capacity close to that in 1972. If energy versus capacity information were available with sufficient precision, changes in operating rates could be accounted for in analysis of energy management results. Unfortunately, few plants have, or can readily obtain, such information except as roughly estimated values.

New Energy Needs: Since 1972 most industries have found it necessary to use more energy to meet new environmental and safety standards, e.g., EPA, FDA, and OSHA regulations. Many of these new facilities have been installed and can be accounted for, and others to be installed in the near future can be estimated quite accurately. Other cases are still unknowns. There are a large number of instances where these installations are profitable such as solvent recovery, and burning wastes for fuel. Nevertheless, they do use energy.

Product Mix: The SEC goals for an establishment inherently are based on a fixed production of specific products, or production from various plant subdivisions. These may vary widely in their own SEC levels. If the mix changes in any given comparative time period, the SEC for that establishment can swing over a wide range. The Manufacturing Chemists Association (MCA) has devised reporting procedures and comparison methods to minimize this effect for its members in reporting their experience.

Economic Criteria and Funds for Investment: Economic criteria for new energy savings projects are by no means uniform within the chemical industry and vary particularly with the size of the company involved. Almost all large companies have well-defined criteria to guide investment or operating expense decisions. They range from simple payout estimates to sophisticated discounted cash flow analysis depending upon the amount, duration, and either specified or implied risks of making a profit from the expenditure.

Most companies permit energy-related projects to use reasonable forecasts of future energy prices. Project life, however, is either explicitly or intuitively kept relatively short. This short life, of course, is normal practice for improvements to old plants, but almost seems incongruous for some cases that were discussed with energy managers in newer plants. It is perhaps a substitute to achieve the same result as a better definition of the risks in the specific project.

Funds for energy-related expenditures compete with available funds for all other uses. Many companies have been permitting slightly longer payout periods, or lower return on investment than normal for energy-related projects of a relatively low investment level. However, large investments require meeting the same criteria as any other corporate investment.

All companies face the high cost of borrowed funds, the uncertainties of energy price forecasts, and the uncertainties of public policies and federal and local regulations as they apply to the duration of benefits to be obtained by energy conservation programs. In some way, each company's economic criteria evaluation take these into account such that relatively quick payout or high return is required.

Perhaps the extremes are exemplified at one end by very small business establishments who require almost instant payout for an energy savings investment; at the other end are public utilities who with their guaranteed return on approved prudent investment, and pass-through of energy cost changes, look at long benefit periods with higher average energy prices, and generally lower cost for borrowed funds.

The large chemical companies fall in between these extremes. The most frequent assessments for other than new plants are found to be on 5-year project life with a few rare cases at 10 years. By contrast, utilities look at 20 to 40 years.

A few of the larger companies whose energy management programs have been in effect for some years, have classified their opportunities and forecast investment requirements into corporate five-year programs. These managers foresee their progress in energy conservation restricted by the amount of funds available rather than by economic criteria, project lead time, or manpower. Others whose long-range plans are less developed, intuitively feel the same way but without supporting information.

ALKALIES AND CHLORINE INDUSTRY

The chemicals in SIC (Product) 2812 include chlorine, caustic soda, sodium bicarbonate, caustic potash, soda ash and miscellaneous alkalies. They are essential to the making of paper, glass, certain plastics, soaps, detergents, petroleum products, textiles and to the treating of municipal water and wastewater.

The distinction between data related to SIC (Product) 2812 and data related to SIC (Establishment) 2812 is essential to an understanding of the estimates that follow. This point will be explored in more detail for SIC 2812 than for any of the other industrial classifications in this volume. It will be done to illustrate the example for this classification alone in the interests of holding this volume to some reasonable size. The reader is referred back to Figure 53 and the related text for a further discussion of this point.

PROCESS TECHNOLOGY INVOLVED

The technologies for manufacturing alkalies and chlorine have historically been based on the processing of common salt—electrolytic manufacture of chlorine caustic and sodium metal from brine or fused salt and the lime and ammonia treatment of brine to produce soda ash.

The basic principle used in the manufacture of caustic and chlorine is electrolysis. When two electrodes are immersed in a brine solution or a bath of molten salt and a source of direct current is attached to the electrodes, sodium ions move toward the negative electrode and chlorine ions move toward the positive electrode. When brine is used, the products are 1 part of chlorine and 1.14 parts of sodium hydroxide (or more commonly caustic soda). When fused sodium chloride is used, the products are 1 part of chlorine and 0.67 part of sodium metal. When a brine of potassium chloride is used or a fused salt of potassium chloride is used, the products are chlorine, caustic potash and potassium metal.

The technologies of electrolysis are of two basic types based on the design of the cell in which the electrolysis takes place, namely, a diaphragm cell which is located between the electrodes preventing the chlorine and caustic liberated at the electrodes from recombining and a mercury cell which has a cathode of mercury metal flowing through the cell. The sodium metal formed in the mercury cell is removed as a sodium amalgam. The amalgam is treated with water outside the cell to produce various concentrations of caustic soda of high purity. For this and other reasons plants using mercury cells will probably continue to operate.

On the other hand, the additional capital for instrumentation and processing to maintain the level of mercury in emissions and effluents does impose an economic penalty on mercury cell operation. Because of the above and because of the potential technological advances in diaphragm cell technology, all new plants in the United States now scheduled for operation prior to January 1, 1980, are expected to be based on diaphragm technology.

A flow sheet of a plant based on diaphragm cell technology prevalent in 1972 is given in Figure 55. Brine from solution mining or brine from the dissolving of rock salt is pumped to a brine purification system. After treatment and heating, the purified brine enters the diaphragm cell. Electrical energy in the form of direct current enters the cell from which chlorine, caustic soda and by-product hydrogen are drawn. The chlorine is cooled, dried and compressed. Some is subsequently liquefied for shipment.

The caustic leaves the cell as about 12% caustic solution with a substantial amount of dissolved salt. Normally the cell liquor is concentrated to about 50% caustic. During concentration most of the salt precipitates and it is removed and recycled. Although not shown in Figure 55, some caustic is further concentrated to 70% solution and also another amount further concentrated to dryness producing solid sodium hydroxide.

The energy values for the various unit operations in Figure 55 are given in the boxes in units that represent millions (10^6) of Btu. Feed materials, products and by-products are underlined with energy values assigned where appropriate. No fuels or electrical energy consumption are itemized in the figure.

Figure 56 is another presentation of the major steps in the chlorine manufacturing process. The process uses diaphragm type electrolytic cells. The major step from an energy consumption viewpoint is the electrolytic separation of the brine into chlorine, hydrogen and cell liquor (dilute caustic solution). This operation accounts for 85 to 90% of the total energy consumption in the electrolytic process.

Figure 57 shows the general arrangement of the chlorine electrolytic cell. Chlorine is evolved from the cell anode while hydrogen and cell liquor come from the cathodic compartment. The cell diaphragm separates the anodic and cathodic compartments.

Figure 58 shows the major steps in the manufacture of caustic soda. This process concentrates the cell liquor from the chlorine diaphragm cell by evaporating water from the dilute caustic solution. Approximately 80% of the total energy consumption occurs in the water evaporation operation which produces 50% caustic soda.

FIGURE 55: PRODUCTION OF CHLORINE AND CAUSTIC SODA (PER NET TON OF CHLORINE)

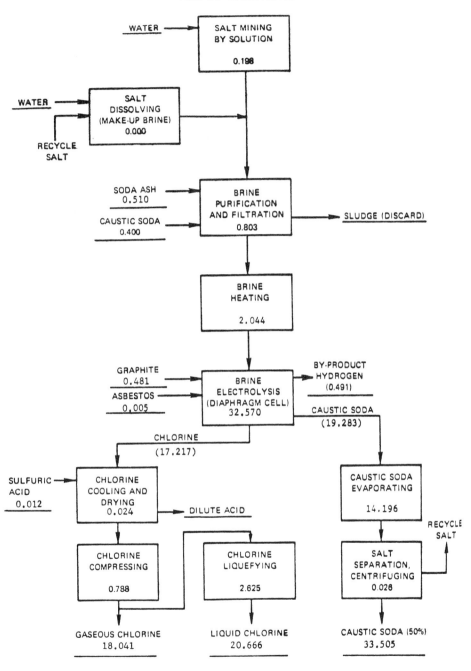

Source: Reference (12)

FIGURE 56: CHLORINE ENERGY CONSUMPTION DIAGRAM

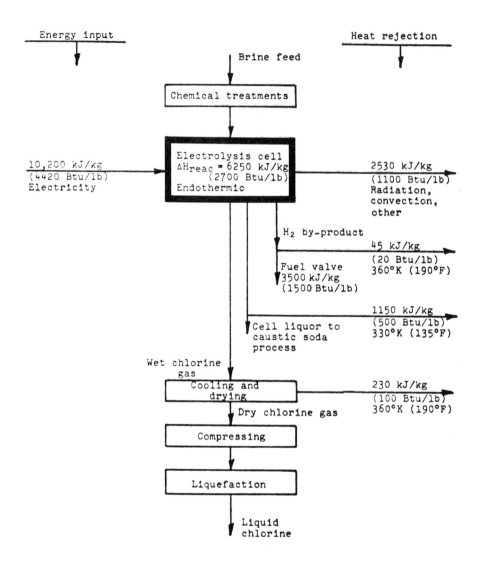

1973 U.S. production: 9.35 x 10^9 kg (20.6 x 10^9 lb)
1973 energy consumption (primarily electricity): 3,300 Mw
    (100 x 10^{12} Btu)
1973 fuel generation (H$_2$): 1,000 Mw (30 x 10^{12} Btu)

Source: Reference (3)

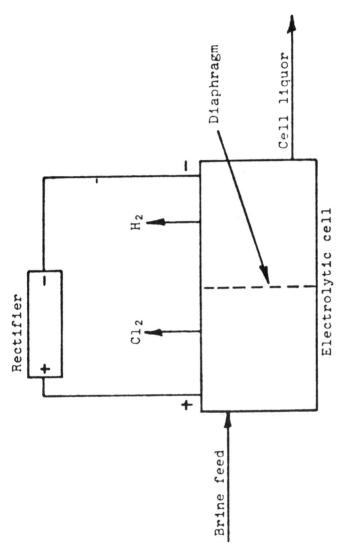

FIGURE 57:   EQUIPMENT DIAGRAM—ELECTROLYTIC CELL

Rejected heat: radiation, convection, other—2,530 kJ/kg (1,100 Btu/lb)
Hot H₂ by-product—45 kJ/kg (20 Btu/lb) at 360°K (190°F)
Hot Cl₂ product—230 kJ/kg (100 Btu/lb) at 360°K (190°F)
Warm cell liquor—1,150 kJ/kg (500 Btu/lb) at 330°K (135°F)

Source: Reference (3)

FIGURE 58: CAUSTIC SODA ENERGY CONSUMPTION DIAGRAM*

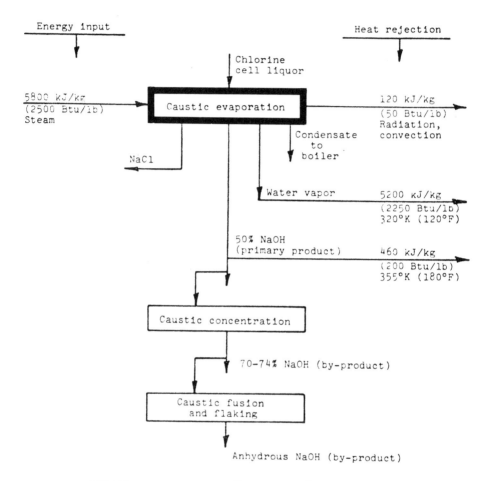

1973 U.S. production: 9.70 x 10⁹ kg (21.4 x 10⁹ lb)
1973 energy consumption (primarily steam): 2,700 Mw (80 x 10¹² Btu)

*Weight of products as 100% caustic soda.
 Energy values are in terms of energy per unit weight of caustic soda
 as 100% caustic soda.

Source: Reference (3)

Figure 59 shows a typical caustic soda evaporation operation. Cell liquor enters the third effect of a three-effect evaporation system and 50% caustic soda leaves the first effect. Sodium chloride in the cell liquor is separated from the caustic soda solution after the solution leaves the first and second effect evaporators. Steam is the source of heat.

FIGURE 59: EQUIPMENT DIAGRAM—CAUSTIC SODA EVAPORATORS

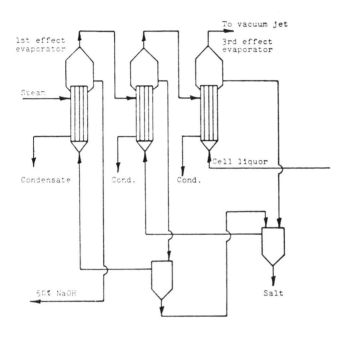

Rejected heat: radiation, convection, other—120 kJ/kg (50 Btu/lb)
Water vapor—5,200 kJ/kg (2,250 Btu/lb) at 320°K (120°F)
Hot product—460 kJ/kg (200 Btu/lb) at 355°K (180°F)

*Energy values are in terms of energy per unit weight of
   caustic soda as 100% caustic soda.

Source: Reference (3)

Technology, by definition, means practical application of science and engineering, which in turn, involves costs. In 1972 a typical plant designed for 500 tons per day of chlorine using metal anodes required a capital cost of a little over $20 million, exclusive of land, site preparation, electric power generation facilities, plant water systems, and auxiliary maintenance, laboratory, and administration buildings. The estimate was based on ENR index on the Gulf Coast = 1800. Inflation in plant construction between 1972 and 1975 has boosted the capital cost in 1975 to over $50 million for the same plant.

The cost of energy in such a plant represents about 35% of the total costs of manufacture including depreciation, and depreciation was estimated at about 17% of total costs of manufacture.

Figure 60 shows the major steps in the synthetic sodium carbonate manufacturing process. The Solvay process is used. Major steps from an energy consumption viewpoint are the lime kiln operation, the compression of carbon dioxide, and the calcining of sodium bicarbonate to sodium carbonate. These steps account for more than 70% of the total energy consumption in the process.

FIGURE 60: SODIUM CARBONATE ENERGY CONSUMPTION DIAGRAM

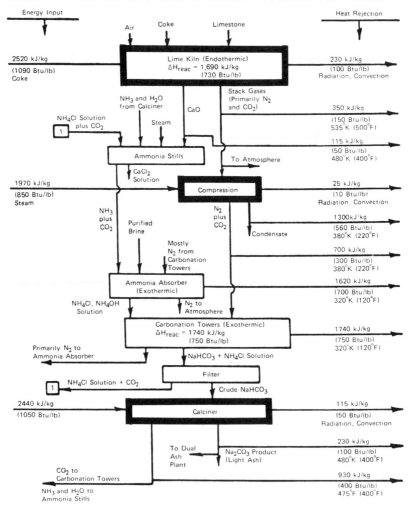

1973 U.S. production: 6.80×10^9 kg (15.0×10^9 lb)
1973 energy consumption (primarily steam, coke): 2,000 Mw (60×10^{12} Btu)

Source: Reference (3)

Figure 61 shows the lime kiln operation. Coke supplies energy to convert calcium carbonate to calcium oxide and carbon dioxide.

FIGURE 61: EQUIPMENT DIAGRAM—LIME KILN

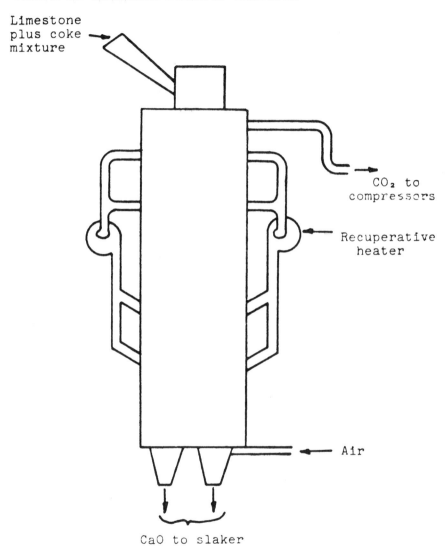

Lime kiln

Rejected heat: radiation convection—230 kJ/kg (100 Btu/lb)
Stack gases—485 kJ/kg (210 Btu/lb) at 535°K (500°F)
Hot lime—115 kJ/kg (50 Btu/lb) at 480°K (400°F)

Source: Reference (3)

Figure 62 shows the compression of carbon dioxide and nitrogen from the lime kiln. Steam is used to provide energy to drive the compressors. Figure 63 shows the calcining of sodium bicarbonate to sodium carbonate. Steam is the heat source.

FIGURE 62: EQUIPMENT DIAGRAM—COMPRESSORS

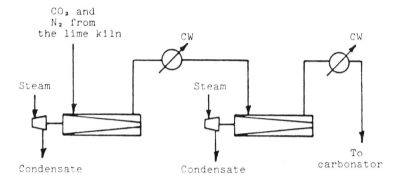

Rejected heat, radiation, convection—25 kJ/kg (10 Btu/lb)
Condensate (vapor)—1,300 kJ/kg (560 Btu/lb) at 380°K (220°F)
Hot compressed CO_2 and N_2—700 kJ/kg (300 Btu/lb) at 380°K (220°F)

FIGURE 63: EQUIPMENT DIAGRAM—CALCINER

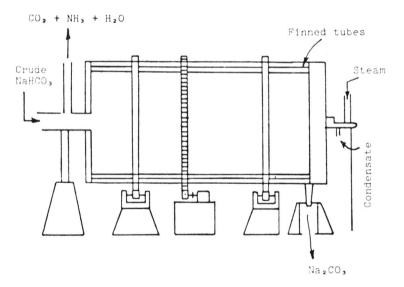

Rotary dryer

Rejected heat: radiation, convection—115 kJ/kg (50 Btu/lb)
Hot product—230 kJ/kg (100 Btu/lb) at 480°K (400°F)

Source: Reference (3)

MAJOR ENERGY CONSERVATION OPTIONS TO 1980

Table 75 indicates the causes of energy losses in the electrolytic brine separation operation. It also shows the approximate magnitude of the losses and possible energy conservation approaches. Table 76 shows the causes of energy losses in the evaporation of water from cell liquor. It also shows the approximate magnitude of the losses and some possible energy conservation approaches.

TABLE 75: CHLORINE ENERGY CONSERVATION APPROACHES

Causes of Energy Losses	Approximate Magnitude of Losses	Energy Conservation Approaches
Electrolysis cell		
Anode overvoltage	460 kJ/kg (200 Btu/lb)	Operation modification (lower current density) Research and development (improve anode material)
Cathode overvoltage	1,400 kJ/kg (600 Btu/lb)	Operation modification (lower current density) Research and development (improve cathode material)
Voltage drop across diaphragm	580 kJ/kg (250 Btu/lb)	Design modification (thinner diaphragm)
Voltage drop in electrolyte	460 kJ/kg (200 Btu/lb)	
Voltage drop in anode-cathode assemblies	280 kJ/kg (120 Btu/lb)	Design modification
Oxygen evolution on anode	230 kJ/kg (100 Btu/lb)	Research and development (improve anode material)
Unaccounted for	580 kJ/kg (250 Btu/lb)	–
Overall process		
Lack of heat recovery from H_2, Cl_2, and cell liquor streams	1,425 kJ/kg (620 Btu/lb)	Design modification (waste heat recovery)
Radiation, convection, other heat losses from electrolysis cell	2,530 kJ/kg (1,100 Btu/lb)	Insulation
Failure to use H_2 by-product as fuel	3,500 kJ/kg 1,500 Btu/lb	Waste utilization

Note: Electrolysis cell losses are electrical. The fuel value of these losses would be approximately three times as large as the values listed.

TABLE 76: CAUSTIC SODA ENERGY CONSERVATION APPROACHES

Causes of Energy Losses	Approximate Magnitude of Losses	Energy Conservation Approaches
Rejected heat		
Radiation, convection	120 kJ/kg (50 Btu/lb)	Insulation Maintenance
Water vapor from last effect	5,200 kJ/kg (2,250 Btu/lb)	Design modification (waste heat recovery)
Hot product	460 kJ/kg (200 Btu/lb)	Design modification (waste heat recovery)

(continued)

TABLE 76: (continued)

Causes of Energy Losses	Approximate Magnitude of Losses	Energy Conservation Approaches
Overall process		
Use of excess stream	1,400 kJ/kg (600 Btu/lb)	Design modification (add additional effect to evaporation operation)
Low NaOH concentration in cell liquor	2,300 kJ/kg (1,000 Btu/lb)	Research and development (ionic membrane diaphragm)
Production of anhydrous NaOH	230 kJ/kg (100 Btu/lb)	Market modification (substitute 70 to 74% NaOH for anhydrous)

Note: Energy values are in terms of energy per unit weight of caustic soda as 100% caustic soda.

Source: Reference (3)

Table 77 shows the causes of energy losses in the lime kiln operation, the compression operation, and the calcining operation. It also shows the approximate magnitude of losses and possible energy conservation approaches.

TABLE 77: SODIUM CARBONATE ENERGY CONSERVATION APPROACHES

Causes of Energy Losses	Approximate Magnitude of Losses	Energy Conservation Approaches
Rejected heat		
Radiation, convection	370 kJ/kg (160 Btu/lb)	Insulation Maintenance
Stack gases	350 kJ/kg (150 Btu/lb)	Design modification (waste heat recovery)
Uncondensed steam from compressor	1,300 kJ/kg (560 Btu/lb)	Process integration
Process streams	5,340 kJ/kg (2,300 Btu/lb)	Design modification (waste heat recovery)
Overall process		
Nonisothermal compression	115 kJ/kg (50 Btu/lb)	–
Nonisentropic compression	115 kJ/kg (50 Btu/lb)	–
Gas for compression is only 40% CO_2	1,150 kJ/kg (500 Btu/lb)	Process modification (use higher oxygen content combustion air)
Heat required to dry limestone and coke	115 kJ/kg (50 Btu/lb)	Design modification (enclosed storage)
Heat lost in heating impurities in limestone	230 kJ/kg (100 Btu/lb)	
High water content in calciner feed	345 kJ/kg (150 Btu/lb)	Design modification

Source: Reference (3)

The energy used for manufacturing alkalies and chlorine in SIC 28 in 1972 has been estimated to be 377×10^{12} Btu as shown in Table 78. This estimate excludes both the energy consumed in the paper industry (SIC 26) for the manufacture of chlorine and caustic for captive use and the energy in mining and processing natural soda ash.

Caustic and chlorine are produced from salt using the principle of electrolysis. It is estimated that 3,425 kwh and 9,680 pounds of steam were required in 1972 to produce one electrochemical unit (e.c.u.) consisting of 1 ton of chlorine, 1.14 tons of caustic soda and 57 pounds of by-product hydrogen. This electrical energy and steam amounts to a total of 34 million Btu per e.c.u. for the average plant which generates half of its electrical energy.

TABLE 78: ESTIMATED ENERGY USAGE IN 1972 FOR MANUFACTURING ALKALIES AND CHLORINE IN SIC 28 ESTABLISHMENTS

Products in SIC 28 Establishments	1972 Production (10^6 short tons)	Energy[a]	
		Specific Energy 10^6 Btu/ton	Total 10^{12} Btu
Chlorine	9.38[b]	34	319
Caustic soda	9.67[c]	--	(d)
Caustic potash	0.19	--	(d)
Soda ash (synth.)	4.31	13	56
Other alkali	0.10	20	2
	23.65		377

(a) Purchased electrical energy was converted to Btu using 3,412 Btu/kwh.
(b) Includes all chlorine from all processes including fused salt processes.
(c) Does not include cell liquor used as such.
(d) Included in electrochemical unit with chlorine.

Source: Reference (5)

In 1972 the mix of electrolytic cells in chlorine and caustic plants was about 23% mercury cells and 77% diaphragm cells. Table 79 presents estimates of the average energy consumption in terms of kwh and pounds of steam for each of the major operations and assuming the average mix of electrolytic cells. The average energy usage per electrochemical unit is 3,425 kwh and 9,680 pounds of steam. Any particular plant may be operating economically as much as 10% above or 10% below the average.

Table 80 presents the estimated total energy usage in terms of 10^6 Btu per e.c.u. in the chlor-alkali industry in 1972. It has been said that about one-half of the electrical energy used in producing chlorine and caustic is generated within the plant and half is purchased. Assuming this to be the case, Table 80 shows the conversion of 3,425 kwh and 9,680 pounds of steam to Btu on a net basis and on a gross basis. Energy use on a net basis is calculated at 3,412 Btu per kwh for half of the electrical energy used and the remainder at 9,500 Btu/kwh. The latter figure assumes the average efficiency achieved in generating electrical energy and steam on site. It was further assumed that in 1972 only 50% of by-product hydrogen was sold as such or burned usefully. On the basis of the estimate presented in Table 79 and Table 80 the specific energy consumption per electrochemical unit in 1972 is estimated to be about 34 million Btu per e.c.u. when calculated on a net basis and 46 million Btu per e.c.u. when calculated on a gross basis.

TABLE 79: AVERAGE ENERGY USAGE PER ELECTROCHEMICAL UNIT
IN CHLOR-ALKALI INDUSTRY IN 1972

	Energy Used, per e.c.u.	
	kwh	Steam, pounds
Salt mining		
Electrical energy	18	–
Brine Purification		
Steam		500
Electrical energy	5	
Brine heating		1400
Electrolysis (AC input)		
Mercury Cells 3700 kwh		
Diaphragm Cells 3050 kwh		
Industry mix of 23% mercury		
cells and 77% diaphragm cells	3200	
Chlorine Cooling and Drying		
Electrical energy	2	
Chlorine Compressing	75	
Liquefaction of 50% of chlorine	125	
Concentration of Caustic		
88% of caustic conc. to 50%*		6100*
8% of caustic conc. to 70%		960
4% of caustic conc. to solid		720
Totals	3425	9680

*Energy shown is for concentration of cell liquor from diaphragm cells.

Source: Reference (5)

TABLE 80: TOTAL ENERGY USAGE PER ELECTROCHEMICAL UNIT IN
CHLOR-ALKALI INDUSTRY IN 1972

	Energy Used 10^6 Btu/e.c.u.	
	Net (a)	Gross (a)
Total Electrical Energy 3425 kwh		
50% generated at 9500 Btu/kwh	16.23	16.23
50% purchased at 3412 Btu/kwh	5.84	
at 10,500 Btu/kwh		17.98
Total pounds of steam 9680 pounds		
50% at 1,250 Btu/lb	6.05	6.05
50% at 1,440 Btu/lb	7.10	7.10
Total Energy Required	35.22	47.36
Hydrogen Credit		
50% sold as hydrogen or burned usefully	-1.20	-1.20
	34.02	46.16

(a) Net and gross in this context refer to the difference in the
factor used for converting purchased electrical energy to
equivalent Btu. The "net energy" is converted using 3412
Btu per kwh. The "gross energy" is converted using 10,500 Btu
per kwh, which is the average energy input to electric
utilities per kwh produced.

Source: Reference (5)

The estimated usage of energy in the production of all alkalies and chlorine
products in 1972 is shown in Table 81. Of the total of 27.72 million short
tons, 85% of 23.65 million short tons are produced in SIC 28.

The paper industry (SIC 26) produces chlorine and caustic for its own captive
use. Natural soda ash is mined in the West (mining operations are not considered in manufacturing), and hence data on such activities are not included in
the Census of Manufactures.

Table 81 shows the total energy consumed in 1972 for producing alkalies and
chlorine, SIC (Product Class) 2812 was 408 x 10^{12} Btu on a net basis and
533 x 10^{12} Btu on a gross basis.

TABLE 81: ESTIMATED ENERGY USAGE IN 1972 BY SIC (PRODUCT CLASS) 2812, ALKALIES AND CHLORINE

Products	1972 Production, 10⁶ short tons	Net Energy[e] SEC 10⁶ Btu per ton	Net Energy[e] Usage 10¹² Btu	Gross Energy[e] SEC 10⁶ Btu per ton	Gross Energy[e] Usage 10¹² Btu
In SIC 28 Establishments					
Chlorine	9.38[a]	34	319	46	431
Caustic	9.67[a]	--	(c)	--	(c)
KOH	0.19	--	(c)	--	(c)
Soda ash (synth.)	4.31	13	56	13	56
Other alkali	0.10	20	2	20	3
Subtotal	23.65		377		490
In Non-SIC 28 Establishments					
Paper Mills					
Chlorine	0.47	34	16	46	22
Caustic	0.54	--	(c)	--	(c)
Mining Operations					
Natural soda ash	3.06	5[d]	15	7[d]	21
Subtotal	4.07		31		43
Total SIC (Product) 2812	27.72		408		533

(a) Includes chlorine from all processes including fused salt processes.
(b) Does not include cell liquor used as such.
(c) Included in the electrochemical unit with chlorine.
(d) Excludes increased energy for transporting natural soda ash from the Green River Basin, Wyoming, and other Western locations to Eastern markets.
(e) Same note as on Table 80.

Source: Reference (5)

Following is a brief summary of options for reducing energy per unit of output and the maximum feasible application of them prior to January 1, 1980.

New Plants: The major contribution to the reduction of energy per e.c.u. will come from new plants placed in operation between 1972 and 1979.5 and the retirement of older facilities. Operating with the best technology available in 1979.5, about 2,896 kwh and 9,150 lb of steam will be required per e.c.u. These quantities of energy are equivalent to about 27 million Btu for the average plant which generates about half of its own electrical energy.

Of course only the capacity installed between 1972 and 1979.5 will operate with this level of energy efficiency. Assuming that 13% of the plants operating in 1972 are retired, and the remaining plants can achieve a 5% reduction in energy usage through improved housekeeping, the practical minimum average energy usage in 1979.5 is estimated to be 31 million Btu per e.c.u.

Housekeeping: Experience has shown that a well-organized and intensive program for improved housekeeping in the chlor-alkali industry can achieve from 4 to 5% savings. A maximum of 5% is a reasonable goal for 1979.5.

Metal Anodes: Conventionally graphite was used as the anode material in both the diaphragm and mercury cells. The most discussed technological change taking place since 1970 has been the use of coated metal anodes. A cell using metal anodes can operate at a lower voltage than a graphite cell. In a number of previous reports it has been asserted that a cell using metal anodes requires 10 to 15% less electrical energy. This level of savings has not been achieved in most plants for economic reasons. The operator of such a cell has the option of operating with reduced energy consumption per e.c.u. or operating the cell with the same energy consumption and achieving an increase in cell output of 10 to 15%.

Cell output is a function of anode current density. The economics favor operation for maximum output of the cell. Although it has been asserted that a plant converted to metal anodes can reduce energy consumption by 15 to 18%, it is not economically feasible in most cases to operate the plant in such a manner. In this analysis it is estimated that no significant savings of energy can be attributed to the replacement of graphite anodes with coated metal anodes.

Modified Diaphragms: The relationship between power consumption in a metal anode cell with a modified diaphragm is very similar to that previously discussed. The economics favor operating cells with metal anodes and modified diaphragms for maximum cell output. In such cases all unit operating costs are reduced except for cell energy. In this analysis, it is estimated that no significant savings in energy will be achieved with use of modified diaphragm as a replacement for asbestos.

Increased Number of Stages in Evaporators: The cost of concentrating caustic in terms of energy is substantial. It is said that most evaporators currently in use are double and triple effect. One quadruple effect evaporator is being installed. In the absence of specific data it can be said categorically that capital invested in advanced design concentrators would save energy. It is unlikely that the energy savings will justify immediate replacement. More likely, as caustic concentrators are replaced, triple and quadruple effect equipment will be installed, depending on the economics of the location.

Hydrogen Utilization: There is little agreement regarding the level of hydrogen utilization in 1972. Some assert that most of it was flared or vented, while others claim that substantial amounts were cooled, cleaned and used beneficially. It is assumed that 50% of the hydrogen in 1972 was so used. By mid-1979, it should be possible to have facilities for recovering all of the by-product hydrogen and using it either as a fuel in the boilers or as a chemical in captive uses or as a product for sale. It has been estimated that the maximum amount of energy that could be recovered from the hydrogen would be about 2.4 million Btu per e.c.u.

In conclusion, by mid-1979 it is estimated that facilities for recovering all of the by-product hydrogen will be in place. This will result in an additional energy credit of 1.2 million Btu per e.c.u., or about 3.5%.

Combined Steam-Electric Generation: About half the chlor-alkali plants generate their own steam and electric power, however, data substantial enough to confirm such a statement is not readily available. Data on the efficiency of existing steam and electric power generation facilities is similarly difficult to obtain.

The most energy efficient facilities would be a gas turbine and steam turbine in sequence, in which the electric power would be generated at about 7,700 Btu per kwh and the steam at 1,250 Btu per pound. But with limitations on gas in new plants, such combined cycle facilities will not be prevalent in new facilities. Facilities already in place for the generation of electric power and steam cannot be readily modified except for their conversion to liquid fuels from gaseous fuels.

It is estimated that no significant increase in efficiency can be achieved by new power plants installed between 1972 and 1979.5 and hence the value of 9,500 Btu/kwh has been used.

Facilities for Handling Liquid Fuels: Most, if not all, plants currently operating on natural gas have installed stand-by facilities for handling liquid fuels. Most of these plants are on the Gulf Coast where the availability and cost of solid fuels precludes their use. The penalties associated with the usage of liquid fuels are small enough to be neglected in this analysis.

Use of Off-Peak Power: Some reports on energy usage in the chlor-alkali industry have pointed up the advantages of using off-peak power. Use of off-peak power means that a chlor-alkali plant would reduce power at certain times when utilities are operating at peak loads. One company estimates that the economics of reducing power by 30% during peak loads would necessitate an 8-mil adjustment in the price of energy. The use of off-peak power does not appreciably affect the energy usage per e.c.u.; it will, of course, help the utility company with its base-load planning and operation.

Energy Required for Meeting OSHA and EPA Standards: No attempt has been made in this short period to develop precise estimates of the additional energy required to meet OSHA and EPA standards. Industry sources indicate that such energy requirements are relatively small in relation to the total for diaphragm cells—of the order of 0.5×10^6 Btu per e.c.u. for diaphragm cells; and, for mercury cells, the energy requirements may be several times that figure.

GOAL YEAR (1980) ENERGY USE TARGET

The previously presented data provide the basis for estimating an energy efficiency goal for SIC (Product Class) 2812. The calculations based on those data and estimates are summarized in Table 82.

This table shows that in 1979.5, with 1972 technology, energy usage would be 491×10^{12} Btu. With maximum energy reduction practices in mid-1979, production of alkalies and chlorine in SIC 28 will require about 451×10^{12} Btu. These estimates of energy usage indicate a maximum energy efficiency goal of 8%.

TABLE 82: ESTIMATED ENERGY USAGE IN SIC (PRODUCT CLASS) 2812
IN 1979.5 IN SIC 28 ESTABLISHMENTS BASED ON MAXIMUM FEASIBLE
ENERGY REDUCTION PRACTICES

	1979.5
Production, 10^6 short tons	30
Total Energy Required, 10^{12} Btu	
Using 1972 Technology	491
Using 1979.5 Technology	451
Specific Energy Consumption, 10^6 Btu per short ton	
Using 1972 Technology	16.4
Using 1979.5 Technology	15.0
Maximum Energy Efficiency Goal, in percent	8

Calculation of Energy Efficiency Goal:

$$\text{Net Goal (including offsets)} = \frac{491-451}{491} = 8 \text{ percent}$$

Source: Reference (5)

Energy usage in 1979.5 will be an average of the energy usage of the newest, most energy-efficient plants and the older plants still in operation. The method for estimating this average is as follows:

(a) Estimate the energy usage in the most energy-efficient plants.
(b) Assume that 13% of the least energy-efficient capacity operating in 1972 is retired. It was assumed that such plants were operating at 10% in excess of the 1972 energy consumption per e.c.u.
(c) Assume that all new plants built between 1972 and 1979.5 are using the most energy-efficient technology.
(d) Estimate that the plants operating in 1972 which are still operating in 1979.5 can save 5% of their previous energy usage through improved housekeeping.

Taking into account the above advances in technology, it is estimated that the best available technology in operation in 1979.5 will use energy as indicated in Table 83. This assumes a cell mix of 20% mercury cells and 80% diaphragm cells. On this basis the estimated energy usage per e.c.u. is 2,896 plus 9,150 pounds of steam.

When these values are summarized, estimated best technology is about 29×10^6 Btu per e.c.u. on a net basis and about 39×10^6 Btu per e.c.u. on a gross basis as shown in Table 84. This includes a hydrogen credit of 2.4×10^6 Btu per e.c.u.

TABLE 83: ESTIMATED BEST TECHNOLOGY, 1979.5

	Estimated Energy Usage, per e.c.u.	
	kwh	Steam
Salt mining Electrical energy	19	
Brine Purification Steam Electrical energy	 5	 450
Brine Heating		1250
Electrolysis (AC input)		
Mercury cells 3450 Diaphragm cells 2475 Industry mix of 20% mercury cells and 80% diaphragm cells	 2670	
Chlorine cooling and drying	2	
Chlorine compressing	75	
Liquefaction of 50% chlorine	125	
Concentration of caustic		
88% of caustic conc. to 50%* 8% of caustic conc. to 70% 4% of caustic conc. to solid		5770* 960 720
Totals	2896	9150

*Energy shown is for concentration of cell liquor for diaphragm cells.

TABLE 84: ESTIMATED BEST TECHNOLOGY, 1975

	Estimated Energy Usage, 10^6 Btu per e.c.u.	
	Net (a)	Gross (a)
Total Electrical Energy 2896 kwh		
50% generated at 9500 Btu per kwh	13.76	13.76
50% purchased at 3412 Btu per kwh	4.94	
at 10,500 Btu per kwh		15.20
Total pounds of steam 9150 pounds		
50% at 1.25 Btu per pound	5.72	5.72
50% at 1.44 Btu per pound	6.59	6.59
	31.01	41.27
Less hydrogen credit with usefully sold or burned in boilers	2.40	2.40
	28.61	38.87

(a) Same note as on Table 80.

Source: Reference (5)

Following the above method, the minimum energy usage in chlorine and caustic plants in 1979.5 was estimated to be 31 million Btu per e.c.u. on a net basis and 41 million Btu per e.c.u. on a gross basis. The calculations for these estimates are given in Table 85.

In summary, the energy usage in chlorine and caustic manufacture in 1972 and 1979.5 are estimated to be as follows:

	Net 10^6 Btu	Gross 10^6 Btu
1972 energy usage per e.c.u.	34	46
1979.5 minimum energy usage per e.c.u.	31	41

TABLE 85: ESTIMATED MINIMUM SPECIFIC ENERGY CONSUMPTION IN CHLOR-ALKALI MANUFACTURE IN 1979.5

	Production, 10^6 short tons	SEC 10^6 Btu/ e.c.u.	Energy Usage		
			Net 10^{12} Btu	10^6 Btu/ e.c.u.	Gross 10^{12} Btu
1972 Plants	9.85	34	335	46	453
Less 13% retired at +10% over average energy	1.28	37	47	51	65
1972 Plant Still in Operation in 1979.5	8.57		288		388
Less 5% improvement in housekeeping			14		19
1972 Plants Operating in 1979.5			274		369
Plants Built and/or Modernized 1972-1979.5	5.73	29	166	39	223
Total	14.30		440		592
Minimum Energy in 1979.5 per e.c.u.			31		41

Source: Reference (5)

The energy usage in 1979.5 was calculated for SIC 28 establishments in 1979.5 on two bases:

(a) Based on 1972 technology and energy management
(b) Based on 1979.5 technology and management

Table 86 shows that the maximum energy efficiency goal on a net basis is about 8% using the 3,412 Btu per kwh conversion factor for purchased electrical energy and about 10% using 10,500 Btu per kwh.

TABLE 86: ESTIMATED ENERGY USAGE IN SIC (PRODUCT CLASS) 2812 IN 1979.5 BASED ON MAXIMUM FEASIBLE ENERGY REDUCTION PRACTICES AND 1972 PRACTICE

Products	Estimated Production, 10^6 short tons	1979.5 With 1972 Technology		1979.5 With Maximum Feasible Energy Reduction	
		Net Energy,[a] 10^{12} Btu	Gross Energy,[a] 10^{12} Btu	Net Energy, 10^{12} Btu	Gross Energy, 10^{12} Btu
In SIC 28 Establishments					
Chlorine	13.6	462	626	422	558
Caustic	14.0	(b)	(b)	(b)	(b)
KOH	0.3	(b)	(b)	(b)	(b)
Soda ash (synthetic)	1.9	25	25	25	25
Other alkalies	0.2	4	4	4	4
	30.0	491	655	451	587

Estimated Goal:

$$\text{Net Basis (a)} \quad \frac{491-451}{491} = 8.1$$

$$\text{Gross Basis (a)} \quad \frac{655-587}{655} = 10.4$$

(a) Same note as on Table 80.
(b) Included in the energy figure for chlorine.

Source: Reference (5)

SOME PROJECTIONS BEYOND 1980 TO 1990

General conservation and efficiency measures are being applied to chemical industries at the two-digit level, but several process specific measures for use in the chlorine-caustic production are identifiable and quantifiable. A projection of energy consumption in chlorine and caustic soda production based on modifications specific to this process is shown in Table 87.

TABLE 87: PROJECTED ENERGY CONSUMPTION IN CHLORINE AND
CAUSTIC PRODUCTION*

	1967	1971	1973	1977	1980	1985	1990
			Total Production (10^6 tons)				
Pre 1975 Plants	16.1	19.0	21.0	19.7	18.8	17.2	13.8
New Plants	--	--	--	4.2	9.7	20.9	37.2
Total	16.1	19.0	21.0	23.9	28.5	38.1	51.0
			Fuels (10^{12} BTU's)				
Pre 1975	172	203	225	211	201	184	30
New Plants	0	0	0	45	86	170	80
Total	172	203	225	256	287	354	110
			Electric (10^9 KWH's)				
	1967	1971	1973	1977	1980	1985	1990
Pre 1975	24.3	28.6	30.5	27.3	24.4	21.6	17.4
New Plants	--	--	--	5.6	12.2	25.8	45.5
Total	24.3	28.6	30.5	32.9	36.6	47.4	62.9

*Process specific conservation measures only

Source: Reference (10)

Modifications used in the projection include:

(a) Retirement of 1.5% of pre-1975 production capacity each year.

(b) 80% conversion to metal anodes by 1980 (12% reduction in electric consumption).

(c) 30% conversion to synthetic diaphragms and adjustable cathodes by 1980. Complete conversion to both by 1985. (5% reduction in electric consumption.)

(d) Use of bipolar geometrics and quadruple effect evaporators in new plants built after 1977. (5% electric reduction; 30% steam reduction).

(e) Industry-wide use of permionic membranes by 1990. (80% steam reduction.)

The opportunities for energy conservation in soda ash synthesis are not as extensive as those for other chemical production industries. One recent process improvement has been the substitution of superheated steam for direct firing in the calcining of bicarbonate. The substitution led to a 10 to 15% reduction in heat used in this step (about a 4% improvement overall).

General measures applicable to soda ash synthesis include those related to heat retention and improved heat efficiency. They would include increased insulation, improved steam system maintenance and perhaps the increased use of solar heat evaporation. With no new soda ash synthesis plants being planned, major process changes are not anticipated. General measures such as these will be accounted for at the two-digit level.

Although overall soda ash production in the U.S. is expected to grow at a rate of 4 to 5% per annum, the use of natural soda ash is expected to grow even faster and synthetic soda ash production is expected to decline gradually through 1980 at about 0.8% per year. The rate of decline based on Department of Commerce data has been extrapolated to 1990. The projection of energy consumption shown in Table 88 is based on that assumption.

TABLE 88: PROJECTION OF SODA ASH SYNTHESIS ENERGY CONSUMPTION

	1967	1971	1977	1980	1985	1990
Production (10^3 tons)	4,849	4,275	3,716	3,627	3,484	3,347
Shipment value (10^6 \$, 1967)	147	130	113	110	106	101
Fuel consumption* (10^{12} Btu)	47.0	41.5	35.3	34.1	32.7	31.5

*Does not include coke consumption.

Source: Reference (10)

INDUSTRIAL GASES INDUSTRY

Industrial gases are defined as the elements obtained from air, hydrogen, helium, carbon dioxide, nitrous oxide, and acetylene. The air-separation products are oxygen, nitrogen, argon, neon, krypton, and xenon. Some important gases not included in SIC 2813 are chlorine (2812), fluorine (2819), ammonia (2873), sulfur dioxide (2819), and natural gas, all other hydrocarbon gases except acetylene and halocarbon gases.

The products of this class had a shipment value (1972) of 660 million dollars, just over 1% of the chemical industry's shipments. Primary energy consumption in SIC 2813 is to produce oxygen and nitrogen because of their large volume production, and hydrogen and acetylene because of their high energy requirement per pound.

These four account for 98% and energy savings opportunities were determined for them and argon. Carbon dioxide energy consumption was determined but the total was so small no savings goals were calculated. Nitrous oxide, helium, and all the other air gases were ignored because their volume was so low. "Other gases" and "gases not specified by kind" were treated as gaseous nitrogen.

Note that certain amounts of industrial gas production are not included in product category 2813 but are included elsewhere. Nitrogen and hydrogen generated and used in ammonia establishments are counted in 2873. Hydrogen generated and used in methanol establishments is in 2869. Carbon dioxide generated and used in establishments making urea is in 2873 and in those making soda ash in 2812. Acetylene made in small generators from calcium carbide in railyards, shipyards, welding shops, etc., is excluded. Hydrogen produced in oil refineries for captive use is also excluded.

Table 89 gives the major uses for the six largest volume gases. The uses are diverse but note that almost all relate to producing some other material, rather than use as part of the final demand. This service role, the low value per pound, and high cost of transportation result in the location of most plants close to

229

the point of use. The extreme example is the captive oxygen plant adjacent to a steel mill.

TABLE 89: MAJOR USES OF SELECTED 2813 CHEMICALS

Oxygen
      Steelmaking
      Foundry metal melting
      Metal fabrication
      Medical
      Aerospace propellant
      Synthetic fuels making (future)
      Waste water treatment (future)
Nitrogen
      Inert gas shield
      Refrigeration including food freezing
      Ammonia making (future)
Hydrogen
      Chemical and refinery use
      Aerospace propellant
      Vegetable oil hardening
      Electric generator cooling
      (Ammonia manufacture, SIC 2873)
      (Methanol manufacture, SIC 2869)
Argon
      Steelmaking
      Inert gas shield
Acetylene
      Chemical intermediate
      Metal fabrication
Carbon dioxide
      Refrigeration
      Beverage carbonation
      Chemical intermediate
      Inert gas shield, including fire fighting
      Propellant and pressure

Source: Reference (5)

PROCESS TECHNOLOGY INVOLVED

Oxygen

Air consists of 78% nitrogen, 21% oxygen, 0.8% argon and small amounts of neon, krypton, xenon, and carbon dioxide. In principle, one cleans and dries an air stream, removes the carbon dioxide, and liquefies the air. Careful continuous fractional distillation of the liquefied air gives separate streams of the component gases if the equipment is designed for total separation.

The xenon and krypton are available from the highest boiling fraction (BP –109°C and –152°C). Next to be removed is oxygen (BP –183°C). In most air separation plants, krypton and xenon are left in the oxygen stream. Argon, which boils near oxygen (BP –185°C), is frequently removed with the nitrogen (BP –196°C) but some plants recover argon.

Neon (BP –249°C) can be obtained by separate redistillation of the nitrogen. The cooling is obtained by compressing the air, cooling it in the heat exchanger, then allowing it to expand, which drops the temperature to –186°C. A double distillation column is used, the lower one separating the air at high pressure into a liquid nitrogen fraction, which is removed at the top, and a still bottoms of liquid nitrogen rich in oxygen.

These streams are introduced into the upper (low pressure) column where the nitrogen and argon are removed at the top in a waste stream containing some oxygen. The still bottoms consists of liquid oxygen, which can be removed as such or allowed to evaporate and be removed from close to the bottom of the still. Boiling of the condensed oxygen at the bottom of the upper still provides the cooling for the nitrogen condensation at the top of the high-pressure still.

Downward flowing condensate cools and liquefies the air admitted at the bottom of the lower still. The cold off gases are led through heat exchangers to cool the incoming air and thus conserve energy. Thus the electricity to power the compressors indirectly cools, liquefies, and separates the gases.

Nitrogen

Nitrogen can be obtained by air liquefaction and distillation by modifying the design shown for production of oxygen so that the off stream of nitrogen is purer. This means use of a distillation column with more theoretical plates and a reflux ratio designed for higher purity of both the oxygen fraction and the nitrogen fraction (at an increased capital cost).

Argon

Argon can be removed from near the center of the low-pressure still in the air liquefaction plant. Because it boils between oxygen and nitrogen it is difficult to remove in pure form. The usual procedure is to remove a stream of argon mixed with oxygen and a bit of nitrogen. Hydrogen is added and passage of the mixture over a platinum catalyst converts the oxygen to water, which can be frozen out. Distillation separates the argon from the excess hydrogen and the residual nitrogen.

Figure 64 shows the major steps in the oxygen/nitrogen manufacturing process. The compression of air accounts for almost 100% of the energy consumption in this process. However, the amount of compression required is dependent on the heat exchange between feed and product streams, and on the design of the distillation column.

Another representation of the oxygen/nitrogen separation process with the emphasis on nitrogen production is shown in Figure 65. The energy values for various unit operations are given in the boxes in units that represent millions (10^6) of Btu. Products, major material inputs, by-products, and waste products are underlined with energy values assigned where appropriate. A summation of all of the values shown on the flowsheet gives the total energy requirement per net ton of primary product and is shown at the bottom of the flowsheet. No fuels or electrical energy consumption are itemized on the flowsheet.

FIGURE 64: OXYGEN/NITROGEN ENERGY CONSUMPTION DIAGRAM*

1973 USA production (oxygen): 14.5×10^9 kg $(31.9 \times 10^9$ lb)
1973 energy consumption (electricity): 530 Mw $(16 \times 10^{12}$ Btu)

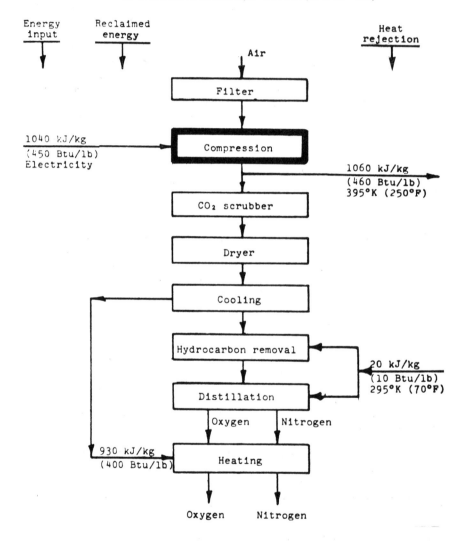

*Energy is expressed in terms of energy per unit weight of oxygen produced.

Source: Reference (3)

FIGURE 65: PRODUCTION OF NITROGEN FROM AIR

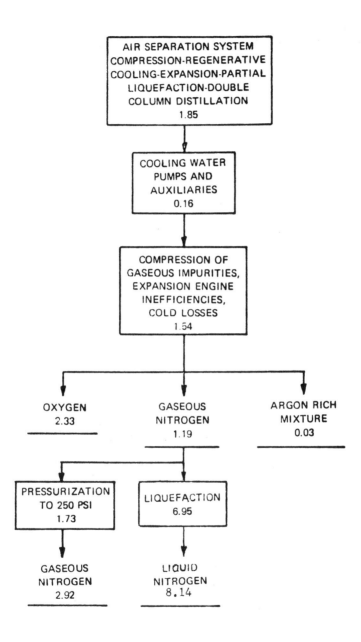

Source: Reference (12)

Figure 66 shows the compression of air before it is cooled by product streams. Electricity is used to drive the compressor.

FIGURE 66: EQUIPMENT DIAGRAM—AIR COMPRESSORS

Rejected heat:
   Hot compressed air—1,150 kJ/kg (500 Btu/lb) at 395°K
      (250°F)

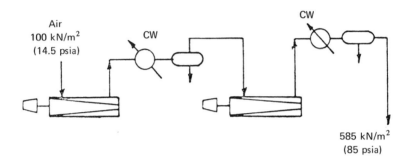

Note: Energy is expressed in terms of energy per unit weight
       of oxygen produced.

Source: Reference (3)

Figure 67 shows the distillation column used in the oxygen/nitrogen process to separate the components in air. The design of this column along with the efficiency of heat exchange between feed and product streams plays a major role in determining the amount of air compression required.

Hydrogen

The principal method of making hydrogen is by reforming natural gas with steam. In this process, shown in Figure 68, the natural gas is heated with steam in an externally heated reformer furnace, giving rise to carbon monoxide and hydrogen.

$$CH_4 + H_2O \longrightarrow CO + 3H_2$$

The CO plus more steam is converted to CO_2 and more hydrogen in a water gas shift reactor.

$$CO + H_2O \longrightarrow CO_2 + H_2$$

The CO_2 is removed by scrubbing with a base such as monoethanolamine, which is regenerated by steam stripping. Second and third stages of shifting and

FIGURE 67: EQUIPMENT DIAGRAM—AIR DISTILLATION COLUMN

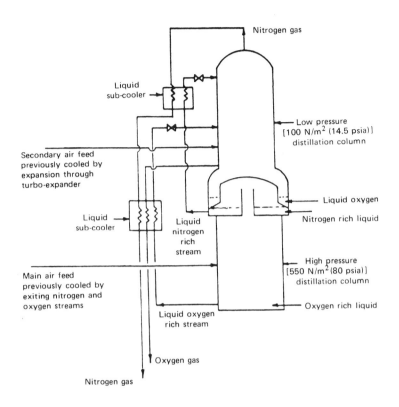

Note: Energy is expressed in terms of energy per unit weight
of oxygen produced.

Source: Reference (3)

scrubbing are frequently used to raise the purity to 99.9%.

Another method of producing hydrogen is partial oxidation, in which the feed
(light oil fractions, heavy oil fractions, or even coal) are mixed with oxygen and
steam. The oxidation produces internally the heat required for hydrogasifica-
tion. The products, as in steam reforming, are principally CO and hydrogen.
Shifting is performed as in the steam reforming. To remove traces of CO,
methane, and other light hydrocarbons, a liquid nitrogen wash is frequently used.
The recovered material can be used as fuel or reformed with steam. This approach is not used in the United States for commercial hydrogen production.

Hydrogen can be prepared by electrolysis of water. Because of the high energy
requirements this is not a major method of production.

FIGURE 68: FLOW DIAGRAM FOR HYDROGEN PRODUCTION

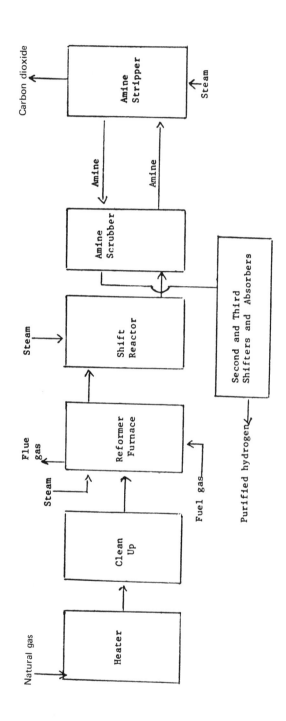

Source: Reference (5)

By-product hydrogen constitutes about 30% of the hydrogen for sale, coming from manufacture of sodium hydroxide, sodium chlorate, sodium perchlorate, and potassium hydroxide as well as surplus refinery gases from cracking and reforming. (Much by-product hydrogen is burned as fuel or used captively in the refinery.)

Acetylene

The principal methods of manufacture of acetylene are by partial oxidation of natural gas and by adding water to calcium carbide. In the first reaction, natural gas plus 95% oxygen combine in a reactor designed to quench the reaction quickly (after 0.01 to 0.001 second). About 30% of the methane forms acetylene, the rest, CO, CO_2, H_2, and water. The CO and H_2 can be recovered and used in methanol or ammonia synthesis.

$$CH_4 + O_2 \longrightarrow C_2H_2, CO, CO_2, H_2, H_2O$$

The process flows are shown in Figure 69. The traditional method of acetylene manufacture is by adding water to calcium carbide.

$$CaC_2 + 2H_2O \longrightarrow C_2H_2 + Ca(OH)_2$$

Because of the low price of natural gas and its derivatives, the rising prices of coke and electricity for making the calcium carbide, and the cost of adding pollution controls on the calcium carbide plants, much of the carbide-based capacity was shut down during the late '60s and early '70s, at a time of declining chemical markets for acetylene. (Ethylene-based chemistry replaced much acetylene usage.) However, calcium carbide is the most convenient method for generating small amounts of acetylene, e.g., cylinder filling and welding.

There are several other methods of producing acetylene, such as the Wulff pyrolysis of hydrocarbons and electric arc cracking of hydrocarbons. These are not important in the United States today.

By-product acetylene can be recovered from the cracking of hydrocarbons for ethylene production. This amounts to about 30 million pounds per year in the continental United States and is expected to grow to 50 million by 1979.5, about 10% of the total then.

Carbon Dioxide

Nearly all of the CO_2 produced in the United States is a by-product that must be removed and is frequently available in concentrated form. Sources are: 62% from ammonia manufacture; natural gas treatment, 30%; petroleum refining, 1.5%; fermentation, 1%; ethylene oxide production, 1%; lime kilns, natural wells, sodium phosphate, and combustion, 1.5%.

MAJOR ENERGY CONSERVATION OPTIONS TO 1980

Production processes involved can be roughly divided into two categories, each with very different energy conservation options. The first consists of purifying, compressing, or refrigerating gases obtained from the air; hydrogen, acetylene,

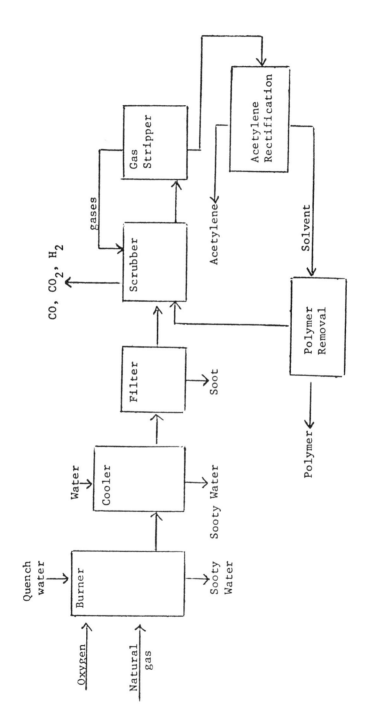

FIGURE 69: FLOW DIAGRAM FOR ACETYLENE PRODUCTION

and carbon dioxide obtained as by-products; and acetylene made by reacting calcium carbide with water. The second consists of hydrogen and acetylene production from natural gas, which involves an energy-intensive reaction, about 5 to 10 times the energy per pound of the first category.

In the purification compression and refrigeration of gases, process energy can be saved by installing more efficient compressors, using heat exchangers and distillation columns having less pressure drop, and using more insulation and more efficient heat exchangers. Since electrical energy has always been the major operating expense, these factors have always been under scrutiny in plant design.

In general, small changes in old plants are possible with today's equipment costs and estimated future fuel costs, leading to perhaps 1 to 2% improvement. Improved cleaning and maintenance and added storage can improve it a further 1 to 2%. In new plants designed to improve power consumption and to limit product purity to specification levels, savings of 10 to 20% can be achieved with increased capital investment. In this study, 10% was used, a consensus from industry representatives that ranged from 3 to 20%. (One involved a 7% saving at a 5% increase in capital cost.)

Oxygen

The energy requirements per pound for oxygen separation will vary with the product mix and purity sought, the production level, and the design characteristics. Numbers from the literature and interviews vary widely, from 450 to 2,750 Btu per pound for gaseous oxygen. The high values apparently double count the electricity and its steam equivalent. If these values are discarded, the range is 450 to 850 Btu per pound. When one looks at nitrogen energy requirements the discrepancies can be explained in part as follows.

In some sources the energy for producing pure oxygen and a waste nitrogen stream is assigned to oxygen. If the plant also produces pure nitrogen, only the incremental energy requirement for purification and final compression of the nitrogen are assigned to that product. This custom parallels the cost accounting system used since in most plants nitrogen is priced as a by-product.

A careful study by Battelle for the Bureau of Mines that used standard 1973 operating practice and typical product purities proportioned the liquefaction energy among the oxygen, nitrogen, and argon according to the national tonnage production.

However, this underestimates the amount required for gaseous oxygen since many oxygen plants make only oxygen. For this study the value of 590 Btu per pound for gaseous oxygen was multiplied by 1.25 to reflect the single-product plants, giving a value of 738 Btu per pound or 432 kwh per ton. For liquid oxygen, the Bureau of Mines number, 1,460 Btu per pound or 852 kwh per ton, was used directly since the liquid plants are more likely to be multiproduct plants.

The savings options for oxygen plants are believed to allow 3% energy saving in old plants and 10% in new ones. However, a word of caution is necessary. The savings depend greatly on the age and history of the plant, its particular product mix, operating rate, and local power costs.

Air-separation plants have a very high proportion of operating costs attributable to power costs (on the order of 65%). For this reason, energy efficiency has been stressed for years. However in the 1960s when many oxygen plants were built for steel mills, the design compromised a bit on energy savings to achieve lower capital costs.

In Europe where fuel costs have always been much higher, specific energy use is about 10% less. However, because the energy inefficiencies are related to large, capital intensive components, modifications to old plants to produce savings of 10% would be prohibitively expensive. Nevertheless, minor changes can be made that will effect small savings. More frequent cleaning, for example, when an unplanned shutdown occurs, can reduce energy usage.

Reducing product purity to no more than the contract level can reduce energy requirements, particularly at high purity. Going from 90 to 95% increases energy use by 5% on an equivalent oxygen basis. Going from 95 to 98% increases energy by 3%, and from 98 to 99.5% by 4%. These changes may save 1 to 21% overall. Housekeeping improvements such as fixing leaks or reducing lighting and office space heat are not really relevant to oxygen plants. (A large oxygen plant has only four workers per shift.)

Better planning and increased storage can result in savings of 1 to 2%, but for merchant plants, careful systems planning has been used for years, with operation of base load units, peaking units, and transportation of liquids between plants.

One reason for good planning is that bringing a plant on line from ambient temperature requires the equivalent of the power for 6 to 8 hours of production. In addition the turn-down ratio possible is only 0.60 to 0.75 (sources disagree) of rated capacity. As one lowers the production rate, energy per pound of product increases in a curvilinear manner.

At 75% of capacity, 9.3% more power per pound of output is required. Systems analysis is used to decide whether to shutdown one of six plants and increase transportation or operate all six at reduced output levels. This sort of systems integration is not possible with captive plants. When the steel mill shuts down, the oxygen plant shuts down. Also, better integration of captive units into merchant production is hampered by contractual arrangements, particularly if the user supplies the power for the facility.

In new plants, larger savings are possible, at an increase in capital costs. One reference point is a recently designed plant that had a 7% reduction in power needs at a 5% increase in capital costs. Thermodynamic analysis shows savings of 30% as the theoretical limit (11). This means complete reversibility—infinitesimal production from infinitely expensive plants.

Savings of 20% are within the realm of technical possibility at very high cost. Savings of 10% are thought by designers and contractors to be reasonable. Some operating companies set the limit as low as 3%. Ten percent has been used in this study for new plants. These savings will be achieved through improved compressor design, careful choice of the outlet pressure, and reduced pressure drop in heat exchangers and in the distillation columns.

No extra energy requirements to meet OSHA or EPA regulations or for fuel switching are expected.

Table 90 shows the causes of energy losses in the compression operation, the distillation operation, the heat exchange operation, and in the turbo-expander operation. It also shows the approximate magnitude of the losses and possible energy conservation approaches.

TABLE 90: OXYGEN/NITROGEN ENERGY CONSERVATION APPROACHES

Causes of Energy Losses	Approximate Magnitude of Losses	Energy Conservation Approaches
Rejected heat in hot air from compressors	1,060 kJ/kg (460 Btu/lb)	Design modification (waste heat recovery)
Overall process		
Nonideal flow volume of liquid down column	45 kJ/kg (20 Btu/lb)	–
Temperature differences in reboiler-condenser and liquid subcoolers	45 kJ/kg (20 Btu/lb)	–
Temperature differences between fluids in main heat exchange equipment	350 kJ/kg (150 Btu/lb)	–
Nonisothermal and nonisentropic compression losses	350 kJ/kg (150 Btu/lb)	Maintenance

Note: Energy is expressed in terms of energy per unit weight of oxygen produced. All overall process losses are electrical. The fuel value of these losses would be approximately three times as large as the values listed.

Source: Reference (3)

Nitrogen

As with oxygen, the energy per pound depends on many design and operating characteristics of the nitrogen production system. The Bureau of Mines number of 475 Btu per pound was multiplied by 1.25 to reflect the fact that some nitrogen plants produce only nitrogen and discharge a waste stream of impure oxygen. Thus the value used is 593 Btu per pound or 278 kwh per ton.

For liquid nitrogen, 1,330 Btu per pound or 777 kwh per ton, from the Bureau of Mines study was used. The discussion of energy savings in oxygen plants applies here. Savings of 3% in old plants and 10% in new plants are believed economically realistic.

Argon

Data on fractions of argon sold as liquid and gas are lacking. The average value of 1,099 Btu per pound or 644 kwh per ton used assumes about 40% is gaseous at 750 Btu per pound and 60% is liquid at 1,330 Btu per pound. The discussion of energy savings in oxygen plants applies here also. Savings of 3% in old plants and 10% in new plants are believed to be economically realistic.

Hydrogen

Hydrogen from natural gas constituted 70% of all production, the balance being by-product sold or shipped between plants, which amounted to 91.2 million pounds. It was assumed that this amount (not percentage) would be constant since petroleum refinery capacity is not growing rapidly and since their internal hydrogen needs are growing. The balance in 1979.5 was presumed to be 56% gaseous, down from an estimated 70% in 1972.

Gaseous hydrogen is reported to require 42,500 Btu per pound. Liquefaction of hydrogen was estimated to require twice the energy per pound of oxygen, 2,370 Btu per pound, or a total of 45,240 Btu per pound.

Old hydrogen plants can achieve a savings of about 1.5% in energy through improved maintenance, avoiding steam leaks, etc. However, this is offset by an expected 1.5% increase due to switching from natural gas to distillate oils for the external firing of the reformer furnace. This energy use constitutes about 35% of the entire fuel and feedstock consumption of natural gas. Thus for gaseous hydrogen production in old plants, no change in specific energy consumption is expected, for liquid hydrogen production savings are expected to be as much as 0.5% on the overall consumption. These estimates for old plants may be conservative since they were received from an industry source.

For new hydrogen plants, planning will allow considerable savings. A major contractor of ammonia plants indicated fuel-plus-feedstock savings of 10% would be reasonable. If so, this could mean as much as 20% on fuel savings. However, a conservative 10% was used on new liquid hydrogen plants, a total of 40,600 Btu per pound. (Because of the current overcapacity of hydrogen plants, all gaseous hydrogen was presumed to be supplied from old plants.) No additional energy use to meet OSHA and EPA regulations is anticipated.

Acetylene

Acetylene made by partial oxidation of natural gas is reported by one industry source to require 52,700 Btu per pound and by a second 40,000, exclusive of feedstock. These are large numbers and reflect the low yield in the reaction. The first number was used because the source supplied savings data that were useful. Savings of about 15% have been achieved by addition of a heat exchanger and an additional saving of 15% is expected. (A survey of all natural-gas-based acetylene plants would be necessary to get more firm data.)

By-product acetylene and acetylene from calcium carbide require energy for collection and compression. (It is sold in cylinders, dissolved in acetone.) This is estimated at 50 Btu per pound and has been ignored. In 1972 it was estimated that 52% of all acetylene was in this category.

In 1979.5 the forecast is for capacity to continue to exceed production by a wide margin. By-product acetylene is forecast to grow from 30 million pounds to

51 million pounds, i.e., at the growth rate projected for ethylene. The balance of the market demand, 48 million pounds, is expected to be supplied half from natural gas and half from calcium carbide, the ratio of the capacity for these forms in 1975.

The estimate for savings in natural-gas-based usage may be optimistic but probably is balanced by the conservative estimate for hydrogen from natural gas. No energy requirements to meet EPA and OSHA regulations are expected.

Carbon Dioxide

Energy usage of 68 Btu per pound and total production are so low that no attempt to develop savings options was made.

GOAL YEAR (1980) ENERGY USE TARGET

Production of industrial gases in 1979.5 is expected to be 100.8 billion pounds, primarily oxygen and nitrogen. As shown in Table 91, the energy requirement will be 1,012 Btu per pound using 1972 technology and 933 Btu per pound using improved technology.

The values reported in the table can be contrasted with estimated 1972 production of 69.2 billion pounds and energy consumption of 78.9×10^{12} Btu. This gives a specific energy consumption of 1,135 Btu per pound. The change from 1972 to 1979.5 is caused primarily by a change in product mix, with a forecast decline in the relative amount of acetylene, which is the most energy intensive gas under study.

TABLE 91: SUMMARY, PRODUCTION, AND ENERGY ESTIMATES— SIC PRODUCT CLASS 2813

	1979.5 All Products SIC 2813
Production, 10^9 pounds	100.8
Total Energy Required, 10^{12} Btu	
1972 Basis	102.0
1979.5 Basis	94.0
Specific Energy Consumption, Btu/lb	
1972 Basis	1,012
1979.5 Basis	933
Preliminary Energy Efficiency Goal,[a]	
Percent	7.8

(a) Calculation for energy efficiency goal:

$$\frac{1012-933}{1012} = 7.8 \text{ percent.}$$

Source: Reference (5)

TABLE 92: PRODUCTION AND ENERGY ENERGY ESTIMATES–SIC 2813

	Estimated Production 10⁹ Pounds		Energy Used		Energy Usage, 1979.5 10⁹ Btu		Energy Efficiency Goal, Percent
	1972	1979.5	10⁹ Btu	Btu Per Pound	1972 Basis	1979.5 Basis	
Oxygen							
Gas	24.5	34.4	18,080	738	25,390	24,120	5.0
Liquid	4.47	6.45	6,520	1,460	9,420	8,930	5.2
Nitrogen							
Gas	9.24	17.6	5,480	593	10,440	9,780	6.3
Liquid	4.74	8.56	6,280	1,330	11,410	10,710	6.1
Hydrogen	0.304	0.422	9,220(a)	43,300(a)	14,340(a)	14,090(a)	1.7(a)
Argon	0.390	0.846	430	1,100	930	870	6.5
Acetylene	0.774	0.532	19,580(a)	52,700(a)	12,670(a)	8,870(a)	30.0(a)
Carbon Dioxide	3.22	2.76	322	100	200	200	--
All Others(b)	21.8	29.1	12,950	593	17,280	16,420	5.0
Total	69.4	100.6	78,900	101,900	102,000	94,000	7.8

(a) Synthesis from natural gas. Energy requirements small for other processes.
(b) Includes amounts for which no production or shipment tonnage are reported.

Source: Reference (5)

The energy requirements with and without energy conservation are listed by compound in Table 92.   Of the gases for which data are available, oxygen has the largest requirement, followed by nitrogen, hydrogen, and acetylene. Acety-lene has the largest requirement per pound and the largest opportunity for savings.

INORGANIC PIGMENTS INDUSTRY

The inorganic pigments industry (SIC 2816) manufactures pigments for paints, fillers in rubber, use in storage batteries, corrosion resistance, printing inks, fillers in fabrics, and a wide range of other usages. The manufacture of pigments in 1972 amounted to 1.7×10^6 short tons of which TiO_2 represents almost 50% of the industry production on a tonnage basis, ZnO about 15%, and no other industry segment amounts to as much as 10%. The smaller segments of the industry include colored pigments, lead pigments, and extender pigments.

The Battelle study for FEA (5) covering TiO_2, ZnO, chrome color, and the iron blue color, includes about two-thirds of the dollar volume of the industry. Lumped into "all others" are lead and iron oxide pigments (both low-value, high-tonnage items). The industry makes a wide variety of other pigments by many processes and even a major effort would obtain data on a relatively small energy usage. This small portion was assumed to display equivalent energy usage on a dollar basis.

PROCESS TECHNOLOGY INVOLVED

Titanium Dioxide

Figure 70 is a simplified flow diagram of the titanium dioxide manufacturing processes. An ilmenite ore can be either beneficiated (by hydrochloric acid leaching) to produce synthetic rutile or processed directly with sulfuric acid. Rutile ore is processed by the chlorination process.

In the sulfate process, the titanium dioxide-bearing material is digested with hot sulfuric acid. Water is added to the mixture to dissolve the soluble constituents, mainly titanyl sulfate and iron sulfate. Scrap iron is added to reduce that portion of the iron that is in the ferric state to the ferrous state to lower consumption of acid and to precipitate iron. After filtering to remove the sludge, the filtrate is concentrated by evaporation and hydrolyzed to precipitate titanium hydrate (metatitanic acid). Rutile seed is added during hydrolysis or prior to

FIGURE 70: SIMPLIFIED FLOW DIAGRAM FOR MANUFACTURE OF TiO$_2$

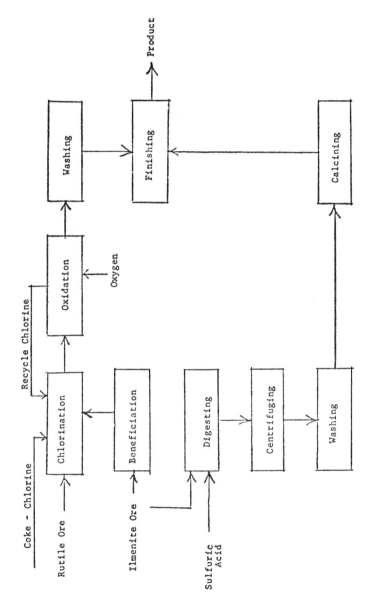

calcining to promote the formation of the rutile rather than the anatase type of titanium dioxide. The hydrate is filtered, repulped, and refiltered to remove impurities, then calcined to the oxide. After finishing treatments, the oxide is ground to the required size.

In the chloride process, rutile is mixed with petroleum coke and chlorinated to produce titanium tetrachloride. After the tetrachloride has been purified by distillation, it is vaporized and converted to the dioxide by burning with oxygen. At the same time, small amounts of metal chlorides or other agents, as well as titanium dioxide seed are injected to induce formation of the rutile crystal form of titanium dioxide rather than the anatase form. The titanium dioxide is collected and treated to remove residual chlorine and other impurities. It is coated with aluminum and silica hydroxides, dried, and ground to size specifications. The energy-intensive operations are the beneficiation, chlorination which uses coke as a fuel, calcining of sulfate TiO_2 and drying and milling the TiO_2. Milling is a particularly energy-intensive process because of the fine size (0.25 micron) required.

Zinc Oxide

Zinc oxide is produced by two major methods, the American process (57%) and the French process (27%), and from secondary materials (16%). The American process follows the same processing steps as production of zinc metal except that when zinc vapor comes from the retort or furnace it is burned to make oxide rather than condensed to make metallic zinc. The French process burns metallic zinc to make the oxide. While the French process appears to be more energy efficient than the American process, the metallic zinc raw material uses the same amount of energy in its manufacture as is used in the American process to make zinc oxide.

Secondary materials may substitute for zinc sinter in making American process oxide or may be burned directly in a French process. Secondary die-casting alloys are frequently burned in the French process. Figure 71 is a flow diagram of zinc oxide manufacture. Different equipment is utilized in each plant, so that each plant has its distinctive process.

The zinc sulfide ore is roasted by burning it in air to form sulfur dioxide and zinc oxide. This is an exothermic reaction and heat is recovered in a waste-heat boiler. The sulfur dioxide is cleaned and processed into sulfuric acid. The roasted ore is mixed with coke, heated, and fumed to remove cadmium vapors. The cadmium fume may be sent to another plant for recovery of cadmium values. The fumed oxide then is mixed with more coal (in some processes all of the coal or coke is added before cadmium removal) and heated (external heat, internal electric heating, or by contact with hot combustion gas depending upon specific equipment) and zinc vapor and carbon monoxide exit from the top of the reactor.

In a zinc refinery, the metal vapors are condensed. To make oxide, air is added and the zinc vapors burn to zinc oxide. The zinc oxide is collected and may be sold as is, or treated further by calcining, depending upon product desired. In the French process, zinc metal is heated and vaporized, then burned in air. The combustion products are frequently used to melt the zinc for energy conservation.

FIGURE 71: FLOW DIAGRAM FOR MANUFACTURE OF ZINC OXIDE

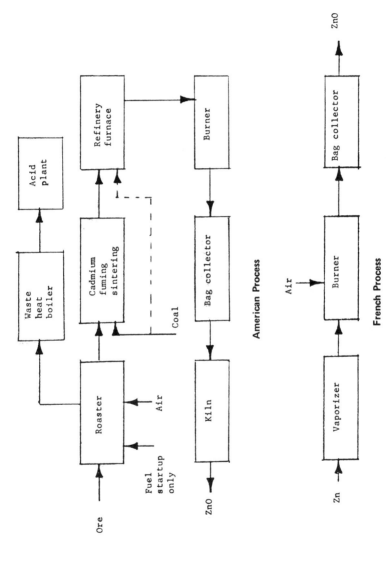

Source: Reference (5)

Other Inorganic Pigments

The other pigments are made by a wide variety of processes. Generally, they include dissolution of two chemicals and then controlled mixing to precipitate the pigment. The pigment is then dried and ground to break up agglomerates formed in drying. Other pigments are natural materials and only require grinding. Some specialized pigments require a sintering process.

MAJOR ENERGY CONSERVATION OPTIONS TO 1980

Energy usage on a unit basis as presented in Table 93 was estimated from a recent Battelle report (12) to the Bureau of Mines which reported values for the total processing energy for TiO_2 and ZnO production, including mining. This data base was modified to insure that processes not practiced in SIC 2816 establishments were deleted. The data on energy required for chrome colors and iron blue obtained from an industry representative are included in "others" to respect proprietary data.

Energy used by the rest of the industry was estimated as being the same on a dollar basis as the known part of the industry. This estimate of total energy is about 10% more than an estimate by the Bureau of Census for 1971, an excellent agreement since 1972 production was about 6% greater than 1971 production.

TABLE 93: UNIT SPECIFIC ENERGY CONSUMPTION IN SIC 2816, 1972 BASIS

Product	Energy Usage, Btu/lb
Chloride TiO_2 from Rutile	28,500
Rutile from Ilmenite	11,000
Sulfate TiO_2	28,900
American Process ZnO	18,900
All Others	16,300

Source: Reference (5)

Interviews with industry representatives indicated that these energy uses were approximately correct; however, the industry was reluctant to present actual energy usage. Generally, they were willing to talk about percentage energy changes over the 1972 to 1980 period. Growth in the industry was assumed to occur utilizing this higher energy consumption rate. Du Pont's new De Lisle plant uses a different process for making chloride TiO_2 from ilmenite; however, energy usage was assumed to be the same. Pigment manufacturers are essentially locked into their energy usage. Major process changes required to substantially reduce usage are generally not available. Most energy savings must come from careful housekeeping, increased insulation, better maintenance and changes in operating procedures. A plant that was operated inefficiently in 1972 can save substantial energy.

The zinc oxide segment of the industry will not be able to save substantial energy, however, because that industry was under severe competitive pressure just before 1972. About one-half of the primary producers of ZnO failed, and only the most efficient producers survived. Therefore, the potential for savings in that segment is very limited.

The inorganic pigments industry has a wide variety of energy intensive processes. Generally, calcining for particle-shape control, removal of impurities, drying, and grinding use major quantities of energy in the pigment plants. Some processes emit hot or combustible gases. Most plants use large quantities of steam. In the short term (before 1980), housekeeping and extra insulation appear to be the major sources of energy savings. On a longer term basis, process-heat recovery may be a viable option. However, some plants are not in a position to use the low-grade energy saved by heat recovery.

Many of the companies in the industry have an energy conservation committee headed by a senior corporate executive. Generally a plant survey has been made under the committee's auspices, to ascertain where energy can be saved in the plant. In most plants, a consulting engineering firm performed the survey. Some of the suggestions from these surveys include:

> Better steam trap and steam line maintenance,
> Return of condensate steam to boiler,
> Better insulation on both equipment and buildings,
> Boiler modifications for better boiler efficiency,
> The use of high-pressure boilers with topping turbines,
> A computer program to select most energy efficient
>   equipment at different levels and modes of operation,
> Recovery of waste process heat,
> CO boilers on some processes,
> Better dewatering before drying pigments,
> Use of more efficient dryers,
> Replace steam jet ejectors with mechanical fans or pumps.

A few plants were using steam-jet ejectors to produce vacuum or for ventilation air in 1972. Mechanical pumps or fans, using only about 5% of the energy of the ejectors, allow for substantial improvement in these operations.

Another energy-saving option that most companies have used is to decrease the comfort heating and cooling in the offices. Generally, the comfort systems are turned down during off hours and office temperatures are higher in the summer and lower in the winter than they were in 1972.

Product Changes: Product changes also have some effect on energy usage.

TiO_2 is purchased because of its hiding power. One company reported that the hiding power of TiO_2 will increase by 30% between 1972 and 1980 (another company said there would be little change); therefore, 30% less pigment is needed to paint the same area. However, the changes increase the production energy on a pound basis by 15%.

Effect of Plant Capacity on Usage: Energy usage in one company was estimated (in terms of energy used at capacity) at 65% at 50% production and at 25% at zero production. The actual numbers will depend upon the particular equipment installed in any plant and will vary substantially from plant to plant. One company estimated they would reduce their energy usage by 1% on a pound basis because of greater production in 1979.5 than 1972.

EPA Compliance: EPA regulations will require increased energy usage of about 7.5% in the 1972 to 1980 period. The sulfate TiO_2 plants may require 15% more energy to neutralize large quantities of waste sulfuric acid. The chloride TiO_2 plants require little additional energy for pollution control. Chromate pigment plants require extensive control to prevent the discharge of hexavalent chromium.

Fuel Switching: Fuel switching has had little effect on the industry as the switch has generally been from gas to oil or propane without change in fuel efficiency. The boiler in one new plant is using coal rather than oil at a 6% energy penalty, primarily because of particulate collection costs.

Housekeeping: Improved steam line and steam trap maintenance have reduced energy usage in the industry by about 5%. (In 1972, 10 to 15% of the steam generated was lost through failed steam traps, uninsulated or poorly insulated steam lines, and steam leaks.) Most of the savings was obtained in 1972. Since a large pigment plant has thousands of traps, and traps fail frequently, a constant surveillance of traps must be maintained. If the traps drain into open sewers, surveillance is easy. However, more typically they drain into closed condensate return lines and special instrumentation is needed to detect leaks.

Steam-line-insulation maintenance is also important. A small break in the insulation can waste large amounts of heat. A comment was made by an industry representative that steam usage previously increased by 10% during a rainstorm because of rainwater striking uninsulated steam pipes.

Insulation: Better insulation on equipment and buildings can save energy by reducing losses. Insulation on equipment will conserve small amounts of energy (1 to 2%) by reducing heat losses. Generally, the heavier insulation is added when the insulation must be replaced for some other reason. The major part of the insulation cost is for installation rather than the material. Building insulation can save significant amounts of energy. Architectural designs used in 1950 to 1970 were energy inefficient. Insulation under a roof is particularly energy conservative. One company reported a 10°F change in room temperature after roof insulation.

Improved Boiler Plant Operation: Boiler modifications and changes in operating procedure can save energy. The energy in the flue gas can be reduced by cooling it and by reducing excess air. Cooling is limited by the dew point of the flue gas and industrial boilers frequently discharge at hundreds of degrees higher. Addition of an economizer or air preheater is usually economic with present fuel prices. Frequently excess air can be reduced by simple operating adjustments. Some excess air is needed to prevent smoking and to completely burn the fuel. Normal excess air is from 10% with gas to as high as 50% with some coals. However many industrial boilers were operated with 100% excess air in 1972.

The production of mechanical or electrical energy by a topping turbine reduces the overall plant energy demand. In a topping turbine, mechanical energy is produced using less than 50% of the energy required by a public utility. However, since purchased electricity is charged at 3,512 Btu/kwh, the use of a topping turbine which requires about 5,000 Btu/kwh appears energy inefficient. When considered from the overall system, however, this is not the case.

One company indicated they reduced energy consumption by 3% by using their most efficient boilers. They also said an off-line computer program is used to select the most efficient equipment for each operating condition. This computer program will go on-line before 1980 to relieve the operators from making equipment selection. The return of condensate steam to the boiler saves water make-up and the condensate steam is hot which conserves some energy.

Recovery of Process Heat: The recovery of waste process heat always appears attractive. In the American process, the zinc oxide-containing gas from the reactor exits at about 1200°C and appears to be a prime target for energy conservation. However, the technology for using this very dusty hot gas is not available and will not be available before 1980. Also, the steam requirements of a zinc oxide plant are minimal and no use for steam generated from this gas is apparent unless it could be sold. No other process-heat-recovery option that was not practiced in 1972 appears practical.

The manufacture of $TiCl_4$ also produces carbon monoxide. This carbon monoxide might be used as a fuel gas. However, it contains chlorine which corrodes boiler tubes. Again technology cannot be developed before 1980 to use this gas for fuel.

Improved Dryer Operation: Pigment drying is energy intensive throughout the industry. Better dewatering of the filter cake before drying might save up to 5% of the plant energy. Some of the dewatering options are: high forming pressures for the cake; air or steam blowing of the cake; and centrifugal dewatering of the cake. Any of these changes might be made before 1980 if a plant determines this is a practical and economical energy-saving option.

Dryer efficiency varies from one type unit to another. If a plant is using an inefficient dryer, a more efficient dryer may save up to 10% of the plant energy. Generally, direct-fired dryers are more efficient than steam dryers because of losses in the steam system.

GOAL YEAR (1980) ENERGY USE TARGET

An energy increase of 1,400 Btu/lb is expected primarily on raw materials substitutions and changes in production technology. Currently rutile ore is utilized in the manufacture of TiO_2 by the chloride process. It is anticipated this pattern will be altered with the use of ilmenite ore for about 50% of the TiO_2 produced by the chloride process. Beneficiation of ilmenite ore to synthetic rutile will require 11,000 Btu/lb of TiO_2 made from synthetic rutile. It is then processed as rutile. The raw material change to ilmenite is necessary because of a shortage of rutile.

Other energy conservation measures throughout the entire industry are expected to conserve about 1,400 Btu/lb. With the 1,400 Btu/lb increase in energy caused by lower grade raw materials and comparable conservation, no net change in the energy per lb can be achieved. These estimates alone do not set an energy efficiency goal. Changes in processes and products between 1972 and 1979.5 also affect the goal.

The pigment industry has been growing at the rate of 4% per year. In the 7.5 years between 1972 and 1979.5, it is expected to grow from 3.4 billion pounds to 4.5 billion pounds. See Table 94. The Department of Commerce estimated inorganic pigments to be a $1.4 billion industry in 1980, a growth rate of 8%. The difference is apparently due to inflation.

TABLE 94: SUMMARY OF PRODUCTION AND ENERGY ESTIMATES, SIC 2816

	All Products, SIC 2816 1979.5
Production, 10^9 lb	4.52
Total Energy Required, 10^{12} Btu	
1972 Basis	88.2
1979.5 Basis	80.8
Specific Energy Consumption, Btu/lb	
1972 Basis	19,500
1979.5 Basis	17,900
Energy Efficiency Goal, (gross) percent	9
Energy Efficiency Goal, (net) percent	0

Calculation of Gross Goal:

$$\text{Goal} = \frac{19,500-17,900}{19,500} = 9 \text{ percent.}$$

Source: Reference (5)

The estimated energy efficiency goal is 9% on an individual establishment basis, but zero on a tonnage basis because of processing changes. The major energy increase due to process change is the use of ilmenite ore in new TiO_2 plants rather than rutile ore. New plants built since 1972 are burdened with this extra energy usage.

New pollution control devices will increase the energy use by 6 to 7% over the entire industry with most of that increase due to water pollution control in the sulfate TiO_2 segment. The chromate pigment segment also has substantial water-pollution problems which will increase energy usage.

Better housekeeping is expected to save 5 to 15% of the energy usage, the amount primarily dependent on the efficiency of operations in 1972. Modification of the dewatering and drying processes can be expected to save another 2 to 7% of the energy. Careful overall energy control will save 0.1% here and 0.5% there, which will aggregate to another percent of the energy used. Product upgrading will perhaps increase energy usage by 1%.

Overall a savings of 9 to 21% can be expected before deducting the 7% loss for pollution control, thus making the savings goal 2 to 14%. The 7% goal is an approximate average. The energy per pound of product in each process is expected to be reduced by about 9%. However between 1972 and 1979.5 the more energy intensive processes are expected to supplant the less energy intensive ones and, on the average, there will be no change per pound of product.

OTHER INDUSTRIAL
INORGANIC CHEMICALS

On a product basis, the SIC 2819 category includes inorganic compounds of aluminum, potassium, sodium, antimony, arsenic, barium, bismuth, cadmium, calcium, chromium, cobalt, copper, iron, lead, magnesium, manganese, mercury, molybdenum, platinum, radium, strontium, tantalum, thallium, tungsten, nickel, phosphorus, rare earths, selenium, silver, sulfur, tin, and zinc, as well as all inorganic acids except nitric and phosphoric, catalysts, reagent grade inorganics, bromine, activated carbon, hydrogen peroxide, iodine, mercury, phosphorus, silica gel, sulfur, inorganic bleaches, and radioactive isotopes and other radioactive materials produced outside AEC plants.

Eight products or groups of closely related products were identified for which production and shipment data were given and the value of shipments of each in 1972 was $50 million or more. The total value of shipments of these eight products was 40.5% of the total value of shipments of all SIC 2819 products by all establishments. The eight chemicals selected to represent SIC 2819 in the Battelle survey for FEA (5) have many uses. Their major uses are shown in Table 95.

TABLE 95: MAJOR USES OF SELECTED CHEMICALS

Sulfuric acid
  Phosphate fertilizers
  Petroleum refining
  Ammonium sulfate
  Alcohols
  Titanium dioxide
  Other chemicals
Hydrofluoric acid
  Fluorocarbons
  Aluminum industry
Aluminum oxide
  Aluminum manufacture
  Aluminum sulfate
  Adsorbents
  Catalysts

Aluminum sulfate
  Pulp and paper
  Water purification
Sodium phosphates
  Detergents and water treatment
  Foods
Calcium phosphates
  Animal feeds
  Dentifrices
Phosphorus
  Phosphoric acid not for fertilizer
Sodium silicates
  Silica gel and catalysts
  Soaps and detergents
  Pigments

Source: Reference (5)

PROCESS TECHNOLOGY INVOLVED

Sulfuric Acid

About 90% of the U.S. production of sulfuric acid is manufactured by burning elemental sulfur in air to give sulfur dioxide (SO_2), oxidizing the SO_2 catalytically to sulfur trioxide (SO_3), and absorbing the SO_3 in dilute sulfuric acid to produce concentrated sulfuric acid. Smaller quantities of sulfuric acid are also produced from smelter gas, refinery acid sludge, hydrogen sulfide, pyrites, and other sulfur-containing materials. These miscellaneous sources combined account for no more than 10% of U.S. production of sulfuric acid, although production from smelter gas could increase significantly in the future, and power plant stack gas is a possibility. As long as elemental sulfur is economically available, it is cheaper and safer to transport sulfur than sulfuric acid, and the sulfur-burning process should dominate the manufacture of sulfuric acid far into the future. Figure 72 shows a generalized flow chart for the production of sulfuric acid by burning sulfur.

FIGURE 72: PRODUCTION OF SULFURIC ACID FROM FRASCH SULFUR

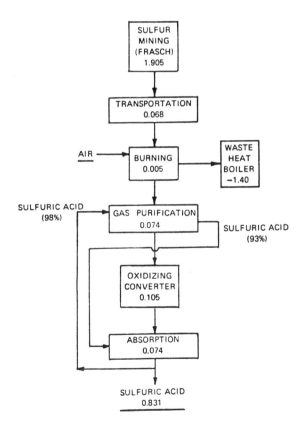

Source: Reference (12)

The energy values for various unit operations are given in the boxes in units that represent millions (10^6) of Btu. Products, major material inputs, by-products, and waste products are underlined with energy values assigned where appropriate. A summation of all of the values shown on the flowsheet gives the total energy requirement per net ton of primary product and is shown at the bottom of the flowsheet. No fuels or electrical energy consumption are itemized on the flowsheet.

This flowsheet represents a compromise between old plants and new. Of about 125 plants in 1975 operating totally on sulfur, about 50% were constructed prior to 1960 and about 30% prior to 1954. Newer plants have been designed to use more steam to operate blowers and pumps, with a concomitant reduction in the usage of electricity. The burning of sulfur and the catalytic oxidation of SO_2 are exothermic reactions, and they generate more energy than is used elsewhere in the process.

Hydrofluoric Acid

Hydrofluoric acid (HF) is produced by treating acid-grade fluorspar with concentrated sulfuric acid in a rotating furnace. Impure HF is evolved from the furnace as a hot gas. The gas passes upward through a packed tower which returns dust and most of the unreacted sulfuric acid to the furnace. A water-cooled precooler removes water and a solution of HF in sulfuric acid from the bulk of the HF gas, which is then condensed by refrigeration to 98.5% hydrofluoric acid. This product can be further purified to at least 99.95% HF by steam-heated distillation.

Hydrofluoric acid is sold in anhydrous (99.95%) and technical (70%) forms. It can also be used for some captive uses without condensation and distillation. The less purification performed, the less energy required. Estimates of energy consumption have ranged from 28 million Btu/ton of anhydrous HF to as little as 7 million Btu for HF used in synthesizing some fluorochemicals. Hydrofluoric acid is a hazardous chemical to discharge from a plant and a valuable chemical to retain. The vent gases pass through both an acid scrubber and a water scrubber. The sulfuric acid in the acid scrubber returns any HF it retains to the process. The aqueous effluent from the water scrubber is neutralized prior to discharge.

Aluminum Oxide

Aluminum oxide (alumina, Al_2O_3) occurs abundantly in nature, generally as impure hydroxides in bauxite and laterite minerals. Alumina is produced commercially in many hydrated and anhydrous forms for a broad variety of uses. However, about 90% of the alumina produced is used as the raw material for the production of aluminum. Only this process will be considered here.

Figure 21 under "Aluminum Industry" shows a generalized flow chart for the production of aluminum oxide in the course of aluminum production and the reader is referred to that figure at this point. The screened and dried bauxite ore is ground with a small amount of lime, then is digested with caustic soda (NaOH) under heat and pressure, forming soluble sodium aluminate. Through a series of pressure-reducing tanks and filter presses, the insoluble impurities are removed as red mud, which is normally impounded. The clarified solution is cooled and seeded with previously formed alumina hydrate. Alumina trihydrate precipitates, leaving soluble caustic soda in solution. The alumina is removed from the slurry by filtration, and the filtrate is concentrated for recovery of caustic soda. About 28% of the energy consumed in the process is used for recovery of caustic soda, but the process would not otherwise be economical.

After the alumina trihydrate is filtered, it is calcined to Al_2O_3, using natural gas heat to drive off water. Dust from the calciner is the only significant air pollutant. By the end of 1976 the installation of electrostatic precipitators and baghouses is expected to eliminate this problem.

Aluminum Sulfate

Aluminum sulfate is produced by reacting sulfuric acid with bauxite in lead-lined steel reaction tanks at 220° to 230°F for 15 to 20 hours. A reducing material, generally barium sulfide, is then added to the slurry to reduce ferric sulfate impurity to colorless ferrous sulfate. The solids are separated from the solution in settling tanks or thickeners, using a coagulant to assist in removal of finely divided particles. The clear, supernatant liquid may be sold as such, or it may be concentrated further by evaporation of water in open, steam-coil-heated lead-lined evaporators. When concentrated to about 61.5°Bé, the solution quickly solidifies on cooling in flat iron pans or on a cooling table. The product may be broken into lumps or ground to a uniform powder for shipment.

Commercial aluminum sulfate contains about 13 or 14 mols of water instead of the theoretical 18 mols. Anhydrous aluminum sulfate can be obtained by further dehydration of the commercial grade. The commercial or technical grade usually contains a maximum of 0.5% iron. An iron-free product (0.005% iron, maximum) can be prepared in the same manner, except that pure aluminum hydrate is used as starting material, instead of bauxite. In this process there are no waste solids to be removed. The pure aluminum hydrate can be derived by using the process for preparing aluminum oxide, stopping short of the calcination step.

Sodium Phosphates

Sodium phosphates are produced and sold in at least seven modifications, as follows: monobasic, NaH_2PO_4; dibasic, Na_2HPO_4; tribasic, Na_3PO_4; meta, $(NaPO_3)_6$; tetrabasic, $Na_4P_2O_7$; acid pyro, $Na_2H_2P_2O_7$; and tripoly, $Na_5P_3O_{10}$. These salts are made by reacting phosphoric acid with sodium carbonate or sodium hydroxide, and processing further the reaction products. Sodium tripolyphosphate accounts for about 80% of the production of all sodium phosphates, and is the only one considered here.

The appropriate quantities of dilute phosphoric acid and dilute sodium carbonate are blended and boiled in a mixing tank, driving off carbon dioxide and forming a mixture of 1 mol of mono- and 2 mols of disodium phosphate (Na_2O/P_2O_5 ratio of 1.67). The solution is filtered hot to remove silica and iron and aluminum phosphates, then is sprayed into a rotary continuous kiln. In the kiln the solvent water is evaporated and additional water is formed by condensation of the phosphate moieties and driven off. The product tripolyphosphate is ground for shipment.

Calcium Phosphates

Calcium superphosphate, triple superphosphate, and other fertilizer phosphates are manufactured from phosphate rock and sulfuric acid, and are included in SIC 2874. Calcium orthophosphates are manufactured from furnace-grade (pure) phosphoric acid and are used primarily for feed and food uses. They are produced in monobasic, dibasic, and tribasic forms. The dibasic form accounts for about 85% of production, and is the only calcium phosphate process considered here.

The chemical process is an extremely simple one. A dilute lime slurry is mixed continuously with dilute phosphoric acid in a Stedman pan mixer. This mixer provides good mixing of the solid and liquid phases, and prevents coating of the solid lime particles by the solid dicalcium phosphate that is formed, thus permitting the reaction to go to completion. The paste that is formed is dried in a tube dryer, and the product is ground prior to bagging for shipment.

If quicklime is used instead of powdered limestone, the product does not need drying prior to grinding. Either defluorinated wet-process phosphoric acid (from phosphate rock and sulfuric acid) or furnace-grade phosphoric acid (from oxidation of elemental phosphorus), which does not require defluorination, can be used.

Phosphorus (Elemental)

Elemental phosphorus is produced by reducing phosphate rock with carbon in an electric-arc furnace. Most of the phosphate rock suitable for this purpose in the United States is mined in Tennessee, Idaho, Montana, Utah, and Wyoming. The silica that is present with the ore is beneficial in forming a fluid calcium silicate slag.

Figure 73 shows a flowsheet for the production of phosphorus. The energy values for various unit operations are given in the boxes in units that represent millions (10^6) of Btu. Products, major material inputs, by-products, and waste products are underlined with energy values assigned where appropriate. A summation of all of the values shown on the flowsheet gives the total energy requirement per net ton of primary product and is shown at the bottom of the flowsheet. No fuels or electrical energy consumption is itemized on the flowsheet.

The ore is agglomerated by nodulizing, in which the phosphorite is heated to incipient fusion at 2200° to 2700°F. After cooling, the large nodules are crushed and screened to a range of sizes that readily permits escape of the gases formed in the furnace. The furnace is charged with nodulized ore, coke breeze, and silica as needed. The charge reaches fusion temperature near the bottom of the furnace, forming a fluid calcium silicate slag and releasing P_2O_5, which reacts with carbon (coke) to form phosphorus vapor and carbon monoxide. Iron impurities in the charge react with some of the phosphorus to form ferrophosphorus. Slag and ferrophosphorus are tapped from the furnace periodically.

Gases leaving the furnace consist of phosphorus and carbon monoxide. Entrained dust is removed in a hot electrostatic precipitator and the phosphorus is condensed by cooling in a water spray tower. The carbon monoxide is drawn off by a vacuum pump and is used as the major source of heat for nodulizing. The condensed phosphorus is stored under water to prevent spontaneous combustion. The water used for condensation of the phosphorus is treated with lime to precipitate fluorides and traces of phosphorus prior to discharge.

FIGURE 73: PRODUCTION OF ELEMENTAL PHOSPHORUS

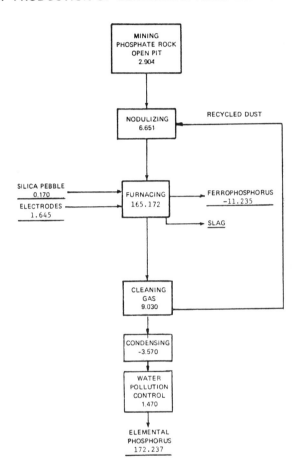

Source: Reference (12)

Sodium Silicates

A variety of sodium silicates, ranging in molar compositions from $Na_2O \cdot 4SiO_2$ to $2Na_2O \cdot SiO_2$, is produced by fusing sand (SiO_2) and soda ash (Na_2CO_3). Colloidal silicates, ranging from $Na_2O \cdot 1.6SiO_2$ to $Na_2O \cdot 4SiO_2$, are sold in aqueous solutions as water glass. The sodium silicates that are higher in SiO_2 content are glasses, important mainly for adhesive and binding properties. The more alkaline silicates, higher in Na_2O content, are used chiefly as cleaners and detergents.

The nature of the product depends on the ratio of reactants charged to the regenerative furnace and the physical form desired. The appropriate mixture of sand and soda ash is heated to 2200° to 2600°F in the furnace. Carbon dioxide is evolved and the melted materials flow slowly through the furnace. The fused melt is run in a thin stream into chilled molds on a conveyor, where it cools to a semitransparent solid.

The hot product will shatter if it is sprayed with cold water. The shattered product can then be fed to grinding and screening equipment for the preparation of granular sodium silicates or can be passed into a rotary dissolver. The silicate is dissolved with superheated steam, then is clarified and adjusted for specific gravity in a settling tank. This product may be sold as water glass or may be forced through fine openings into an air-swept solidification chamber, where the water is removed. This procedure gives a powdered form of sodium silicate.

Uranium Enrichment

The enrichment of uranium for use in weapons, power plants, and medical applications is very energy intensive and nearly all of its consumption is in electricity. Of the electricity consumed by SIC 2819 industries in 1967, 50% of it was due to uranium enrichment but less than 5% of the total SIC 2819 shipments were enriched uranium. (Exact shipments in 1967, military and nonmilitary, are still classified.) With the tremendous growth rate of over 18%/yr through 1985 (according to AEC estimates), uranium enrichment will soon dominate both total and electrical energy consumption in this 4-digit group, however.

As it occurs in nature, uranium consists of 99.3% of the U-238 isotope and only 0.7% of the fissionable isotope U-235. In order to sustain the chain reaction in present nuclear reactors the concentration of U-235 must be increased to about 2 to 4%.

Presently, the enrichment of ore in the United States is carried out through a gaseous diffusion technique. Raw ore is first upgraded to about 85% U_3O_8 by conventional hydrometallurgical methods. This U_3O_8 is then converted to UF_6 either by solvent extraction purification followed by complete fluorination or by first fluorinating the ore and then purifying by a distillation method. UF_6 must be used in the gaseous diffusion enrichment as it is the only known suitable volatile compound of uranium.

The diffusion enrichment is based on the fact that the average velocity of gas molecules are dependent on their masses. The U-235 isotope is lighter and on the average will travel faster than the U-238. In the gaseous diffusion process the container wall is the inside of a porous barrier through which diffusion is accomplished at very high pressures. Because the weight ratio of the two fluorinated isotopes is so close to unity, the degree of separation achieved in a single stage is very small. Upgrading the U-235 content from 0.7 to 4.0% requires about 1,200 diffusion stages in series.

The gaseous diffusion process consumes a tremendous amount of electrical energy in the recompression of the UF_6 gas between stages. In 1972, even at an electrical cost of only 5.5 mills/kwh, the power cost was about 60% of the total production cost for uranium. The electric consumption is so great that in 1956, when the diffusion plants were operating at peak capacity for defense purposes, they consumed in excess of 50 billion kwh/yr or almost one-tenth of the electrical energy of the entire U.S. at that time. The level of energy consumption (and production) was gradually reduced after 1956, reaching a low of 16.3 billion kwh in 1970. Although a large stockpile was established during the sixties, production is again on the increase in anticipation of the large demand for nuclear-based electrical power in the seventies and eighties. Power consumption was 25.1 billion kwh in 1972 and 28.1 billion kwh in 1973.

The first step in the production expansion will be to gradually restore the existing plants to the high level of operation of the 1950s. As production begins to approach full capacity, the system will be slightly less energy efficient than at present. The second phase of the projected capacity expansion involves modification of process equipment with emphasis on the installation of improved diffusion barriers and compressors. These modifications should provide substantial increases in output without accompanying increases in power consumption. The so-called Cascade Improvement Program (CIP) will involve installation of the improved hardware over the period 1976 to 1980.

Additional productivity gains are also available by increasing the UF_6 mass flows beyond those achievable with present drive systems. To do this both the electric power and cooling systems must be uprated. The Cascade Uprating Program (CUP) is expected to increase the production capacity of the existing diffusion plants by 25% and should have no substantial effect on the unit energy efficiency of the enrichment process. Expansion of production in this manner will begin in 1979 and be completed in the mid-80s.

MAJOR ENERGY CONSERVATION OPTIONS TO 1980

In general, there do not appear to be significant energy conservation options for this 4-digit SIC category. Most of the processes are relatively simple and are not highly energy intensive. Furthermore, most of the processes are well-established, and most engineering improvements have already been made. Many of the plants may be relatively old and may not incorporate the most modern improvements, but they are generally amortized and are not economically replaceable. The newest engineering features are likely to be placed only in new construction to meet growth in demand, which is expected to be only about 2.3%/yr for the category as a whole.

Improved housekeeping and maintenance should enable modest energy savings in some processes, particularly those employing steam. Increased utilization of waste energy is possible in some processes that require furnaces or kilns and in some processes that include highly exothermic reactions. However, this option is viable only if there is a use in adjacent facilities for the energy saved.

The consumption of fuels and electricity by SIC 2819 establishments in 1971 based on the product make-up of this SIC group in 1971 is reported in a special report by the Census of Manufactures. The quantity of natural gas used as fuel (not feedstock) in ammonia systhesis in 1971 (220×10^9 ft^3) and the electricity used for uranium enrichment in 1971 (12.2×10^9 kwh) was subtracted from the total energy usage to determine the energy used by privately owned and operated SIC 2819 establishments in 1971 on a 1972 product basis.

The value of fuels and of electricity purchased in 1972 by SIC 2819 establishments is reported in the Census of Manufactures. Assuming a 10% increase in the prices of all fuels and electricity between 1971 and 1972, and assuming the same ratios of fuels applied to both years, the quantities of fuels and electricity were calculated for 1972. The quantity of electricity used for uranium enrichment in 1972 (25.1 x 10^9 kwh) was subtracted from the total to obtain the amount of energy used by privately owned and operated SIC 2819 establishments in 1972.

On the assumption that the coverage ratio of 79% and specialization ratio of 89% applies to production as well as to value of shipments, it was estimated that the total energy required by all establishments to produce all SIC 2819 products in 1972 was 312.32 x 10^{12} Btu. The eight chemicals that were selected to represent SIC 2819 products varied widely in their energy requirements in 1972, from 23,790 Btu/lb for elemental phosphorus to –980 Btu/lb for sulfuric acid.

Sulfuric Acid

Sulfuric acid is a special case in that it requires only modest electrical energy for moving gas and liquid streams, and it generates considerable energy in the oxidation of sulfur and SO_2, and the absorption of SO_3 in dilute acid. This excess energy is used to generate steam, only part of which can be used in the sulfuric acid plant. In most installations, the acid plant is part of a larger complex, and the excess steam can be used elsewhere, or sold.

In the more modern plants, more of the steam is used to power blowers and pumps, thus replacing more electric power. However, many plants were constructed prior to 1960 and may not have been modernized for effective use of steam. Most will continue to operate as long as they can do so economically. Battelle estimates that by 1979.5 the displacement of some old plants and other new construction, coupled with better housekeeping in older plants, will produce additional energy equivalent to about 25% of the electrical energy used now, or about 20,000 Btu/ton of sulfuric acid.

A complicating factor for sulfuric acid plants was introduced in 1971 with the necessity to reduce emissions of SO_2 to the atmosphere. Dual absorption, in which the exit gases are reheated to initiation temperature and passed again through a catalytic converter, is the most common method of meeting EPA standards. This technology requires additional energy to heat and move the gases. On the assumption that all plants will meet the EPA standard by 1979.5, Battelle estimates that about 540,000 Btu/ton of sulfuric acid will be required for this purpose. The net energy saving for sulfuric acid using 1979.5 technology will thus be –520,000 Btu/ton. This number looms large when it is considered that sulfuric acid accounts for nearly 46% of the total production of SIC 2819 chemicals by all establishments.

Phosphorus

Phosphorus, which consumes considerable energy per ton, is also a special case. Virtually all of the energy purchased is electrical. Coke breeze is purchased as a feedstock, not a fuel. Its purpose is to chemically reduce P_2O_5 to P_4. In doing so, the coke is partially oxidized to CO, which is subsequently burned to CO_2 as a fuel in the nodulizer. It does not seem likely that energy can be saved in this process. The energy required for water-pollution control, however, is already considered in the process.

Uranium Enrichment

Even with existing plant production capacity increased by virtue of the planned improvement and uprating programs and with the utilization of substantial enriched uranium stock-piles, it is anticipated that new plant enrichment capabilities will be required by about 1980.

New plants built in the early 1980s are all likely to continue to use the gas diffusion enrichment techniques. Projected improvements in diffusion process technology include more efficient barriers and compressors, and the introduction of power recovery systems. These changes will markedly improve the energy efficiency. A new plant is expected to consume only two-thirds as much electricity per unit work as a present diffusion plant.

The Atomic Energy Commission is presently developing a new method for U-235 enrichment based on the use of a high-speed centrifuge. The mixture of gaseous U-235 and U-238 in the hexafluoride form is fed to the centrifuge. The heavier U-238 molecules tend to concentrate at the outer walls of the centrifuge and the lighter U-235 enriched gas is then removed from the center of the centrifuge. As with gaseous diffusion the efficiency of each individual separation unit is very low and hundreds of centrifuges in series must be operated to achieve the desired enrichment. The gas centrifuge is more energy efficient than the present gaseous diffusion techniques and is expected to require only one-tenth as much electricity.

Other Products

Modest energy savings may be possible in the production of hydrofluoric acid, alumina, and sodium phosphates. These processes use considerable quantities of steam and also require furnaces or kilns. With improved housekeeping, waste-heat boilers, and other energy economizers, it may be possible to save 1.35 million Btu/ton of hydrofluoric acid, 1.15 million Btu/ton of alumina, and 200,000 Btu/ton of sodium phosphates.

Prospects seem poor for saving much energy in the remaining three processes. It may be possible to save as much as 80,000 Btu/ton on the production of sodium silicates through better utilization of excess furnace heat. Perhaps 20,000 Btu/ton can be saved on the production of aluminum sulfates through better housekeeping. It does not seem likely that any energy can be saved in the production of calcium phosphates. Neither of these processes uses much energy to start with.

Energy consumption for all other SIC 2819 products was estimated as follows. The energy used for production of each of the selected eight products by all establishments in 1972 was determined by multiplying the production figure by the energy per ton. The sum of these numbers was subtracted from the 312.32×10^{12} Btu estimated earlier as the total energy required by all establishments to produce all SIC 2819 products in 1972. This left 209.89×10^{12} Btu as the energy required to produce an estimated 25.766 million tons of all other SIC 2819 products, or an average of 8.146 million Btu/ton in 1972. On the assumption that 7% of the energy used in producing these "all other" products could be saved, a saving of 570,000 Btu/ton could be realized for the "all other" products by 1979.5.

GOAL YEAR (1980) ENERGY USE TARGET

Production quantities for the eight selected chemicals were projected individually to 1979.5 on the basis of the growth histories of these chemicals during the past 5 to 8 years. It was assumed that production of the "all other" products, most of which are produced in much smaller quantities than are the selected eight, would collectively increase at about 4%/yr. Estimated production of all SIC 2819 products by all establishments in 1972 and 1979.5 is shown in Table 96.

TABLE 96: SUMMARY OF PRODUCTION AND ENERGY ESTIMATES, SIC 2819 PRODUCTS

	All Products [a]	
	1972	1979.5
Production, 10^9 pounds	136.6	162.0
Total Energy Required, 10^{12} Btu		
1972 Basis	312.3	376.5
1979.5 Basis	n.a.	367.5
Specific Energy Consumption, Btu/pound		
1972 Basis	2,287	2,324
1979.5 Basis	n.a.	2,269
Energy Efficiency Goal, percent [b]	n.a.	2.4

(a) All SIC 2819 products manufactured by all establishments.
(b) Calculation of energy efficiency goal:

$$\text{Goal} = \frac{2324-2269}{2324} = 2.4 \text{ percent}$$

n.a. = not applicable

Source: Reference (5)

The total energy used by SIC 2819 establishments in 1972 was estimated on the basis of energy consumption in 1971 minus the electricity used by government-owned, privately operated establishments and the natural gas used as fuel in the synthesis of ammonia. This usage of energy on an establishment basis was converted to a product basis on the assumption that the coverage ratio and specialization ratio apply to production as well as to value of shipments. The energy usage for 1972 production of the eight selected chemicals was subtracted from the total usage to estimate the energy used in producing all other SIC 2819

TABLE 97: PRODUCTION AND ENERGY ESTIMATES, SIC 2819 CHEMICALS

	Estimated Production, 10^6 pounds		Energy Used, 1972		Energy Usage, 10^12 Btu, 1979.5		Energy Efficiency Goal, Percent
	1972	1979.5	Total, 10^12 Btu	Btu per Pound	1972 Basis	1979.5 Basis	
Sulfuric acid	62,368	70,252	-61.12	- 980	-68.85	-50.58	-27.6
Hydrofluoric acid	518	866	3.46	6,680	5.78	5.20	10.0
Aluminum oxide	12,408	12,880	98.52	7,940	102.27	94.86	7.2
Aluminum sulfates	2,790	3,664	6.75	2,420	8.87	8.83	0.5
Sodium phosphates	2,596	2,596	15.45	5,950	15.45	15.19	1.7
Calcium phosphates	1,418	2,290	4.25	3,000	6.87	6.87	0.0
Phosphorus	1,082	1,082	25.74	23,790	25.74	25.74	0.0
Sodium silicates	1,876	2,082	9.38	5,000	10.41	10.33	0.8
All other 2819 chemicals	51,532	66,282	209.89	4,073	269.97	251.07	7.0
Total	136,588	161,994	312.32	2,287	376.51	367.51	2.4

Source: Reference (5)

products. The total energy required to produce all SIC 2819 products in 1972 and the average energy consumption per pound are shown in Table 96.

The anticipated energy saving in 1979.5 using 1979.5 technology amounts to only 2.4% for the overall spectrum of SIC 2819 products. This appears to be a small but achievable goal overall.

Potential energy usage for each of the eight selected chemicals and the composite all other products was calculated using the estimated 1979.5 unit energy consumption figures. Table 97 shows these results, with tons converted to pounds.

SOME PROJECTIONS BEYOND 1980 TO 1990

Although the Battelle study for FEA (5) excludes Government-owned (albeit privately operated) establishments, the utilization of power in the enrichment of uranium is, as already pointed out, large and growing. Further as discussed in the Project Independence Blueprint (10), major energy savings are in view for uranium enrichment beyond 1980.

The Atomic Energy Commission is anticipating an 18.5% per year growth in the demand for enriched uranium between 1975 and 1985, with the use of nuclear fuels in electric power plants growing at approximately the same rate.

Through most of the seventies, production will exceed demand in that year, but by the end of 1982 the stockpiles will be gone and new enrichment plants will be required if demand is to be met.

Table 98 gives a projection of enrichment production and demand through 1990. It has been assumed that centrifuge enrichment will become commercially available in 1985 and completely capture the new plant market because of its superior energy efficiency.

A projection of the electric energy consumption due to uranium enrichment in U.S. plants is given in Table 99. The consumption levels in existing plants are based on 1973 estimates by the AEC.

New gas diffusion plants built after 1980 are assumed to be 35% more efficient than present-day diffusion plants and the centrifuge plants are assumed to be ten times more energy efficient than present gas diffusion plants.

Table 100 combines the sales data in Table 98 with the energy use data in Table 99 to give energy consumption ratios for uranium enrichment through 1990. Notice that the 1967 energy consumption ratio is about nine times larger than the 1971 value; this is because the demand level was comparatively low. One reason for this might be that the level of military shipments in 1967 was high (they have not been released to the public). It might also be the case that a large amount of stockpiling was done in 1967.

Much of the fluctuation in the coefficients between 1975 and 1985 is due to the adding to or taking from stockpiles in any given year.

TABLE 98:　PROJECTED URANIUM ENRICHMENT PRODUCTION THROUGH 1990 (MILLIONS OF SEPARATIVE WORK UNITS*)

	1967	1971	1975	1977	1980	1985	1990
Uranium consumption (10^6 work units)	1.0**	6.5	8.3	16.3	27.7	45.5	78.7
Value of shipments (million 1967 $)	26	169	216	424	720	1,183	2,046
. . . . Enriched Uranium Production, 10^6 Work Units							
Existing plants	8.5	7.0	14.2	18.0	25.4	27.7	27.7
Part due CIP program				1.9	5.4	5.8	5.8
Part due CUP program					2.9	4.7	4.7
New plants***						18.9	52.3
Part due to gas diffusion						18.9	18.9
Part due to centrifuge							33.4
Total production	8.5	7.0	14.2	18.0	25.4	46.6	80.0

*Separative work units with degree of ^{235}U enrichment (3% ^{235}U requires 4.3 units per kilogram).
**In 1967 figures do not include military consumption.
***Assumes centrifuge takes over all new demand from 1985 to 1990.

CIP = Cascade Improvement Program
CUP = Cascade Uprating Program

TABLE 99:　URANIUM ENRICHMENT—PROJECTED ELECTRICITY CONSUMPTION, 10^9 kwh

	1967	1971	1975	1977	1980	1985	1990
Existing plants	24.9	19.2	39.9	48.8	59.5	64.6	64.6
New gas diffusion plants*						38.7	38.7
New centrifuge plants**							10.0
Total	24.9	19.2	39.9	48.8	59.5	103.3	113.3

*2.05 x 10^9 kwh per million separative work units.
**0.3 x 10^9 kwh per million separative work units.

TABLE 100:　ENERGY-PRODUCTION COEFFICIENTS FOR URANIUM ENRICHMENT

	1967	1971	1975	1977	1980	1985	1990
Required supply (millions of 1967 $)	26*	169	216	424	720	1,183	2,046
Total electricity consumption (billion kwh)	24.9	19.2	39.9	48.8	59.5	103.3	113.3
Energy coefficient (kwh/$)	958	114	185	115	82.6	87.3	55.4

*1967 figure does not include military sales.

Source:　Reference (10)

PLASTIC MATERIALS INDUSTRY

Plastic materials, synthetic resins and nonvulcanizable elastomers (SIC 2821) are thermoplastic and thermosetting polymeric materials, and thermoplastic elastomers obtained via the chemical reaction of monomers and oligomeric polymer pressures. The important products of this industry are low-density polyethylene, high-density polyethylene, polystyrene, polypropylene, polyvinyl chloride, phenolic resins, acrylic resins, polyester and alkyd resins, cellulose plastic materials, amino resins, hydrocarbon resins and petroleum polymer resins.

The products of this class had a shipment value of about $4.5 million in 1972, slightly less than 7% of the value of the whole chemical industry's shipments. The energy consumption is primarily due to production of six general types of polymers, namely low-density polyethylene, polyvinyl chloride, high-density polyethylene, polystyrene, polypropylene and phenolic resins.

These six products account for 88% of the energy consumption and 80% of the SIC 2821 production. Polyester and alkyd resins, amino formaldehyde resins and acrylic resins account for another 3.5% of the required energy for this industry. For the six major types of resins, energy savings opportunities were estimated. Other polymers were treated as a mix of condensation and addition type polymerization processes and savings estimated by extrapolation from data obtained from the six major polymers' processes.

The fuel equivalent of the monomer feed stocks for polymerization processes are given in the third column of Table 101. The feed stocks for polymerization probably represent 80 to 90% of the feed stock to the monomer production and thus the fuel equivalent of raw materials for the plastics industry is a significant and often unaccounted-for portion of the energy use in plastics production. The feed stocks for any industry should be counted at their actual energy value so that accurate analyses of fuel utilization can be made and all resources can be fully utilized. This is particularly important in industries which use fuels, fuel derivatives or potential fuels as raw materials. This is the case for plastics where the feed stocks are largely derived from petroleum and hence, compete with fuel consumers for petroleum.

Since there is very little waste of feed stock in the polymerization processes, the fuel values of the current monomer inputs to polymerization processes are very nearly the minimum energy requirements for the four plastics considered.

TABLE 101: FUEL USE FOR PLASTICS PER TON OF PRODUCT

	Total Industry Fuel Used[a]	Fuel for Polymerization[b]	Fuel Value of Monomer[c]
	$(10^6$ Btu/ton of product)		
Low-density polyethylene	45.8	15.3	44.8
High-density polyethylene	37.2	13.2	45.7
Polystyrene	18.5	8.0	37.3
Polyvinyl chloride	23.8	7.1	17.9

(a) Includes monomer production (but not fuel equivalent of feed stocks) and inefficiencies in steam and electrical generation.
(b) Fuel equivalent for steam and electricity required for polymerization process analyzed.
(c) Fuel equivalent of monomer feed stock, not included in Columns 1 and 2.

Source: Reference (8)

PROCESS TECHNOLOGY INVOLVED

Low-Density Polyethylene (LDPE)

Low-density polyethylene is manufactured by a number of high-pressure processes. Two different types of reactors are used for these processes. The tubular process unit is generally considered to require about 14% less energy than the autoclave process. The normal conversion of ethylene to LDPE is only 15 to 25% per pass to prevent the buildup of a highly viscous reaction system difficult to transfer and separate. This low conversion requires separating and recycling large quantities of ethylene in heated separators. Generally, two levels of pressure are used in the separation step, and monomer recompression is necessary for recycle. Figure 74 shows a generalized flow sheet for producing LDPE. Specific operating conditions are generally considered to be proprietary; hence, available process details for U.S. production are quite limited. Energy requirements, however, have been reported for several processes and are summarized in Table 102.

For purposes of this study, the Project Independence value (Gulf Oil Process— 1975) was assumed to be realistic for overall 1973 LDPE production and an effective savings of 7.5% was assumed between 1972 and 1973 in estimating a specific energy value for 1972. The various values for specific energy show the wide variation in process technology. The potential for energy savings through building plants based on new processes is evident.

FIGURE 74: GENERALIZED FLOW SHEET FOR LOW-DENSITY POLY-ETHYLENE

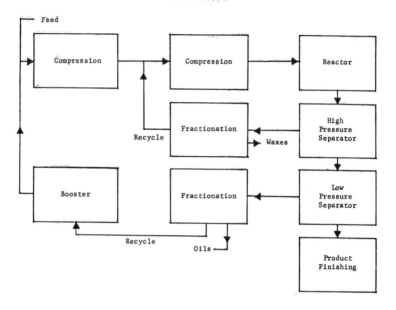

TABLE 102: SUMMARY OF LDPE PROCESS ENERGY REQUIREMENTS

Process	Reactor	Net Specific Energy Requirements (Btu/lb x 10^3)
Ato Chimie	Tubular	2.66 (1973)
Ato Chimie	Tubular	2.39 (1975)
Imhausen International	Tubular	2.49 (1975)
Gulf Oil Chemicals	Stirred	3.56 (1973)
Gulf Oil Chemicals	Stirred	6.98 (1975)

Source: Reference (5)

The energy and availability summary for low-density polyethylene (LDPE) is given in Table 103. The input values for steam and electrical power inputs do not reflect their generation and transmission losses. The most important energy and availability input to the polymerization process is the raw material, which provides 44.8 x 10^6 Btu/ton of LDPE. The energy of the LDPE output is 40.5 x 10^6 Btu/ton and most of the difference between these energies is the energy of polymerization, -3.3 x 10^6 Btu/ton of LDPE. The energy and availability of polymerization is apparently completely wasted, although low pressure steam and thus electricity could be generated. The reactor coolant leaves the reactor at 380°F or slightly higher, and low pressure, superheated steam could be generated at these temperatures. The boilers, turbines and condensers would be large and expensive, but this system could be designed for a high effectiveness and thus much of the availability decrease of polymerization could be recovered as electricity. Up to 5% of the fuel equivalent of the feed stock could be converted to electricity in this way.

Another input to LDPE polymerization is the electrical input to vapor compression. This input is shown in Table 103 as a direct input of 1.3 x 10^6 Btu/ton of LDPE. When generation and transmission losses are considered, this becomes a fuel equivalent of 4.0 x 10^6 Btu/ton of LDPE. The compression work per ton of LDPE is very high because the yield from the reactor is low, 10 to 25%, so that 4 to 10 tons of ethylene is compressed for each ton of LDPE produced. The conversion ratio or reactor yield is controlled by catalyst injections to relatively low levels so that the heat of polymerization can be adsorbed by the feed injections and by the cooling capabilities of the heat exchangers.

The use of the heat of polymerization to reduce or eliminate the need for electricity purchase for ethylene compression is technically feasible. The economics will have a strong influence on the extent to which waste heat will be used for power generation.

TABLE 103: ENERGY AND AVAILABILITY SUMMARY FOR THE LOW-DENSITY POLYETHYLENE INDUSTRY

	Energy*	Availability*
Input		
Ethylene	44.83	42.31
Steam	1.73	0.59
Cooling water	0.60	0.00
Electricity (direct equivalent)	1.32	1.32
Total	48.48	44.22
Output		
Polyethylene (low density)	40.46	39.27
Steam	0.52	0.12
Cooling water	5.15	0.14
Ethylene	1.75	1.64
Losses	0.60	–
Total	48.48	41.17

*10^6 Btu/ton LDPE

Source: Reference (8)

High-Density Polyethylene (HDPE)

High-density polyethylene is manufactured by one of several low-pressure processes which are completely different from LDPE processes. Consequently, the plants are not adaptable to production of conventional LDPE. However, a similarity is found in some older HDPE and polypropylene plants such that either product can be made.

HDPE is generally made in hydrocarbon solvent or diluent. Several types of highly reactive catalysts are employed, and the reaction is characterized by nearly complete conversion of ethylene in a single pass. Regardless of the catalyst used, three other major process variations exist. Initially, HDPE was made by solution polymerization processes in which quantity of solvent and reaction temperature were both significantly high to keep the polymer in solution. Next introduced were slurry polymerization processes in which quantities of solvent and lower reaction temperature were such that the polymer formed as a dispersed solid or slurry. Finally, gas-phase processes that do not require solvent were developed.

Specific details on the various processes used in the U.S. production of HDPE are proprietary. Several processes and specific energy requirements, however, are described in the literature, and were used as background in this analysis. Generalized flow sheets for the three processes are shown in Figure 75. Published specific energy requirements for several processes are summarized in Table 104.

TABLE 104: SUMMARY OF AVAILABLE HDPE PROCESS ENERGY REQUIREMENTS

Process	Net Specific Energy Requirements, Btu/pound x 10^3
Solvay and Cie	2.74 (1973)
Solvay and Cie	2.50 (1975)
Stamicarbon BV	3.14 (1973)
Stamicarbon BV	1.93 (1975)
Union Carbide	0.68 (1973)
Union Carbide	0.40 (1975)
Montedison SPA	4.17 (1973)
Friedrich Uhde GmbH	1.87 (1973)
Mitsubishi Chemical Industries	2.88 (1973)
Snam Progetti	5.35 (1975)
Hoechst	1.31 (1975)
Veba-Chemie AG	2.07 (1975)
	3.67 (1973)

Source: Reference (5)

The specific energy values in Table 104 show the broad spread in energy requirements from process to process, and also indicate the possible reduction in specific energy consumption as new plants are built. For this study, the Project Independence (Veba-Chemie—1973) value was assumed to represent valid data for 1973. Energy usage in 1972 was based on an adjustment of this figure.

The energy and availability summary for the polymerization of High-Density Polyethylene (HDPE) by the Montecatini process is given in Table 105. The steam and electrical power inputs listed are direct inputs and do not reflect generation losses. When considering the losses in utilities generation, the overall efficiency drops from 75.1%, given in Table 105, to 73.4%. In contrast to the LDPE production, the HDPE process has a high yield and no compression

FIGURE 75: GENERALIZED FLOW SHEET FOR HIGH-DENSITY POLY-
ETHYLENE

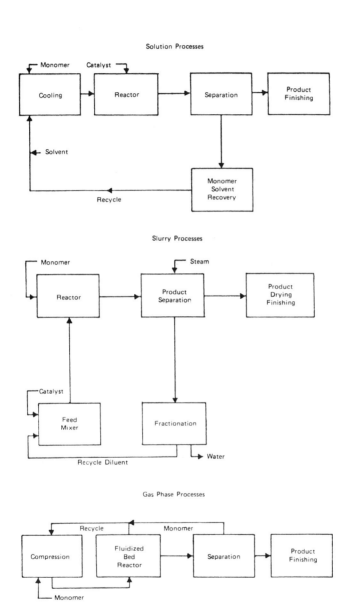

Source: Reference (5)

requirement. The electrical input to the HDPE process analyzed was primarily
for mixture agitation and materials handling; thus, significant reductions in electricity use are not likely. The process temperatures are less than 100°C, and
may approach 50°C; thus, the cooling water for this process has very little availability and waste heat recovery is not feasible.

TABLE 105: ENERGY AND AVAILABILITY SUMMARY FOR THE HIGH
DENSITY POLYETHYLENE INDUSTRY

	Energy	Availability
	(10[6] Btu/ton HDPE)	
Input		
Ethylene	45.69	43.12
Steam	6.24	2.13
Cooling water	1.06	0.00
Electricity (direct equivalent)	0.62	0.62
Total	53.61	45.87
Output		
Polyethylene (high density)	40.28	39.18
Steam	1.87	0.43
Cooling water	10.82	0.27
Ethylene	0.44	0.42
Losses	0.20	-
Total	53.61	40.30

Source: Reference (8)

There do not appear to be any large availability destructions in the process analyzed. The internal details of the process could not be determined with sufficient accuracy to allow an analysis of specific steam uses. It does appear that
the primary steam usage is for process heating and drying; thus, the pressure
and temperature levels are quite modest as compared to power plant practice.
This suggests that the present process steam generation equipment be replaced
with high pressure and temperature boilers and turbine generators. The turbines
would be run so that their exit steam would be at the desired process conditions. Such installations require capital investment and higher fuel consumption,
but these costs are offset by the electrical output. The exact improvement possible with this step could be evaluated accurately if further process information
were available.

Polyvinyl Chloride (PVC)

Polyvinyl chloride (PVC) resin is the second largest family of high polymers,
based on amount produced, in the United States. This class of material (PVC
homopolymer and copolymers consisting primarily of vinyl chloride) accounted
for nearly 17% of all plastics produced.

PVC resins are produced by suspension, emulsion, bulk and solution processes.
Although historically, the emulsion process was first used commercially, advantages

of the aqueous suspension process make it the most widely used process at present. Accordingly, suspension polymerization accounts for an estimated 85% of the U.S. production. Generally suspension process facilities, with the exception of polymer separation, can be used for emulsion polymerization. The emulsion process is commercially used for nominal production of specialty systems including plastisols, organisols and latexes (isolation of the polymer is not required in the last application).

The bulk process which has been developed produces PVC of comparable properties with lower capital investment and energy requirements. The bulk process is reported to require about 750 kwh per metric ton of product compared to 1,940 kwh for the suspension process. Accordingly, several U.S. companies have licensed the bulk process.

Despite the large volume production, 2,171,000 metric tons in 1973, very little specific process data for PVC manufacture has been published. Undoubtedly, such engineering information is protected to maintain a competitive production cost position. The exception to this is the data released on the Pechiney-Saint Gobain bulk polymerization process which is of considerable interest because of the reportedly reduced demands on energy and investment capital. Because of these advantages, publication is undoubtedly used to stimulate licensing possibilities. Accordingly, there were 20 commercial installations using this mass process in 1971, including three in the United States.

The energy and availability summary for polyvinyl chloride (PVC) production is given in Table 106. As with HDPE and PS, the primary energy and availability input is with the feed stock which is largely passed on to the product.

TABLE 106: ENERGY AND AVAILABILITY SUMMARY FOR THE POLYVINYL CHLORIDE INDUSTRY

	Energy	Availability
	(10^6 Btu/ton PVC)	
Input		
Vinyl chloride	17.87	17.09
Process water	-0.07	0.00
Cooling water	0.16	0.00
Steam	5.53	1.50
Electricity (direct equivalent)	0.09	0.09
Total	23.58	18.68
Output		
Polyvinyl chloride	15.45	15.50
Process water	4.17	0.64
Cooling water	1.76	0.04
Steam	1.10	0.19
Vinyl chloride	0.85	0.51
Losses	0.25	–
Total	23.58	16.88

Source: Reference (8)

The most important input stream from which dissipation occurs is the process steam. The steam is primarily used for drying the large quantities of process water from the polymer. Other processes are available or under development which would eliminate the high requirement for process water and the attendant steam for drying. These could not be analyzed due to a lack of detailed process information. They may show significant reductions in fuel requirements; although, they do require stirring of the reactor charge and this would increase the fuel requirements. A significant improvement in the present processes could be obtained by combining electrical power generation with the raising of process steam.

Polystyrene

Styrenics are the third most important class of high polymers, based on the amount produced, in the United States. Polystyrene homopolymer accounted for slightly more than 12% of all 1971 plastics production, including thermoplastics and thermosets. Rubber-modified polystyrene accounted for 2.1%, and styrene copolymers such as ABS (acrylonitrile-butadiene-styrene) and SAN (styrene-acrylonitrile) accounted for another 3.7% of the total resin production. In addition, there are a number of specialty copolymers containing styrene which are produced in limited quantities. These are principally block and graft copolymers of styrene and butadiene.

Continuous mass polymerization is the most commonly used process for making polystyrene. Energy requirements for this process are projected at around 1,000 kwh per metric ton of polystyrene. The suspension process used for certain grades of polystyrene requires much larger energy requirements and capital investment. Further, the mass polymerization has a somewhat better overall conversion efficiency, about 97.5% compared to 92.5% for the suspension process.

The energy and availability summary for polystyrene (PS) production is given in Table 107. The results of the polystyrene analysis are quite similar to the results of the HDPE analysis given above.

TABLE 107: ENERGY AND AVAILABILITY SUMMARY FOR THE POLYSTYRENE INDUSTRY

	Energy	Availability
	(10^6 Btu/ton polystyrene)	
Input		
Styrene	37.25	36.49
Process water	0.00	0.00
Cooling water	0.06	0.00
Steam	5.87	1.68
Electricity (direct equivalent)	0.14	0.14
Total	43.32	38.31

(continued)

TABLE 107: (continued)

Output	Energy Availability (10⁶ Btu/ton polystyrene)	
Polystyrene	35.84	35.53
Process water	4.52	0.86
Cooling water	0.66	0.01
Steam	1.27	0.23
Styrene	0.91	0.89
Losses	0.12	–
Total	43.32	37.52

Source: Reference (8)

The main energy and availability inputs to the process are in the feed stock and very little of these is lost during polymerization. The second largest input to polymerization is process steam for mixture preheating and drying. The steam and process water exit temperatures are low and little availability is lost in these streams. The raising of process steam if combined with electrical power generation would offer an opportunity for process improvement and more effective use of fuel.

MAJOR ENERGY CONSERVATION OPTIONS TO 1980

The processes involved in the manufacture of the six major types of polymers can be divided into two major categories, addition and condensation polymerization. Five of the polymers, polyvinyl chloride, low-density polyethylene, polypropylene, polystyrene and high-density polyethylene, are prepared via addition polymerization techniques. Phenolic resins are prepared by a condensation process with formaldehyde. Addition polymerization processes are subdivided into two major subcategories, aqueous dispersion (suspension and emulsion) solution and bulk polymerization.

Since all of the major polymers are used at 100% solids, a major step with respect to the energy requirement of each is the removal via evaporation of unreacted monomer and solvent or diluent. Major process research is under way in the polymer industry to cement major solution process to essentially bulk polymerization and to increase the conversion of monomer to polymer to minimize the energy required for product isolation. Processes for low-density polyethylene have been developed using the latter concept (higher conversion reactors) that are 35% more efficient than 1972 reactors.

Processes for the synthesis of high-density polyethylene have been developed that are gas phase polymerizations instead of the more conventional slurry or solution process. This reduces the energy requirement to produce this product by about 25%. This technology should be applicable to production of polypropylene as well.

Polyvinyl chloride production is predominantly an aqueous suspension process and no immediate change to a diluent having a lower heat of vaporization is anticipated. Further, a bulk polymerization process has been marketed which effects high conversion to minimize monomer recovery. In addition, the vinyl-chloride polymerization process has become more energy intensive because of OSHA regulations.

Not all of the above technical changes can be built into present and new facilities by January 1, 1980. Most new plants will incorporate them; old plants may or may not—being operated as is, closed or modernized.

The savings achievable will vary from plant to plant but are related to better housekeeping, better utilization of polymerization heat in the addition process and developing high-conversion, high-total-solids processes with the ultimate being a high-conversion bulk polymerization for major SIC 2821 products.

Through housekeeping the industry has reduced from 7 to 20% the energy required to produce a pound of SIC 2821 products. For this study a housekeeping savings of 15% was used. This includes maximizing the utilization of steam systems, insulation of plant and process lines, reuse of hot water previously sewered, establishing a routine heat exchanger cleaning program, improved solvent recovery, revised drying methods, improved turbines, burning waste in boilers and installation of heat-recovery equipment. In projecting the energy savings for new plants the savings was calculated as a sum of housekeeping and new process reductions. Old plants were credited only for housekeeping improvements in efficiency.

GOAL YEAR (1980) ENERGY USE TARGET

Production of plastic materials, synthetic resins and nonvulcanizable elastomers in 1979.5 is expected to be 44.1 x 10^{12} lb, primarily five products, low-density polyethylene, high-density polyethylene, polystyrene, polyvinyl chloride and polypropylene. The projected SIC 2821 totals are presented in Table 108. The energy requirement will be 4,573 Btu per pound using 1972 technology and 3,735 Btu per pound using the maximum improvements in technology projected for 1979.5. The preliminary energy efficiency goal then is estimated to be 17.7%.

The values reported in Table 108 for 1979.5 can be contrasted with those of 1972, 25.9 x 10^{12} pounds and energy consumption of 118.5 x 10^{12} Btu. This gives a specific energy consumption of 4,573 Btu per pound. The decrease going from 1972 to 1979.5 is due to a combination of housekeeping-type improvements coupled with new polymerization processes eliminating or minimizing solvent evaporation steps in the process.

Only one of the major products is involved with a significant energy penalty as the result of new environmental regulations. The energy demand in the production of polyvinyl chloride increased about 5% as the result of new legislation regarding the permissable levels of airborne vinyl chloride monomers in the plant area.

Energy usage and savings opportunities are difficult to define for SIC 2821 products. Individual companies are reluctant to divulge sensitive data required to give a precise analysis of the total industry on a product-by-product basis. A reasonable level of energy conservation for existing plants will effect energy savings of 15% on all new plants constructed since 1972 for the following products: high-density polyethylene, low-density polyethylene, polypropylene. For the remainder, assumed process improvement research will produce only a 7% savings. Similarly, 25% conservation should be achieved in new plants. These judgments were made on the basis of interviews with industry representatives and available new process literature.

TABLE 108: SUMMARY OF PRODUCTION AND ENERGY ESTIMATES, SIC 2821

	All Products SIC 2821	
	1972	1980
Production, 10^9 pounds	25.92	44.09
Total Energy required, 10^{12} Btu.		
1972 Basis	118.5	200.2
1980 Basis	—	164.7
Specific Energy Consumption, Btu/pound		
1972 Basis		4573
1980 Basis		3735
Energy Efficiency Goal, percent		17.7

Source: Reference (5)

In calculation, the same ratio of production to capacity that existed in 1972 was assumed for 1979.5. The energy requirement for the "All Other" resin group was assumed to be the arithmetical average of the requirements for acrylics, phenolics and polyesters. The results of applying these factors to production forecasts are shown in Table 109.

SOME PROJECTIONS BEYOND 1980 TO 1990

Table 110 shows estimates of energy use coefficients for present and future plastics manufacturing. The new plant figures for LDPE reflect the efficiency of the Ato Chimie process for 1974 to 1985 and the 1985 to 1990 figures reflect savings through radiation-induced polymerization or a comparably advanced technology. The new plant HDPE figures for 1974 to 1985 are representative of the consumption levels in new plants. The 1985 to 1990 new HDPE plant energy use figures reflect the application of dry polymerization technology.

New plant consumption figures for polypropylene are intended to reflect the application of advanced technologies similar to those for HDPE.   ꞓ consumption figures for polystyrene and polyvinyl chloride reflect the appꞒication of

TABLE 109: PRODUCTION AND ENERGY ESTIMATES—SIC 2821

| | Estimated Production 10^9 Pounds | | Energy Used 1972 | | Energy Usage 1979.5, 10^12 Btu | | Energy Efficiency Target |
	1972(a)	1979.5(b)	Total 10^12 Btu	Btu/Pound	1972 Basis	1979.5 Basis	Percent
Low-density Polyethylene	5.36	8.82	40.47	7550(c)	66.59	52.60	21.0
Polyvinylchloride	5.12	8.38	30.82	6020(c)	50.45	45.40	10.0
Polystyrene	4.89	6.83	11.0	2250(c)	15.47	12.77	16.9
High-Density Polyethylene	2.30	4.41	9.13	3970(c)	17.51	13.6	22.2
Polypropylene	1.73	5.07	5.09	2940(c)	14.91	11.19	24.9
Phenolic	1.44	2.21	7.79	5411(d)	11.96	9.87	17.5
Polyester	0.93	1.53	1.09	1176(d)	1.80	1.5	17.8
Amino	0.93	1.53	5.03	5411(d)	8.28	6.80	17.8
Acrylic	0.76	1.25	1.71	2250(e)	2.81	2.3	17.8
Alkyds	0.63	1.04	0.74	1176(d)	1.22	1.99	17.8
All Other	1.83	3.02	5.65	3085(f)	9.32	7.7	17.8
Total	25.92	44.09	118.52		200.32	166.72	
Average				4573			Goal 17.7

(a) United States Tariff Commission, Synthetic Organic Chemicals, U.S. Production and Sales, 1972.
(b) Modern Plastics, pp. 40-50, October, 1975.
(c) Projected from Project Independence data for 1973; 7.5% savings assumed.
(d) Hydrocarbon Processing, Petrochemical Issue, November, 1975.
(e) Assumed comparable to polystyrene.
(f) Arithmetic average using specific energy values for phenolics, polyesters and acrylics, and applying this to the remainder.

Source: Reference (5)

improved operating procedures, energy conservation hardware innovations, and increased energy economies of scale, as well as increased use of the mass process for polyvinyl chloride and the continuous process for polystyrene.

Based on these crude but representative assumptions about future energy requirements for plastics production, the overall projection in Table 111 was generated. These projections do not include the impact of energy conservation efforts at existing plants.

TABLE 110: BASIS OF ENERGY PROJECTIONS FOR PLASTICS (PER POUND)

| | .....Pre-1974..... | | New 1974-1985 | | New 1985-1990 | |
	Fuel*	Electric**	Fuel	Electric	Fuel	Electric
LDPE	5.0	0.58	1.0	0.58	0.50	0.44
HDPE	2.3	0.40	2.3	0.20	0.30	0.20
PP	1.7	0.30	1.7	0.15	0.26	0.15
PS	1.5	0.17	1.2	0.14	1.05	0.12
PVC	2.7	0.84	1.4	0.42	1.40	0.42

*10^3 Btu
**kwh

TABLE 111: FUTURE ENERGY CONSUMPTION FOR THERMOPLASTIC PRODUCTION

| Plants | ....Electricity (10^9 kwh).............. | | | | | |
	1967	1973	1977	1980	1985	1990
Pre-1974*	5.20	9.73	9.14	8.64	7.96	7.29
New	–	–	2.83	5.59	11.61	19.31
Total	5.20	9.73	11.97	14.23	19.57	26.60
	................Fuels (10^{12} Btu)................					
Pre-1974	28.4	58.6	55.0	52.1	47.9	43.9
New	–	–	12.7	25.3	53.8	73.6
Total	28.4	58.6	67.7	77.4	101.7	117.5

*Decreases after 1973 due to retirement at 1.5%/yr.

Source: Reference (10)

SYNTHETIC RUBBER INDUSTRY

SIC 2822 materials are synthetic rubbers made by polymerization and copolymerization of specific monomers such that the final product contains reactive functionality that permits further polymerization (vulcanization) when desired. The major products in this industry—styrene-butadiene and polybutadiene—account for 70% of the total produced in 1972.

Other important polymers in SIC 2822 are butyl rubber, nitrile rubber, polyisoprene, and ethylpropylene rubber. The products of this class have a shipment value (1972) of $1,089 million, over 1.5% of the whole chemical industry shipments. The energy consumption is primarily due to the production of SBR because of the large volume of production and butyl, nitrile, butadiene, isoprene, and ethylene-propylene rubbers because of their high-energy requirement per pound. These six account for 93% of the energy consumption. For these materials savings opportunities were determined.

PROCESS TECHNOLOGY INVOLVED

Two types of processes are currently utilized in the production of various rubbers. These are emulsion processes used to produce about 90% of the SBR, nitrile, and neoprene rubbers. Solution processes are used to produce butyl, polybutadiene, polyisoprene, EP rubber, and about 10% of the SBR. It is expedient, therefore, to discuss the two processes rather than production of various products.

Emulsion Process

In emulsion polymerization, the monomers are mixed in an emulsion reaction with water, catalyst, emulsifier, and molecular-weight modifiers. A simplified flow sheet is shown in Figure 76.

Two significantly different emulsion process modifications are used in producing SBR, nitrile, and neoprene rubber. SBR is cited in the description. In the older or hot rubber process, the polymerization is conducted at an elevated temperature.

FIGURE 76: GENERALIZED FLOW SHEET FOR EMULSION POLYMERIZATION OF RUBBER

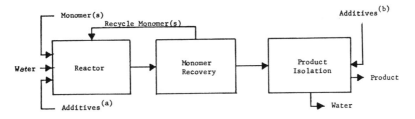

(a) Catalyst, emulsifiers, molecular weight controls.

(b) Oils, carbon black, coagulants

Source: Reference (5)

Steam heats the mixture to reaction temperature (122°F for SBR production). The reaction is exothermic and cooling is required to hold the reaction at temperature. The reaction is carried to a suitable conversion and then stopped by addition of a shortstopping agent. If the conversion limit is exceeded, (72% for SBR) an insoluble gel is formed. The hot process accounts for about 17% of all SBR production.

The cold reaction uses more effective catalysts and activators such that heating is not required to initiate the reaction. However, the polymerization is exothermic and cooling is required to maintain the reaction temperature (41°F for SBR). This reaction, also, must be stopped to prevent gel formation. (For SBR, this conversion limit is 60%.) Cold SBR has superior properties, so this process accounts for about 75% of the SBR production. The balance of SBR is produced by solution polymerization.

Unreacted monomer(s) are removed from the latex by vacuum and/or steam stripping and recycled after any necessary purification. The product rubber is recovered from the latex by coagulation after blending and additive steps are completed. These are primarily mechanical steps. Process heat is required only for drying the crumb after mechanically separating it from the water.

Solution Process

These processes are conducted either in an inert solvent or are carried to a limited conversion such that the product remains partially or totally dissolved in the monomer. Several types of catalysts are used in solution polymerization of various monomers. These, however, are water sensitive such that the general reaction cannot be adapted to an emulsion technology.

The generalized flow sheet for solution polymerization is shown in Figure 77. Conversion levels are lower for solution reactions than for emulsion processes, consequently a greater amount of energy is needed to recover and purify the monomer/solvent.

FIGURE 77: GENERALIZED FLOWSHEET FOR SOLUTION POLYMERIZA-
TION OF RUBBER

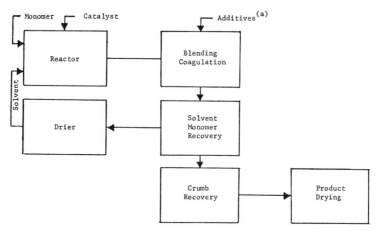

(a) Short stop, antioxidants, water.

Source: Reference (5)

Product properties are controlled to a great extent by the choice of catalyst.
The catalyst used has a significant influence on the specific energy requirements
of the process. As an example, net specific energy requirements for polybutadiene production range from about 12,600 Btu/lb to 22,800 Btu/lb depending
on the catalyst used. These effects are discussed further in a following section.

MAJOR ENERGY CONSERVATION OPTIONS TO 1980

The processes involved in making SIC 2822 products can be divided into two
major categories, solution and emulsion (SBR) polymerization processes. The
products are shipped as 100% solids; hence, each process involves a polymer
purification step. The emulsion polymerization product is heated to remove
unconverted monomer, then the latex is coagulated and most of the water is
removed as liquid with the last traces removed by heating. The solution products on the other hand are submitted to solvent and monomer stripping operations from solutions containing as little as 12% polymer.

The major potential for significant energy conservation ultimately exists in the
development of less energy-intensive processes. This could be achieved by
developing technology to supplant solution production of the various rubbers
with slurry formation, or possibly bulk reaction.

For example, the emulsion process for SBR requires much less energy than the
solution processes for SBR and other elastomers. Solution processes require a
major expenditure of energy in recovery of product by volatilization of the solvent whether inert or monomer. Presently, many of the new elastomers with

superior properties are made by solution polymerization and technological innovations will be required to develop satisfactory nonsolution type processes.

No innovations have been documented to significantly reduce the energy requirements of solution polymerization processes by conversion to slurry or bulk polymerization methods for elastomers. Accordingly, the potential savings of 50% in this category cannot be projected by 1979.5. However, the study estimated that a 5% energy reduction will be achieved in this area by selecting solvents that will permit higher conversion (e.g., low viscosity at 18% solids for polyisoprene than currently obtained at 12% solids in heptane) and solvents that will require less energy to evaporate.

Improved housekeeping and maintenance, better process control and utilization of process heat and better energy balancing within the process or plant have achieved significant energy savings within the industry. These savings are estimated to be 15%. In projecting the energy savings for new plants, a sum of housekeeping and process improvement was assumed. Old plants were given credit only for housekeeping improvements in efficiency.

Plants constructed prior to and around 1972 were based upon the ready availability of low-cost fuels and feedstocks. In this environment, plants were designed for minimum capital and operating costs. This generally neglected consideration of the optimum use of feedstocks, by-products, and process steam. The advent of fuel and feedstock short falls has necessitated reassessing the efficiency of their utilization–on both process level and plant level.

Conservation on the process level includes selection of new reactions with obvious energy savings and incorporating process innovations to minimize energy input requirements. Savings on the process level relate primarily to new plant design and construction. Such new plant energy savings have been estimated to range from 15 to 60% with an industry average generally projected around 25%.

The potentials for such savings are demonstrated in data compiled for specific energy requirements of several rubber processes. Table 112 is from several sources. In the absence of a broad base of specific industrial production/energy requirement data, the data in Table 112 were used as the basis for projecting the 1972 specific energy requirements.

Observations worth noting in Table 112 are the energy savings possible in the SBR emulsion processes. In processes where the use of expander dryers are utilized a specific energy savings of nearly 33% can be anticipated over conventional thermal drying.

Catalyst selection and reactivity also can markedly affect process energy requirements. This is shown for polybutadiene where the minimum specific energy value is 12,600 Btu/lb compared to a maximum value of 22,800 Btu/lb. This represents a 45% savings for the n-butyl lithium process providing the product properties are satisfactory for the projected market.

For example, polymerization of 1,3-butadiene with cobalt-type or titanium-type coordination catalysts produces high-molecular weight rubbers for tire application.

On the other hand, polymerization of 1,3-butadiene with the nickel-type catalyst produces a low-molecular-weight polymer used in varnishes for the air-dry-
ing properties.

TABLE 112: ENERGY REQUIREMENTS OF RUBBER SYNTHESIS TECHNOLOGY

Rubber-Type Process	. . . . Net Specific Energy (Btu/lb)			
	Estimate*	Estimate**	Estimate*** 1973	Estimate† 1980
SBR††		4,680	4,113	3,192
Emulsion A	3,750			
Emulsion B	2,520			
Solution	7,250			
Butyl	18,200	19,900	19,030	14,675
Nitrile	3,135†††		3,135	2,627
Polybutadiene		19,300	18,340	13,375
Cobalt complex	14,400			
Iodine, Ziegler	22,800			
Nickel complex	20,000			
n-Butyl lithium	12,600			
Polyisoprene		17,100	16,240	12,079
Ziegler	16,500			
n-Butyl lithium	14,300			
EP Rubber	18,000		18,000	14,318
Neoprene (and others assumed)			3,520§	3,164
Acetylene	5,140			
Butadiene	1,900			

*Estimate calculated from Project Independence Data (10).

**Estimate calculated from *Industrial Energy Study of the Plastics and Rubber Industries* (23).

***Estimate based on an average between * and **. Value for * was weighted average for SBR 90% by emulsion-10% by solution, other averages arithmetic since weighting factors were not known.

†Specific energy savings of 15% assumed on old plants, i.e., 1972 capacity and specific energy values. Specific energy values for new plant construction assumed to be comparable to technology reported in *Hydrocarbon Process-ing* (Nov. 1973). In the absence of specific data for a product, a 25% energy saving over the 1972 value was used. The composite value is based on a relative weighting of old and new plants operating at the 1972 level.

††A without expander dryer; B with expander dryer.

†††Average value of SBR technology assumed.

§Average value of two processes assumed in the absence of weighting data.

Source: Reference (5)

Other areas with potential energy savings may include mechanically assisted
fractionation or hydrocarbon recovery using waste-heat sources rather than
process steam. Numerous routes to improved plant energy balancing have been
utilized to reduce energy consumption.

Older established plants are constrained in process manipulation and modifications which are practical and economically viable. One type of process manipulation used effectively is production scheduling. The process is operated at near capacity to obtain a favorable energy efficiency, and then placed on stand-by for scheduled maintenance and improvements or operated on a related product. This avoids the frequent loss in energy efficiency encountered in operating continuously at a lower utilization level.

Housekeeping, improved steam-trap maintenance, hot-pipe insulation, improved instrumentation for tighter process and combustion control, replacement of inefficient motors with more efficient drivers have been implemented to various degrees to effect energy conservation.

GOAL YEAR (1980) ENERGY USE TARGET

Production of SIC 2822 products in 1979.5 is expected to be 7.15×10^{12} lb (see Table 113). In 1979.5, the rubbers other than SBR will total 63% of the total vulcanizable elastomer products. These products are made by processes that are more energy intensive than SBR and this is reflected in the specific average energy requirement of 9,265 Btu per pound of product for this SIC category in 1979.5 even when using 1979.5 technology. This specific energy requirement using 1972 technology and the projected 1979.5 product mix is 11,518 Btu per pound. Using these data the preliminary efficiency goal is 19.6 for SIC 2822.

TABLE 113: SUMMARY OF PRODUCTION AND ENERGY ESTIMATES, SIC 2822

	All Products, SIC 2822	
	1972	1979.5
Production, 10^9 lb	4.91	7.154
Total Energy Required, 10^{12} Btu		
1972 Basis	41.77	82.40
1979.5 Basis	--	66.28
Specific Energy Consumption, Btu/lb		
1972 Basis	8,508	11,518
1979.5 Basis	--	9,265
Preliminary Energy Efficiency Goal,[a] percent	--	19.6

(a) Calculation for energy efficiency goal:
$$\frac{82.40 - 62.28}{82.40} = 19.6 \text{ percent}$$

Source: Reference (5)

The values reported in Table 113 for 1972 for production and energy utilization reflect a product mix that is 53% SBR. This product is one-third as energy intensive as the rubbers prepared via solution processes and as mentioned earlier, the average Btu required per pound of product in 1972 was only 8,508. Of the total energy used, 74% was required for the production of 47% of the total SIC 2822 products in 1972.

The estimated potential for savings in the SIC 2822 industry is summarized in Table 114. Comparing the estimated energy requirements for the projected 1979.5 mix and production to the energy requirements using 1972 technology for the 1979.5 projected mix and production indicates an energy efficiency target of 19.6%.

This value reflects the effect of changing product mix which is shifting to rubbers with improved properties, but which are more energy intensive. If the energy savings projection is made on 1979.5 production and technology but using the 1972 mix, the projected energy savings is estimated to be 25.2%.

TABLE 114: PRODUCTION AND ENERGY ESTIMATES SIC 2822 MATERIALS

	Estimated Production, 10^9 lb		Energy Used 1972		Energy Usage 1979.5, 10^{12} Btu		Energy Efficiency Target, percent
	1972(a)	1980(b)	Total, 10^{12} Btu	Btu/lb(c)	1972 Basis	1979.5 Basis	
Styrene Butadiene	2.610	2.590	11.273	4319	11.186	9.51	15.0
Butyl	.290	.400	5.796	19985	7.994	6.57	17.8
Nitrile	.160	.255	.526	3292	0.839	0.68	18.6
Polybutadiene	.670	1.680	12.906	19262	32.360	25.9	21.0
Polyisoprene	.290	.880	4.944	17048	15.002	11.73	21.8
EP Rubber	.200	.660	3.780	18900	12.474	9.73	22.0
All Other	.690	.689	2.550	3696	2.547	2.16	15.0
Total	4.910	7.154	41.775		82.402	66.28	
Average				8507			19.6

Sources: (a) United States Tariff Commission, Synthetic Organic Chemicals U. S. Production and Sales 1972.

(b) Project Independence growth factor (5 percent/year) used in projecting capacity. Production estimated for 1980 at the same level of utilization as 1972.

(c) Specific energy requirements projected from Table 112 (Estimated 1973) by multiplying by 1.05 (equivalent to a savings about 4.8 percent for 1973 on 1972 basis).

Source: Reference (5)

CELLULOSIC MANMADE FIBER INDUSTRY

Fibers made in the form of yarn, textile monofilament, staple and tow from cellulosic polymers constitute SIC 2823. The polymers involved are rayon and cellulose acetates. The products of this class had a shipment value of 627.9 million dollars in 1972. The energy consumption is primarily to produce rayon; it accounts for 69% of the production in SIC 2823 and the two classes of fiber, rayon and acetate account for the total production in this category.

PROCESS TECHNOLOGY INVOLVED

In the manufacture of SIC 2823 fibers the starting raw material is wood pulp characterized by high content of α-cellulose. In rayon the cellulose is solubilized via xanthation reactions and the solution passed through spinerets. In the manufacture of acetate fiber the cellulose is reacted with acetic anhydride and dissolved in acetone for spinning. These two processes are described briefly in the following sections. A second method for preparing rayon is known as the cuprammonium process but this process accounts for less than 10% of the rayon made and is not discussed.

Acetate

Most cellulose acetate used for textile purposes is made by the solution-type, sulfuric-acetic acid process for converting purified cellulose, from wood pulp or cotton linters, into cellulose acetate. The process of next importance is the methylene chloride process. In this process, methylene chloride replaces all or part of the acetic acid and performs as a solvent for the resulting triacetate. These two processes make either primary or secondary acetates.

A third process is the fibrous process. In this process, an inert organic liquid, such as benzene, is introduced to prevent solution of the acetylated cellulose as it is forming. The product is fibrous, resembling the texture of the original cellulose. This process is seldom used because of some inherent disadvantages; the cellulose acetate must be recovered and dissolved before it can be hydrolyzed.

291

In the sulfuric-acetic acid cellulose acetate manufacturing process, first, acetylation grade cellulose is shredded, then it is charged into an acetylator containing most of the acetic acid along with a small quantity of catalyst. Here the cellulose is agitated at 100°F for about 1 hour or until it has become thoroughly activated. Cooled acetic anhydride is mixed with the activated cellulose and the whole is cooled to 30° to 40°F. On reaching this temperature range, the rest of the catalyst is added in a small quantity of acetic acid.

The acetic anhydride and the water introduced in the cellulose react causing a rise in temperature. Thereupon, the cellulose begins to acetylate and the temperature is reduced to control the reaction. As the reaction progresses, the material changes from a soupy to a doughy and then to a clear viscous condition. The acetylation reaction is allowed to continue at a temperature of 100° to 110°F until no more fibers remain and the desired viscosity has been obtained. At this point, weak acid (which furnishes the water for the hydrolysis step) is added slowly while the solution is vigorously agitated and the temperature brought to 100°F.

The solution is then charged to the hydrolysis vessel where it is held at constant temperature until the desired acetyl content has been reached. At this point, the solution is pumped into another vessel together with enough 10 to 14% acetic acid to bring the acetic acid content of the mixture to 23 to 35%. The mixture is agitated violently which causes the acetate to separate from solution as a flaky solid. The resulting slurry is poured into a preliminary washer where the acid is drained off and returned to the recovery system. The cellulose acetate is then washed until the acidity is reduced to about 0.1% acetic acid. The washings are used in the preceding precipitation step.

The cellulose acetate, which is now in a water slurry, is given a final washing to render it acid free. Stabilizing salts are added; these may be hydroxides or carbonates of sodium, magnesium, or calcium. The slurry is dehydrated to a moisture content of 50 to 70%, shredded to break up lumps, and air dried at 225° to 275°F to reduce the moisture content to below 1%, after which it is stored and packaged.

The methylene chloride process is similar to the one just described except that methylene chloride replaces all or part of the acetic acid. The methylene chloride is removed after hydrolysis by adding weak acid at the beginning of the hydrolysis, boiling it off, and then recovering it for reuse in the acetylation step.

The acid drained from the precipitation vessel is recovered as glacial acetic acid for reuse in the process, or for reconversion into acetic anhydride. The economy of the process depends on an efficient recovery of this material. The resulting cellulose acetate product is carefully blended to close tolerances of acetyl content and viscosity and then filtered. The cellulose acetate is then dissolved in acetone and spun into fiber using dry-spinning techniques.

Rayon

Wood pulp is reacted with a solution containing an excess of rayon-grade sodium hydroxide to extract alkali-soluble impurities and produce alkali cellulose.

Unreacted sodium hydroxide (steeping liquor) is pressed out of the cellulose and the alkali cellulose is shredded and aged under carefully controlled conditions. When the product has aged sufficiently, it is reacted with carbon disulfide to form sodium cellulose xanthate. The cellulose xanthate is dissolved in cooled dilute sodium hydroxide, using high-shear mixing for about 2 hours to produce the clear, honey-like solution called viscose.

The viscose is blended, filtered, and deaerated according to a carefully controlled schedule to take advantage of several chemical changes that accompany the process. In this process, called ripening, rearrangement of the xanthate groups takes place, accompanied by partial dexanthation.

When the viscose has ripened sufficiently, pigments, dulling agents and other additives can be added. Then it is extruded through spinnerets into the spinning bath containing sulfuric acid, sodium sulfate, and zinc sulfate. The acidity, as well as the sodium sulfate and zinc sulfate concentration, and the temperature of the spinbath control the coagulation of the dissolved xanthate and the regeneration of the cellulose. Often more than one bath is used. Viscose brings new sodium hydroxide constantly into the bath, rapidly consuming the sulfuric acid in the bath in the formation of sodium sulfate; therefore, the acidity and salt concentration of the spinbath must be maintained within desired limits. The regenerated cellulose filaments that emerge from the spinning bath are either twisted into yarn or cut into staple of uniform fiber length.

Staple is produced in thicknesses ranging from 1.5 to 15 denier; the fiber can be made bright or dull, can be spundyed (a dyeing process in which dyes are added to the viscose), and it may be crimped or uncrimped. Filament textile yarn is mostly produced in the 2 to 3 denier range and, typically, with a conditioned tenacity of 1.7 to 2.1 grams per denier. It is collected in cakes and subsequently purified, finished, dried, and rewound.

MAJOR ENERGY CONSERVATION OPTIONS TO 1980

Energy savings opportunities are more difficult to define for SIC 2823 than for other areas of the chemical industry. The processes are old and the demand for both rayon and acetate fiber is decreasing. As a result the major energy saving opportunity for this group is improved housekeeping.

No new plants are anticipated in this industry. In the existing plants improved housekeeping, better planning and scheduling and small additions to plant capital can save 10% of the energy per pound required for production. Housekeeping improvement includes elimination of steam leaks due to faulty traps, control of space heating and cooling, and regular maintenance of heat exchangers. Meeting OSHA and EPA regulations is not expected to add large energy increases for the period 1972 to 1979.5.

GOAL YEAR (1980) ENERGY USE TARGET

The cellulosic manmade fiber industry is expected to produce about 0.97 billion pounds of product in 1979.5 (Table 115) a decrease of 6.4% per year.

No significant change in product mix is predicted. Energy usage will be 63,000 Btu per pound with 1972 technology and 57,000 with improved technology and operating procedures. The total energy required in 1979.5 on a 1972 basis is 61.4 x 10^{12} Btu and forecast as 55.26 x 10^{12} using improved technology. The 10% savings estimated is a lower goal than for other polymers (SIC 2821, 2823, and 2824); it was established lower because production of the SIC 2823 materials is decreasing significantly with capacity remaining constant.

TABLE 115: SUMMARY OF PRODUCTION AND ENERGY ESTIMATES FOR SIC 2823

	All Products SIC 2823	
	1972	1979.5
Production, 10^9 Pounds	1.394	0.97
Total Energy Required, 10^{12} Btu		
1972 Basis	88.3	61.4
1979.5 Basis	--	55.26
Specific Energy Consumption, Btu/lb		
1972 Basis	63,330	63,330
1979.5 Basis	--	57,000
Energy Efficiency Goal, [a] percent		10

(a) Calculation for energy efficiency goal: $\dfrac{61.4 - 55.26}{61.4} =$ 10 percent.

Source: Reference (5)

SYNTHETIC NONCELLULOSIC FIBER INDUSTRY

Fibers made in the form of yarn, textile monofilament, staple and tow from synthetic polymers constitute SIC 2824. The polymers involved are principally terephthalate polyesters, nylon-6, nylon-6,6, polypropylene, acrylics and moda-crylics. Other fibers, which comprise less than 2% of production, include urethane (Spandex), anidex, vinyon, fluorocarbon and reconstituted protein fibers. Glass fibers are classified as SIC 3229 products.

Since many fiber-making establishments start with simple organic chemicals and make the resin from which the fiber is spun, captive resin making is included in this discussion.

In 1967, the energy consumption in fiber manufacturing was greater than that for plaster or synthetic rubber. As in the case of these other industries, synthetic fiber production depends on steam and electricity for its energy needs. The ratio of steam to electricity used is well suited to the dual generation of steam and electricity through the use of a high-pressure boiler and a back-pressure turbine. It is not surprising, therefore, to find that 50% of the electricity used in fiber manufacturing was self-generated.

The energy to shipment value ratio for synthetic fibers decreased gradually between 1958 and 1967 by 42%. This reduction was due to improvements in manufacturing techniques but also to a shift away from energy-intensive cellulosic fibers.

PROCESS TECHNOLOGY INVOLVED

In fiber making simple building blocks (monomers) are reacted to form long chains of repeating units (polymers), then the polymer is spun into fiber, drawn and wound. Usually the polymer is made and spun in the same establishment; thus, the polymerization technology is described where polymer resin is usually purchased by the fiber maker. (A description of polypropylene resin manufacture is included in the discussion of SIC 2821).

Polyester Fibers

Dimethyl terephthalate (DMT) or very pure terephthalic acid can be reacted with ethylene glycol or certain other diols to give polyester resin that can be spun into fibers, as shown in Figure 78.

FIGURE 78: PREPARATION OF POLYESTER FIBER

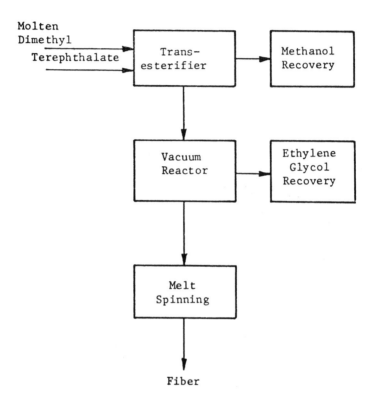

Source: Reference (5)

Usually DMT is heated with 2 mols of ethylene glycol to form diethylene glycol terephthalate (DGT) and methanol. The methanol is recovered and either sold or used in making more DMT. The DGT is heated in several stages so as to have the molecules of DGT react by releasing ethylene glycol and forming increasingly long polymer chains.

The ethylene glycol is continuously removed by vacuum and recovered. The final molten material can be melt spun immediately or cooled, granulated and stored. The melt spinning process is described in a subsequent section.

Nylon Fiber

Nylon, or polyamide, polymer has a repeating unit involving an amide linkage, the same linkage found in wool. The number of carbon atoms in the monomers can vary but the most important have six carbon atoms. There are two types of nylon made with six carbons, nylon-6 made from caprolactam, and nylon-6,6 made from hexamethylene diamine and adipic acid.

In making nylon-6, molten caprolactam is mixed with small amounts of water, catalysts, stabilizer and delusterant and heated in a reactor at 500°F (Figure 79). The reaction is slightly exothermic; excess heat is removed by a coolant.

FIGURE 79: PREPARATION OF NYLON-6

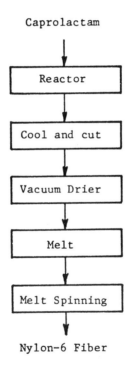

Caprolactam

Reactor

Cool and cut

Vacuum Drier

Melt

Melt Spinning

Nylon-6 Fiber

Source: Reference (5)

Unreacted monomer is removed, either by distilling it away or by casting the resin, quenching, and drying it. The polymer is then melted and spun as described in a subsequent section.

Nylon-6,6 is prepared by the polymerization of the salt formed by mixing hexamethylene diamine and adipic acid in water (nylon salt). Nylon salt solution is fed to a thin film evaporator at about 230°F to remove excess water. This is followed by heating in a second thin film evaporator at 450°F under pressure, where the loss of water from the molecules of acid and amine gives rise to the condensation product desired. This moisture is removed by "flashing" at atmospheric pressure. The hot molten polymer then can be melt spun directly.

Acrylic Fibers

Acrylonitrile, mixed with comonomers, is polymerized in an aqueous system using an oxidation/reduction initiator, as shown in Figure 80.

$$nH_2C=CH-CN \longrightarrow -CH_2-\underset{\underset{CN}{|}}{CH}(-CH_2-\underset{\underset{CN}{|}}{CH})_{n-1}$$

FIGURE 80: PRODUCTION OF ACRYLIC FIBERS

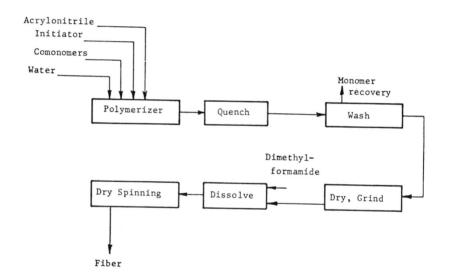

Source: Reference (5)

The polymer is insoluble in water and is removed by centrifuging. Excess monomer is removed by washing with water, from which it can be recovered. The polymer after drying and grinding is ready for spinning. Since the polymer decomposes at its melting point, it is dissolved in dimethylformamide and either dry spun or wet spun to give acrylic fiber. A description of these solvent spinning processes is in the following section.

Manufacture of man-made fibers requires conversion of bulk polymer into fibers, using processes that involve solution or melting of the polymer followed by extrusion through a spinneret. There are three types of commercial spinning technologies: melt spinning, dry spinning, and wet spinning, of which melt spinning is the most common and wet spinning the least.

Melt Spinning: If a polymer can be melted at a reasonable temperature without degradation, the simplest and least energy-intensive process is melt spinning. Polymer is melted by contact with a hot grid in the form of steel tubing. The molten polymer is pumped under high pressure through a spinneret, a plate containing many small holes.

The liquid polymer streams emerge into air where they cool and solidify, then are wound together into a thread and taken up on bobbins. A drawing step orients the fibers and improves their physical properties. Energy is required, of course, to melt the fibers, meter them precisely through the spinneret, and drive the guide rolls and other equipment. Since the molten polymer is frequently very viscous, booster pumps may be required to pump the melt through the spinneret. Nylon, polyester, and polyolefin fibers are produced by melt spinning.

Dry and Wet Spinning: If the polymer cannot be converted into the desired fiber by melt spinning, wet or dry spinning is performed. In these processes the polymer is first dissolved in a solvent, invariably a different solvent than the one in which the polymer was prepared. After extrusion through a spinneret, the solvent is either removed by evaporation (dry spinning) into a hot-air bath or by leaching out into a coagulating bath of a liquid which is a solvent for the spinning solvent but immiscible for the fiber. Dry spinning for the polymer is used for most acrylic fibers and for polyvinyl chloride fibers. Wet spinning is used for some acrylic fibers.

Drawing: Following extrusion by any of these spinning processes, the fibers are invariably drawn and annealed to develop desired physical properties. To accomplish this the driving rolls (which pull the fiber continuously through the spinneret and air or solvent bath) rotate slower than the final take-up rolls, permitting the fiber to be stretched.

This stretching, as much as ten times the original fiber length, allows the polymer molecules to align themselves. The annealing process consists of holding the fiber at an optimum temperature for a finite interval while the final orientation of structure occurs. The drawing and annealing processes increase the strength of the fiber, typically three to ten times its unoriented strength and may often be performed as a separate processing step. Other important fiber properties, including dyeability, extensibility, and dimensional stability are also related to drawing.

MAJOR ENERGY CONSERVATION OPTIONS TO 1980

Energy usage and savings opportunities are difficult to define for man-made, noncellulosic fibers, as they are for most of the chemical industry. The type and subtype of fiber, proprietary processes, special treatments (e.g., for curl, dye acceptance), degree of integration in the establishment, age of the plant and many other factors make the numbers imprecise. In addition individual companies are reluctant to divulge sensitive data. The best estimates of the energy per pound are in the FEA study of 1974 (23).

Table 116 shows average energy consumption in the manufacture of these six major synthetic fibers for 1971 and 1973. Energy required per pound of product varies by a factor of three between the most energy intensive fiber (acetate) and the least energy intensive (polyester).

TABLE 116: COMPARATIVE ENERGY CONSUMPTION IN MANUFACTURE OF SYNTHETIC FIBERS IN 1971 AND 1973

Fiber	Energy Consumption (per pound)		Overall Change, percent
	1971	1973	
Rayon	61.9	50.4	-18.6
Acetate	67.2	63.0	-6.3
Polyester	19.9	17.3	-13.1
Nylon	25.0	20.6	-17.6
Acrylic-modacrylic	47.1	43.4	-7.9
Olefin	24.6	19.0	-20.8
Average drop			-14

Source: Reference (10)

The savings achievable will vary from plant to plant, but are related to better housekeeping, improved systems analysis of operating procedures, and the addition of some heat exchangers, turbogenerators for steam reduction valves, etc. About one-fourth of the industry's electrical energy is internally generated. In older plants there are abundant opportunities to improve efficiency through balancing high- and low-pressure steam use, minimizing excess oxygen used in the boilers, and increasing the use of air preheaters.

In a new plant careful planning can allow better efficiency by recovering waste heat and arranging equipment adjacent where the recovered heat can be used. Capital equipment can be purchased with energy conservation in mind, steam lines can be planned for length as short as possible and turbogenerator units utilized.

The largest user of energy per pound of production is acrylic fiber. If research could develop a technique for melt spinning acrylics, a large additional saving would be possible. However, this has been a goal of research for many years without success. No such breakthrough is incorporated in the goal established here. Energy consumption to meet OSHA and EPA goals in this product area is small. For example, to meet the required water-quality goals by 1977, less than 1% additional energy is required.

GOAL YEAR (1980) ENERGY USE TARGET

The noncellulosic fiber industry is expected to produce about 8.3 billion pounds of product in 1979.5 (Table 117), a growth of 6% per year over the 5.4 billion produced in 1972. Polyesters and polyolefin fibers will grow more rapidly than nylon and the acrylic fibers.

TABLE 117: SUMMARY OF PRODUCTION AND ENERGY ESTIMATES—
SIC 2824

	All Products, SIC 2824 1979.5
Production, 10^9 pounds	8.290
Total Energy Required, 10^{12} Btu	
1972 Basis	200.5
1979.5 Basis	160.8
Specific Energy Consumption, Btu/pound	
1972 Basis	24,200
1979.5 Basis	19,400
Preliminary Energy Efficiency Goal, percent	19.8

Calculation of energy efficiency goal:
$$\frac{200.5 - 160.8}{200.5} = 19.8 \text{ percent}$$

Source: Reference (5)

Energy usage will be 24,200 Btu per pound with 1972 technology and 19,400 Btu per pound with improved technology and operating procedures in 1979.5. These estimates of energy consumption indicate that energy usage in 1979.5 with 1972 technology would be 200.5 x 10^{12} Btu and with the maximum reduction of energy usage by 1979.5, 160.8 x 10^{12} Btu. Of course the product mix will have changed, with a larger growth forecast for polyester fibers, which have a lower energy requirement.

In old plants improved housekeeping, better planning and scheduling, and small additions to plant capital can save 15% of the energy per pound. Housekeeping improvement includes elimination of steam leaks due to faulty traps, control of space heating and cooling, and regular maintenance of heat exchangers.

Systems changes involve scheduling operations so as to optimize energy use. In some cases replacement of pressure reducing valves with turbogenerators, adding new or improved heat exchangers, and preheating combustion air will be economical.

In new plants 30% savings are possible with better design and layout as well as the purchase of more efficient pumps, heat exchangers, etc.

The results of applying these factors to the production forecasts are shown in Table 118. In the calculation, the 83% ratio of production to capacity existing in 1972 was assumed for 1979.5. The energy requirement for "other fibers" was assumed to be the same as for nylon.

TABLE 118: PROJECTED PRODUCTION, ENERGY USAGE AND ENERGY EFFICIENCY GOAL FOR SELECTED PRODUCTS

Fiber	Estimated Production, 10^9 lb		Energy Used, 1972		Energy Usage, 1979.5, 10^{12} Btu		Energy Efficiency Goal, Percent
	1972	1980	Total 10^{12} Btu	Btu per lb	1972 Basis	1979.5 Basis	
Polyester	2327	4444	46.2	19,900	88.9	69.2	22.2
Nylon	1974	2309	49.4	25,500	57.7	47.5	17.7
Acrylic	626	714	29.5	47,100	33.6	27.9	17.0
Polyolefin	347	667	8.5	24,600	16.4	13.1	20.1
Other	82	156	2.1	25,000	3.9	3.1	20.5
Total	5356	8290	135.7	25,336	200.5	160.8	19.8

Source: Reference (5)

SOME PROJECTIONS BEYOND 1980 TO 1990

In the Project Independence Blueprint (10), fiber production is assumed to grow at a rate of 7.7% per year between now and 1980, and will slow to 6% during the eighties. An important factor in the future consumption of energy by the fiber manufacturing industry is the mix of synthetics produced by the industry. As an estimate of this effect, the relative growth of each type of fiber in recent years was extended to 1980. Table 119 shows the likely shift in the production by 1980.

TABLE 119: CHANGE IN MIX OF FIBER PRODUCTION

	Relative Portion of the Market (pounds produced)	
	1973	1980
Rayon	0.118	0.036
Acetate	0.061	0.018
Nylon	0.228	0.116
Acrylics	0.107	0.071
Polyester	0.415	0.677
Olefin	0.070	0.081

Source: Reference (10)

This shift, separate from any potential changes in technology, will reduce average energy consumed per pound of product by 34%. The shift will also reduce product value on a mass basis by 3.5%. Overall the energy required per dollar of shipment drops by 32%. Table 120 shows a projection of energy consumption in fiber manufacturing based on this shift in production and the growth assumptions cited above. The mix of products is held constant between 1980 and 1990. The decreases in consumption due to product mix change are evenly distributed between electricity and fuels.

TABLE 120: PROJECTED ENERGY CONSUMPTION IN MAN-MADE FIBER PRODUCTION*

	1967	1973	1977	1980	1985	1990
Production (10⁶ pounds)	4,264	7,102	10,233	12,402	16,717	22,519
Value (10⁹ $, 1967)	2.91	4.63	6.54	7.80	10.51	14.16
. . . . Electricity (10⁹ kwh)						
Old facilities	3.13	5.37	4.03	3.28	3.01	2.73
New facilities	–	–	2.16	3.13	5.63	8.92
Total	3.13	5.37	6.19	6.41	8.64	11.65

(continued)

TABLE 120: (continued)

	1967	1973	1977	1980	1985	1990
	. . . . .Fuel (10^{12} Btu).					
Old facilities	190	216	162	132	121	110
New facilities	–	–	87	126	226	358
Total	190	216	249	258	347	468

*Reflects only the change in product mix and anticipated increases in production of the six fibers analyzed.

Source: Reference (10)

DRUG INDUSTRY

The drug industry is characterized by a complex spectrum of products. Not only are the various materials produced by a specific manufacturer quite different in nature, e.g., insulin vs vitamins vs cough remedies, but within a particular product category there is a wide variety of dosage forms and packaging. This range of materials is commonly grouped under Standard Industrial Classification (SIC) No. 283, and is subdivided into Biological Products (SIC 2831), Medicinal Chemicals and Botanical Products (SIC 2833) and Pharmaceutical Preparations (SIC 2834).

Energy conservation goals for the drug industry have been estimated for only the three-digit industry classification. This was necessitated by the large number and diverse nature of products in each of the four-digit SIC industries which make up the drug industry and the prevalence of multiproduct manufacturing plants which commonly produce drug products which are classified in more than one of the four-digit SIC drug industries. Drug company plants often comprise more than one million square feet in area, wherein as many as 200 separate products are produced during a single year.

Energy consumption is monitored for the plant as a whole, and identification for separate products is a formidable problem not yet solved by any of the companies in the industry. The establishments included in each of the four-digit SIC industries of the drug industry are briefly reviewed below and discussed in more detail later on.

The biological products industry (SIC 2831) includes establishments primarily engaged in the production of bacterial and virus vaccines, toxoids and analogous products (such as allergenic extracts), serums, plasmas, and other blood derivatives for human or veterinary use. The establishments engaged in this industry accounted for 360.0 million dollars of the 8,018.5 million dollars of drug industry value of shipments in 1972.

The medicinal chemicals and botanical products industry (SIC 2833) includes establishments which are primarily engaged in [1] manufacturing bulk organic and inorganic medicinal chemicals and their derivatives; and [2] processing (grading, grinding and milling) bulk botanical drugs and herbs. Establishments primarily engaged in manufacturing agar-agar and similar products of natural origin, endocrine products, manufacturing or isolating basic vitamins, and isolating active medicinal principals such as alkaloids from botanical drugs and herbs are also included in this industry. The establishments in this industry accounted for 509.0 million dollars of the 8,018.5 million dollars drug industry value of shipments in 1972.

Establishments in the pharmaceutical preparations industry (SIC 2834) are primarily engaged in manufacturing, fabricating, or processing drugs in pharmaceutical preparation for human or veterinary use. The greater part of the products of these establishments are finished in the form intended for final consumption, such as ampuls, tablets, capsules, vials, ointments, medicinal powders, solutions, and suspensions.

Products of this industry consist of two important lines, namely: [1] pharmaceutical preparations promoted primarily to the dental, medical, or veterinary profession; and [2] pharmaceutical preparations promoted primarily to the public. The pharmaceutical preparations industry comprises the largest segment of the drug industry group, both from the point of view of value of shipments and employees. It accounted for 7,149.5 million dollars of the 8,018.5 million dollars of drug industry value of shipments on an establishment basis in 1972.

PROCESS TECHNOLOGY INVOLVED—BIOLOGICAL PRODUCTS—SIC 2831

This segment of the drug industry includes establishments primarily engaged in the production of the following products:

Agar culture media
Aggressins
Allergenic extracts
Allergens
Antigens
Anti-hog-cholera serums
Antiserums
Antitoxins
Antivenom
Bacterial vaccines
Bacterins
Bacteriological media
Biological and allied products:
  antitoxins, bacterins, vaccines,
  viruses
Blood derivatives, for human or
  veterinary use

Culture media or concentrates
Diagnostic agents, biological
Diphtheria toxin
Plasmas
Pollen extracts
Serobacterins
Serums
Toxins
Toxoids
Tuberculins
Vaccines
Venoms
Viruses

The biological products industry accounted for 360.0 million dollars of the 8,015.5 million dollars of drug industry value of shipments in 1972. Of the 360.0 million dollar biologicals industry value of shipments, 82% was actually biological products. Secondary products shipped by this industry were mainly comprised of surgical and medical instruments. The biologicals industry accounted for only 60% of the total biological product shipments of 495.2 million dollars in 1972. Most other biologicals were shipped by the pharmaceutical preparations industry (SIC 2834).

Much of the discussion which follows, dealing with process descriptions, is based on information contained in a Versar, Inc. report to FEA (25). Biological products for human and veterinary use include three major classes:

(1) Blood and serum derivatives; prepared by refinement.
(2) Vaccines, bacterins, and antigens; prepared by biosynthesis.
(3) Diagnostic substances, a miscellaneous group of specialty items for clinical identification; prepared by biosynthesis or extraction.

Blood processing and fractionation are conducted on a scale similar to that for pharmaceutical chemicals manufacture. Extensive refrigeration is required and also the products are dried while frozen under high vacuum. These operations employ primarily electricity rather than steam. On the other hand, the preparation of vaccines, antigens, and diagnostics is done essentially on a laboratory scale. For these latter processes some steam may be used for sterilization, but most of the energy consumption is electrical energy for conditioning or purifying the air in process areas.

Blood Derivatives

Since whole blood has limited shelf life, much of it is centrifuged in the cold (40° to 45°F) to produce a clear plasma, which is then freeze-dried under vacuum to provide a stable product for blood transfusions.

More elaborate processing is required to obtain individual protein fractions from whole serum. These include fibrinogen, gamma globulin, and serum albumin. All refining steps are conducted in the cold. The operations involve successive precipitations with increasing concentrations of ethanol and inorganic salt solutions. Intermediate fractions are centrifuged and freeze-dried, although some portions are redissolved and reprecipitated. Each precipitation may take from 5 to 20 hours so that the overall processing of a batch lasts about 2 weeks. There must be precise control of the temperature as well as the concentrations and times for each precipitation.

The set of operations shown in Figure 81 is repeated many times to isolate individual blood fractions. A typical plant might have a floor area of 20,000 square feet with a tank capacity of 10,000 gallons to produce 750 pounds of gamma globulin in 2 weeks.

Vaccines, Toxins and Antigens

Vaccines, toxins and antitoxins, antigens and bacterins are prepared by cultivating viruses, or in some cases bacteria, in appropriate culture media and then isolating the biological agents by centrifugation. In the case of viruses, the culture medium must consist of living tissue, either egg embryos, or cultivated mammalian cells such as kidney.

By inoculating foreign proteins such as viruses or bacteria into experimental animals and then later withdrawing the blood, antiserums are prepared. Such serums may be used directly for treatment or may also be used for diagnostic purposes in clinical or research laboratories. Figure 82 shows the steps in preparing a viral antiserum.

FIGURE 81: BLOOD FRACTIONATION

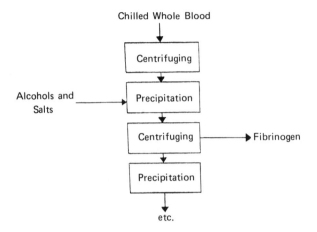

FIGURE 82: PREPARATION OF ANTISERUM

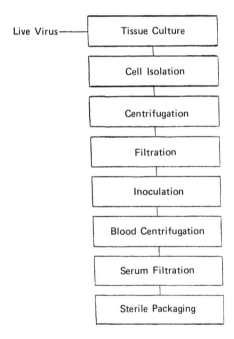

Source: Reference (5)(25)

Diagnostic Substances

Diagnostic agents include test reagents of various kinds such as allergens, blood typing reagents, sera and antisera. A typical procedure might involve the solvent extraction of rabbit brains with chloroform under controlled conditions. To the extract is added 1% of fine silica, and the resulting reagent is used for testing factors involved in blood clotting.

Antigens and allergens prepared by alcoholic or other extraction procedures are often purified by chromatography using laboratory or bench-scale packed columns. Many new techniques of protein purification are used to provide reagents which are as free as possible from side-reactions that might make the tests unreliable.

In all three types of products or processes used by this industry the energy requirements should be roughly the same. Because most operations are conducted on a laboratory scale, and because conditioning of the air and refrigeration are so important, it would seem logical that: [1] energy consumption should correlate best with floor space or cubic feet of space; and [2] a relatively high percent of the total energy should be electrical.

PROCESS TECHNOLOGY INVOLVED—MEDICINAL CHEMICALS AND BIOLOGICAL PRODUCTS—SIC 2833

Manufacturers in this segment of the drug industry are primarily engaged in the production of the following products:

Adrenal derivatives: bulk, uncompounded
Agar-agar (ground)
Alkaloids and salts
Atropine and derivatives
Barbituric acid and derivatives: bulk, uncompounded
Botanical products, medicinal: ground
Caffeine and derivatives
Chemicals, medicinal: organic and inorganic—bulk, uncompounded
Cinchona and derivatives
Cocaine and derivatives
Codeine and derivatives
Digitoxin
Drug grading, grinding, and milling
Endocrine products
Ephedrine and derivatives
Ergot alkaloids
Fish liver oils, refined and concentrated for medicinal use
Gland derivatives: bulk, uncompounded
Herb grinding, grading and milling
Hormones and derivatives
Insulin: bulk, uncompounded
Kelp plants
Mercury chlorides, U.S.P.

Mercury compounds, medicinal: organic and inorganic
Morphine and derivatives
N-methylpiperazine
Oils, vegetables and animal: medicinal grade-refined and concentrated
Opium derivatives
Ox bile salts and derivatives: bulk, uncompounded
Penicillin: bulk, uncompounded
Physostigmine and derivatives
Pituitary gland derivatives: bulk, uncompounded
Procaine and derivatives: bulk, uncompounded
Quinine and derivatives
Reserpines
Salicyclic acid derivatives medicinal grade
Strychnine and derivatives
Sulfa drugs
Sulfonamides
Theobromine
Vegetable gelatin (agar-agar)
Vegetable oils, medicinal grade: refined and concentrated
Vitamins, natural and synthetic: bulk, uncompounded

This industry accounted for 509.0 million dollars of the 8,018.5 million dollar drug industry value of shipments in 1972. The 140 medicinals and botanicals establishments in the U.S. represent plants primarily engaged in production of bulk antibiotics and other bulk synthetic organic medicinal chemicals (SIC 28331); bulk botanicals, bulk naturally occurring vitamins, bulk drugs of animal origin and bulk inorganic medicinal chemicals (SIC 28332). All products are bulk quantity intermediates and are primarily shipped to establishments in the pharmaceutical preparations industry (SIC 2834).

Of the 509.0 million dollars of medicinals and botanicals industry value of shipments in 1972, 79% was actually medicinal and botanical products. Secondary products shipped by this industry are mostly industrial organic chemicals. Many pharmaceutical preparation establishments produce their own bulk medicinals and chemicals; these are never shipped, but are used at the same plant in the manufacture of pharmaceuticals. In the medicinals and botanicals industry the value added as a percentage of value of shipments is relatively low at 51% compared with 76% for the drug industry, 57% for chemicals and allied products (SIC 28).

While the listing of plant and animal extracts is long, fermentation and chemical synthesis account for the most pounds and most dollar sales of finished drugs. Thus, antibiotics alone comprise about 15% of all dollar sales and constitute the largest therapeutic class in the industry. Furthermore, many of the crude plant or animal extracts are also modified either by chemical treatment (e.g., morphine derivatives) or by fermentation (e.g., steroid transformations).

Only three or four of the fifty top-selling ethical drugs are obtained by extraction. Also included in the SIC 2833 category are oils, kelp products and a few others obtained by a process of refinement rather than extraction. These are relatively unimportant in terms of the total industry and therefore their process of manufacture can be considered essentially a subprocess of extraction.

Figure 83 shows a flow chart for a typical fermentation process as used in the manufacture of an antibiotic like penicillin. Nutrient media are weighed, mixed, and pumped to the fermentor which is a batch tank of perhaps 50,000 gallons capacity. The dilute aqueous nutrient solution must be heated to 250°F and held for about an hour to ensure sterilization. The contents are then cooled, inoculated with about 10% by volume of active culture from a seed tank and then aerated and agitated intensely for 3 to 7 days in order to develop a high titer of desired product.

At the end of fermentation, the mold is filtered from the beer in continuous rotary vacuum precoat filters. Solids disposal requires hauling to some site where it can be spread on the land; alternatively, it can be incinerated or in a few instances economically dried for sale as an animal feed. To dry or even to incinerate requires more energy than to haul away but may be better ecologically.

Clarified fermentation beer is processed in continuous, not batch, equipment because of the large volumes to be handled and the short holding times at each step. The concentration and purification typically involves successive extractions using centrifuges and mixers. The volume of liquid is reduced by factors of 50 or 100 so that the final steps of bacteriological filtration and vacuum dehydration are done batchwise in smaller equipment where close quality control can be exercised.

FIGURE 83: TYPICAL PROCESS FLOW SHEET FOR FERMENTATION (ENERGY VALUES BASED ON 1,000 LB OF ANTIBIOTIC)

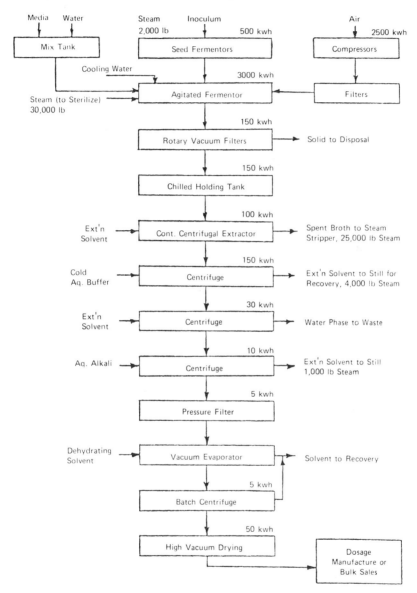

Source: Reference (5)(25)

Besides the mechanical energy needed to drive the filters, centrifuges and refrigeration equipment, steam requirements are high for stripping aqueous wastes and for distillation of solvents. The major energy consumption, however, is in the power to drive the agitators and to compress the air during fermentation.

Fermentation is one of the most energy intensive steps in the drug industry. The main reason is that each kilogram of antibiotic requires the processing of at least several hundred liters, and often several thousand liters of dilute beer. To produce one pound of a high-yielding antibiotic such as penicillin would require 60 to 70 pounds of steam and 6 to 7 kwh of electricity. Similar figures would apply to streptomycin and the tetracyclines. For erythromycin, kanamycin and other antibiotics whose concentration in the beer is not so high, energy costs per kilogram would be even higher.

Even though fermentation is relatively energy intensive as regards the drug industry, it must be pointed out that all utilities comprise only a small fraction of the total cost of manufacture. Even though the basic raw materials are cheap, and the equipment is relatively simple, so much is needed to produce a few kilograms that fixed costs and raw materials account for 70 to 75% of manufacturing costs. Labor and overhead account for another 25 to 30%. Only 2 to 5% of the cost is for utilities.

Chemical synthesis accounts for most of the pharmaceutical manufacture under the SIC 2833 classification. The number and types of chemical transformation are so varied, however, that it is exceedingly difficult to assign values for energy consumption even in a general way. Substantially all are batch operations.

Many drugs are manufactured from relatively inexpensive starting materials in just a few steps. These normally cost less than $10 per kilogram. These include the synthetic organic chemicals itemized in Table 121 below, most of which are the high-volume commodities of the industry.

TABLE 121: ORGANIC CHEMICAL PRODUCTS FROM THE SYNTHETIC CHEMICALS SEGMENT OF THE DRUG INDUSTRY

Analgesics, Antipyretics:	Asprin Dipyrone Aminopyrine
Anesthetics:	Chloroform Ether
Antiinfective agents:	Hexylresorcinol Piperazine Arsenicals
Cardiovascular drugs:	Amyl nitrate Glyceryl nitrate
Dermatologicals:	Salicylic acid
Central nervous system stimulants:	Caffeine
Sulfonamides:	Sulfadiazine and most others
Tranquilizers:	Meprobamate
Vitamins, therapeutic nutrients:	Ascorbic acid Nicotinic acid Nicotinamide Pantothenate Choline

Source: Reference (5)(25)

Vitamins, such as vitamins A, B or riboflavin, and more sophisticated drugs in all of the above classes cost more than $10 per kilogram; they all involve many steps in synthesis, or starting materials that are expensive. The diversity of unit operations that may be employed in each synthetic step is also noteworthy. However, pumping, stirring or filtering require little energy as compared to refluxing, distillation, vacuum evaporation or refrigeration. A typical two-step process, the manufacture of acetylsalicylic acid (aspirin) is illustrated in Figures 84 and 85.

FIGURE 84: MANUFACTURE OF SALICYLIC ACID (ENERGY FIGURES BASED ON 1 TON OF PRODUCT)

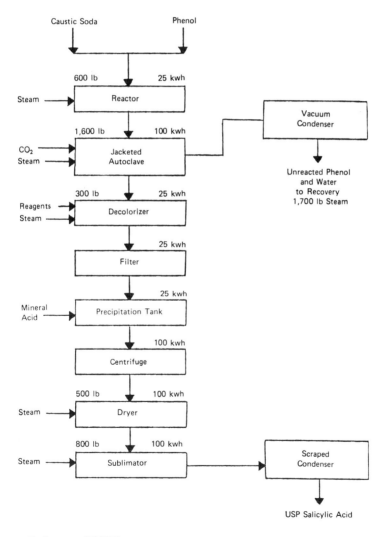

Source: Reference (5)(25)

FIGURE 85: MANUFACTURE OF ASPIRIN FROM SALICYLIC ACID (ENERGY FIGURES BASED ON 1 TON OF PRODUCT)

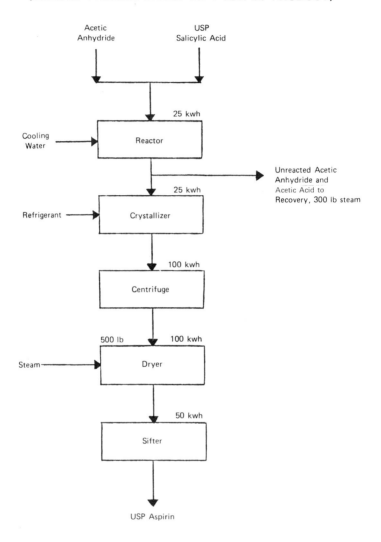

Source: Reference (5)(25)

The first step involves synthesis of salicylic acid. Phenol is reacted with hot 50% caustic and then heated to 130°C and evaporated to dryness in an auto-clave. Either a stirred jacketed pressure vessel or a gas-fired ball mill can be used. The sodium phenolate must be dried and ground up so that it reacts as completely as possible when dry carbon dioxide is introduced. The carbonation reaction occurs at 100°C and 100 to 120 pounds pressure. After CO_2 absorption ceases, the charge is held for several hours at 150° to 170°C. Unreacted phenol is evaporated off under vacuum. The crude product is cooled and dumped to a decolorizer where it is treated with activated carbon and other bleaching agents. Steam is blown directly into the batch so as to remove impurities also. After filtering off the carbon in a press, the free acid is precipitated with sulfuric acid, centrifuged, and dried in a rotating drum to give technical grade crystalline salicylic acid. This must be purified further by sublimation, i.e., vaporization from the solid state, which occurs at 76°C.

Salicylic acid has medicinal use directly or is converted in another step to the methyl ester for liniments and skin ointments. Two-thirds of the production, however, is converted to aspirin, or related analgesics.

Acetic anhydride and salicylic acid are allowed to react at about 90°C with cooling, for 2 to 3 hours in a glass-lined reactor at atmospheric pressure. When reaction is complete, the batch is cooled, filtered to remove a small amount of waste solids, and pumped to a crystallizer. There the temperature is brought slowly to 0°C so as to form good crystals.

These are recovered in basket centrifugals and then dried in a flash or rotary dryer. After cooling, milling and sifting, the product is bagged, or sent to tabletting. If the grade of salicylic acid is too impure, the product may have to be recrystallized from hot ethanol.

The direct steam consumption per ton of aspirin produced is 6,300 pounds, apportioned among the individual operations as shown on the flow sheet. It is interesting that about 90% of the steam is consumed in the manufacture of salicylic acid while only 10% is needed for conversion to aspirin. By contrast the total electrical consumption of 800 kwh per ton of product is more evenly distributed between the two steps. An estimated 500 kwh is used to manufacture salicylic acid; 300 kwh is used to convert to aspirin, per ton of the product.

For the two steps cited here as illustrative, about one-third of the total energy is used in recovery and recycle of reagents (phenol, acetic anhydride) or byproducts (acetic acid). This large fraction is typical of the fine chemicals or pharmaceutical industry as a whole.

For the example cited, total energy used as steam and electricity accounts for 10% of the estimated cost of production exclusive of raw materials cost, but only 3% of direct manufacturing costs including the raw materials. The costs are about equally divided between steam and electricity, and one can assume that the dollar value of the energy portion of the final sales price for such bulk synthetic medicinals is only about 1.5% of the sales value.

The assessment of energy consumption for manufacture of the more expensive synthetic chemicals is difficult. These comprise only a third of the pounds produced, but 70 to 80% of the dollar value, according to the U.S. Tariff Commission in 1972. A typical example would be the chemical synthesis of clinical grade riboflavin.

In contrast to the manufacture of aspirin, this process involves some nine individual steps, starting with xylene, corn sugars and barbituric acid as principal raw materials. Each of the nine steps involves numerous individual operations just as in the case of aspirin manufacture. The overall yield of riboflavin based on input raw materials is only about 15% of the overall yield for aspirin because of losses around each step. Its selling price is about 20 times that of aspirin, attributable in part to the more expensive raw materials and in part to the smaller scale of manufacture, well under a million pounds in 1972 as contrasted with 46 million pounds of salicylate medicinals.

Medicinal products obtained from plants or animals may simply consist of dried ground tissue, but more often the tissues are extracted with water or solvents and these extracts are then concentrated and refined. The operations resemble those used in antibiotic purification but are on a much smaller scale.

Combinations of liquid-liquid extraction, precipitation, evaporation and crystallization may be used. The same equipment may be employed to extract and refine several different plant or animal tissues, with only minor modifications. The flow diagram shown in Figure 86 illustrates a general process which might be typical for the extraction of insulin from pancreas or an alkaloid from plant material.

Either a shredder-chopper (plant material) or roto-cut grinder (soft animal tissue) would be used to reduce the particle size of the biological raw material. Then it would be contacted for a period of hours or even days in a tissue extractor. This equipment might consist of a large percolator, with provision for reboiling and condensing a fresh stream of hot solvent to drop through the packed tissue.

Alternatively, it might be merely a series of leaching or soaking tanks. Water, alcohols, or other petrochemicals such as toluene might be used to extract the active principles. Further extraction into a different solvent and/or precipitation by adding acid or alkali would be combined with filtration and evaporation to further fractionate and concentrate the desired product.

Finally a salting-out or crystallization would be provided and after filtration or centrifugation of the solids they would be dried in trays in a warm-air oven. Most of the operations are conducted batchwise in glass-lined or stainless-steel equipment. The energy requirements for this process are similar to those for a simple one-step or two-step chemical synthesis. Even less electrical energy would normally be required but steam consumption would be comparable because of solvent evaporation and recovery of solvents by distillation. For some drugs such as the steroids or cocaine, additional chemical modifications are made after the biological extracts have been refined.

The production of medicinal chemicals (SIC 2833) really should comprise the major industry of the drug group of industries. The fact that it comprises less than 10% of the total value of products shipped is an artificial result of the method of reporting. After synthesis or extraction the medicinals are made into various pharmaceutical preparations, often at the same plant, and hence are classified in the SIC 2834 rather than SIC 2833 category. Certainly, in terms of energy usage, the manufacture of medicinals and botanicals rather than their compounding into final dosage forms is of predominant importance.

As previously discussed, energy consumption in the drug industry is related to dollar values rather than physical units. The rationale for expressing the energy used to manufacture medicinals in terms of dollar value rather than per pound is easy to justify. Many more chemical steps, and hence more units of steam and electricity, are needed to produce a pound of an expensive vitamin than to produce a pound of some relatively cheap analgesic.

FIGURE 86: TYPICAL EXTRACTION PROCESS

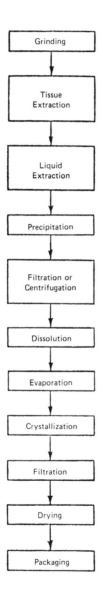

Source: Reference (5)(25)

Likewise an antibiotic produced by fermentation requires expensive concentration and refinement before it can be sold even as a bulk medicinal. A higher dollar value reflects the increased labor, capital and fuel needed in manufacture.

Therefore, one can anticipate that Btu per dollar value should be fairly constant and fairly characteristic for a given type of process in this industry. It should be noted that this is not generally true when considering all of the four-digit SIC drug industries together as a group (of which more later).

PROCESS TECHNOLOGY INVOLVED–PHARMACEUTICAL PREPARATIONS– SIC 2834

In 1972 there were 743 pharmaceutical preparation establishments in the U.S. representing plants primarily engaged in production of dosage form drugs except biologicals. Products in this segment of the drug industry are either ethical (prescription) or proprietary (nonprescription or over-the-counter), and include those shown in the following table, Table 122.

Of the 7,149.5 million dollars of the pharmaceutical preparations industry value of shipments in 1972, only 85% were actually pharmaceutical preparations. Secondary products shipped by this industry are mainly toilet preparations and medicinals and botanicals. The pharmaceutical preparations industry accounted for 95% of the total pharmaceutical preparation products in 1972, as reported in the 1972 Census of Manufactures.

In the pharmaceutical preparations industry the value added as a percentage of value of shipments is very high, 79% vs 76% for the drug industry (SIC 283); 57% for chemicals and allied products; and 47% for all U.S. operating manufacturing establishments. The primary reason for the high value added, is that high research and development costs must be covered by sales.

Figure 87 depicts a general flow chart for the manufacture of dry pharmaceutical preparations in the form of powders, capsules, tablets, or pills. Bulk chemicals are normally received in fiber drums with disposable liners and are handled by battery-powered trucks on skids or pallets by the larger manufacturers.

Separate ingredients are weighed into fiber drums or hoppers, often in isolated weighing rooms with filtered air-conditioning for both temperature and humidity. Some raw materials are screened for removal of contaminates or for control of particle size prior to blending in twin-cone or similar types of blenders. Quality control (QC) is necessary throughout the processing.

The dry blending is also done in separate small rooms with filtered conditioned air. Loading is frequently manual and dust control is a major problem both for protection of employees and prevention of cross-contamination of products. All these operations are, of course, performed batchwise, and the same equipment is used for a variety of different pharmaceutical prepartions.

Particle size is important so that milling of coarse fractions is usually necessary. Some products such as effervescent salts are packaged directly. However, most preparations (vitamins, antibiotics, antacids, etc.) are either encapsulated or formed into tablets. Tablets may be prepared either directly, or after dry or wet granulation, depending on the particular physical characteristics of the individual blends of powders.

TABLE 122: TYPICAL PHARMACEUTICAL PREPARATIONS—SIC 2834

Adrenal pharmaceutical preparations
Analgesics
Anesthetics, packaged
Antacids
Anthelmintics
Antibiotics, packaged
Antihistamine preparations
Antipyretics
Antiseptics, medicinal
Astringents, medicinal
Barbituric acid pharmaceutical
  preparations
Belladonna pharmaceutical
  preparations
Botanical extracts: powdered,
  pillular, solid and fluid
Chapsticks
Chlorination tablets and kits
  (water purification)
Cold remedies
Cough medicines
Cyclopropane for anesthetic use
  (U.S.P. par N.F.), packaged
Dextrose and sodium chloride
  injection mixed
Dextrose injection
Digitalis pharmaceutical preparations
Diuretics
Druggists' preparations (pharma-
  ceuticals)
Effervescent salts
Emulsifiers, fluorescent inspection
Emulsions, pharmaceutical
Ether for anesthetic use
Fever remedies
Galenical
Hormone preparations
Insulin preparations
Intravenous solutions
Iodine, tincture of
Laxatives

Liniments
Lozenges, pharmaceutical
Medicines, capsuled or ampuled
Nitrofuran preparations
Nitrous oxide for anesthetic
  use
Ointments
Parenteral solutions
Penicillin preparations
Pharmaceuticals
Pills, pharmaceutical
Pituitary gland pharmaceutical
  preparations
Poultry and animal remedies
Powders, pharmaceutical
Procaine pharmaceutical
  preparations
Proprietary drug products
Remedies, human and animal
Syrups, pharmaceutical
Sodium chloride solution for
  injection U.S.P.
Sodium salicylate tablets
Solutions, pharmaceutical
Spirits, pharmaceutical
Suppositories
Tablets, pharmaceutical
Thyroid preparations
Tinctures, pharmaceutical
Tranquilizers and mental drug
  preparations
Vermifuges
Veterinary pharmaceutical
  preparations
Vitamin preparations
Water decontamination or puri-
  fication tablets
Water, sterile: for injec-
  tions
Zinc ointment

Source: Reference (5)

Dry granulation (Tablets #1 in Figure 87) involves a preliminary compaction followed by crushing or breaking so as to form small hoppers into the tabletting machines. Some formulations do not lend themselves to dry granulation and so must be granulated wet (Tablets #2 in Figure 87).

FIGURE 87: GENERAL FLOW CHART FOR DRY PHARMACEUTICALS
PREPARATIONS

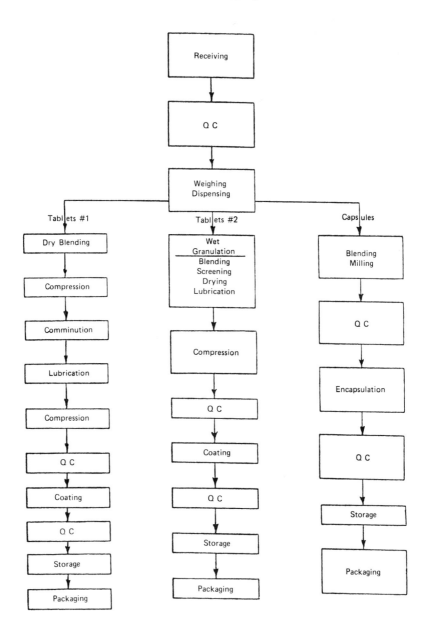

Source: Reference (5)(25)

This involves moistening while mixing in ribbon or Sigma-type blenders and then tray drying or drying in a fluidized bed. Dyes are often admixed during wet granulation so as to obtain distinctive identification of particular products. Pills or tablets are often coated with sugar syrups or other liquid films using sprays and rotating pans. Several distinct coating operations are usually employed, and at this stage trade names may be printed onto the individual pills. Wax coatings and polishing steps make a more attractive and durable product.

Separate rooms are required for each of the following steps:

> Weighing and dispensing
> Blending
> Milling or comminution
> Encapsulation
> Dry-granulation
> Wet-granulation
> Compaction
> Coating and polishing
> Printing
> Filling and packaging

Control of temperature and humidity is necessary in all such areas and each is normally provided with a separate system. Since dusting is a problem in all areas except printing and packaging, high air flows must be maintained; local flows of 500 cubic feet per minute are not uncommon. Many plants use 100% fresh air with discharge to atmosphere, rather than a recirculating air system. The latter requires more expensive and elaborate filtration in order to avoid cross-contamination, but the energy consumption for air-conditioning is only about one-third as great as when fresh air is used.

It is estimated that 60 to 70% of the energy is used for conditioning and circulating the air, 20 to 30% is used for drying after wet granulation, and only 5 to 10% is used in all the mechanical operations of blending, granulating, tabletting and packaging.

Liquid preparations involve simpler processing than dry preparations. The liquid ingredients are received in drums and carboys, or may be transferred directly from dissolution tanks where solutions are made up from dry ingredients. Either gravity flow, air pressure, or pumps can be used for transfer. Measuring is done by metering pumps or other volumetric devices. Such solvent vehicles as glycerol or alcohol are blended in using only mild agitation with propellors or turbines. The liquids are clarified by filtering under pressure using filter presses with disposable pads. They are stored in bulk quarantine until all quality controls signify that they meet specifications and can be packaged.

Semisolid ointments or salves, and emulsions or cremes require more powerful methods of dispersion. Even before blending with melted petroleum or oils, the solids may be milled. The blending is done when the mixture is heated and hence not especially viscous or difficult to mix. To provide a stable suspension of solids, however, roller or colloid mills are used on the cooled product.

Liquid emulsions sold as cremes are homogenized either ultrasonically or by passing through fine orifices under pressure. As in the manufacture of dry formulations, individual operations are conducted in separate air-conditioned rooms, but dusting is not a problem.

Drugs for injection require special handling. Sterile filtration can be followed by filling into sterile ampoules in special sterile rooms. Some heat-stable types can be sterilized by steam after filling. Still other liquids may be lyophilized (i.e., dried to solid form by subliming off the water from a frozen state under vacuum). Because of the expense, both in capital costs and energy, this latter procedure is now less widely used for parenteral products than heretofore.

From 75 to 85% of the total energy consumption for liquid preparations would be attributable to air-conditioning. The electrical power need for mixing and packaging normally comprises the balance. Preparation of emulsions or ointments would, of course, consume more power per unit of product, but any effect on total power consumption would be small. The preparation of parenteral solutions requires relatively more energy; sealing of ampoules (natural gas), steam or hot air sterilization, larger quantities of conditioned air per unit of material makes this an energy-intensive operation. It is estimated that three or four times as much mechanical energy is used per unit, if a liquid preparation is packaged in aerosol form.

In sharp contrast to other industries of the drug group, where plant variations are not very important, the manufacture of pharmaceutical preparations in SIC 2834 is done in two distinct types of establishments. The large companies are few in number but they account for most of the shipments of the industry. Their operations are integrated, so that substantially all of the preparations shipped come from bulk medicinals which they themselves manufacture. Accordingly, their use of energy is more characteristic of bulk manufacture.

MAJOR ENERGY CONSERVATION OPTIONS TO 1980

Efforts Toward Energy Reduction (1972-75)

Apparently every major firm in the drug industry has an energy coordinator who works with total cooperation throughout his firm. Quite often, a committee is appointed to help coordinate and disseminate energy policy throughout the company. Changes made within companies vary widely as a function of their nature and geographic locations.

Improved housekeeping is one means employed by all companies in effecting a reduction in energy use. Themostats in offices are commonly fixed at 68°F maximum during the winter months and do not go below 78° to 80°F during the summer months. Invariably, lights, heating and air-conditioning are turned off during periods when the areas are not in use. Lighting fixtures have been converted to lower wattage bulbs and about half of the fluorescent light tubes in most offices and hallways have been permanently removed.

A significant effort has been directed toward steam distribution systems to minimize the loss of condensed steam. Steam traps have been evaluated to determine their efficiency, and temperature measuring devices have been installed so that traps could be monitored on a routine basis. Consideration has been given to the development of hot water run around systems in order to use low pressure steam to generate water. Waste steam is also used in place of electricity to run turbines. Further, boiler firing controls have been upgraded to improve their efficiency. Finally, steam has been conserved through use of more efficient equipment, such as triple effect evaporators.

Improved building insulation has been of major importance, and some firms have had aerial IR scans of their entire plant site in order to locate losses in heat greater than 40°F. Computer systems are also being introduced in order to maintain maximum energy conservation. By monitoring all systems, it is possible that a 5 to 10% additional energy savings is feasible in a new modern plant.

Considerable attention has been focused on waste disposal systems. Waste gases from production operations are being sent to the boiler plant as fuel to produce process steam. Moreover, in areas where access to a municipal sewage disposal system is not possible, less energy intensive methods of waste treatment have been developed. For example, excess capacity in the biological degradation system at one plant site could be used as an alternative to evaporation and incineration of waste from fermentation processes at another. Use of waste solids for landfill is another alternative to incineration.

Research facilities, a major segment of any drug company, can consume far more energy than production facilities. Laboratories have all kinds of environmental controls depending upon the studies being done. Computer systems have also been introduced here to monitor the air flow, humidity and fan motors throughout the installation, and especially in the areas comprising the animal rooms.

Energy Savings (1972-1975)

Energy consumption in the drug industry in the 1972-1975 period declined almost imperceptibly from 89.3×10^{12} Btu to 88.8×10^{12} Btu (Table 123). However, there was a shift away from natural gas usage toward that of fuel oil. Accordingly, consumption of natural gas declined about 7.5×10^{12} Btu from 32.36×10^{12} to 24.83×10^{12} Btu.

The gain in fuel oil use (5.62×10^{12} Btu) accounted for 75% of this transfer. Electricity and coal additionally gained in the shift away from natural gas with respective gains of 0.93×10^{12} and 0.43×10^{12} Btu in this period. These quantities of energy can be readily translated to actual fuel consumption in barrels, M scf, etc., by employment of the conversion factors given as footnotes in Table 123. The actual shift in terms of percentage distribution can be discerned from Table 124.

The data in Table 124 illustrates that the position of natural gas as the largest energy source for the drug industry in 1972 (36.2%) has been supplanted by that of fuel oil (38.0%) in 1975. No significant trends toward other sources (gasoline, diesel fuel or kerosene) are apparent.

The complex diversity of products indigenous to the drug industry was explored in preceding sections. The vast number of products and formulations has discouraged the industry from placing their energy usage on any weight or volume basis. Most products are simply not sold in this manner, which makes such a method of energy measurement meaningless. A few firms have considered the measurement of energy consumption on a unit basis.

TABLE 123: ENERGY CONSUMPTION—DRUG INDUSTRY (SIC 283) (MM Btu)

	1972	1975
Fuel Oil (1)	28,146,200	33,775,200
Natural Gas (2)	32,364,700	24,834,300
Propane (3)	137,300	192,000
Butane (4)	500	0
Electricity (5)	16,212,500	17,140,900
Coal (6)	12,450,900	12,876,300
Gasoline (7)	8,600	9,200
Diesel (8)	100	1,100
Kerosene (9)	0	6,800
Totals	89,320,800	88,835,800

(1)	6287 M Btu/Bbl	(6)	26,200 M Btu/short ton
(2)	1032 M Btu/MSCF	(7)	5253 M Btu/Bbl
(3)	4011 M Btu/Bbl	(8)	6093 M Btu/Bbl
(4)	4011 M Btu/Bbl	(9)	5670 M Btu/Bbl
(5)	3.41 M Btu/KWH		

TABLE 124: ENERGY CONSUMPTION—DRUG INDUSTRY (SIC 283) PERCENTAGE DISTRIBUTION

	1972	1975
Fuel Oil	31.5	38.0
Natural Gas	36.2	28.0
Electricity	18.1	19.3
Coal	13.9	14.5
All Others	0.3	0.2
Totals	100.0	100.0

Source: Reference (5)

This latter technique requires that the total number of such units as penicillin and laxatives are totalled, and the total quantity of energy is divided by this base. Such a base is misleading, owing to substantial divergence in energy use of products in the industry. The few firms that have endeavored to employ this technique have found wild swings in energy consumption. This has led them to doubt their own results.

The most generally accepted technique (adopted by the Pharmaceutical Manufacturers Association) has been to use a constant dollar value of sales as the basis for measurement. This mechanism sets 1972 as the base sales year (index = 100) and divides total energy consumption by the sales volume of that year. Succeeding years simply apply the approximate index relative to 1972.

The pharmaceutical industry has found that the impact of inflation has not been as great a factor in their operations as it has for industry at large. This is apparent from the data in Table 125 and also the data in Table 126.

Domestic sales are adjusted to 1972 constant dollar basis and are shown in Table 127. These are related to energy consumption in Table 128. The data in Table 128 indicate a decline in specific energy usage (Btu/$) of 18.1% in the 1972-75 period.

TABLE 125: CHANGES IN SELECTED WHOLESALE PRICE INDICES* (WPI)

Year	All WPI	All Industrials	Pharmaceuticals (BLS)
1949	78.7	75.3	-
1965	96.6	96.4	103.2
1974	160.0	153.8	104.2
1975	174.9	171.5	113.2

* 1967 = 100

TABLE 126: INDEX OF MANUFACTURERS' PRICES TO RETAILERS FOR ETHICAL PHARMACEUTICALS

Year	Index
1967	100.0
1968	99.1
1969	100.1
1970	101.0
1971	102.9
1972	102.4
1973	102.7
1974	109.3
1975	116.2

TABLE 127: U.S. DRUG SALES

Year	Domestic Sales, MM $	Index	Sales in 1972 Dollars (MM $)
1972	5375	100.0	5375
1973	5913	100.3	5895
1974	6552	106.7	6141
1975	7404	113.5	6523

Note: The sales figures above were obtained from the PMA's Annual Survey Report, but are not the same as those used by the PMA's Energy Conservation and Management Program in their estimates of specific energy consumption in the manufacture of ethical pharmaceuticals.

TABLE 128: 1972-75 ENERGY SAVINGS

Year	MM Btu	10^3 Btu/Dollar of Sales	Improvement Percent
1972	89,320,800	16.62	-
1975	88,835,800	13.62	18.1

Source: Reference (5)

In an independent study, the PMA's Energy Conservation Management Program reported a decline of 27.6% in the same time period. The primary distinction in the results of the two studies is the estimation of U.S. domestic sales which was employed. There is a basic uncertainty regarding the composition of the sales figures which were employed by the PMA.

The sales figures used in this analysis are also from PMA sources, as noted in Table 127 but represent domestic sales over the period on a consistent basis. In any event, there is no doubt that the drug industry had achieved a significant reduction in energy consumption in the 1972-75 period. This reduction appears to have been at least 18.1% and could have been as great as 27%. The value of shipments data for SIC 283 cited earlier includes foreign sales, and therefore is significantly greater than those employed in Tables 127 and 128.

A precautionary note against the use of sales value as the basis for measuring energy consumption in SIC 283 is that some products have relatively low value and are highly energy intensive while others are the converse. Furthermore, the lifespan of products in this industry is relatively short, and significant change-overs in product spectrum could greatly alter the apparent energy consumption per dollar. It is plausible to assume, however, that on an industry basis, significant changes in product mix by one firm are counterbalanced by opposite changes in another during the period from 1972 to January 1, 1980 which is used in this study.

Constraints on Future Savings

The drug industry is subject to a number of constraints that will adversely affect continued energy savings. Such constraints are induced by regulations introduced by such Federal agencies as EPA, OSHA and particularly FDA. With respect to the latter, the drug industry is subject to Good Manufacturing Practices (GMP) which form a part of the FDA regulations.

Such GMP requirements are energy consumptive. The regulations impose standards of temperature and humidity for the storage of intermediates and finished products. There are also standards to prevent cross-contamination by use of special air filtration systems for portions of the plant. In addition, separate formulation, filling and packaging areas are required for many products.

Such regulations not only increase the energy requirements for the operations directly, but also indirectly. In the latter instance, a larger physical plant is required for many operations which had previously been achieved in a smaller space. Newer regulations also require that products be stored at a greater minimum distance (2 feet) from walls than they had been previously. This has the effect of reducing warehouse capacity and therefore the overall attendant energy costs for storage per se. Also, animals used in research and in testing procedures must be maintained under certain specified conditions of humidity and temperature.

Actual energy consumption in a facility is often a function of the configuration of the various process units contained therein. The more encapsulating machines or pill packaging machines there are in a room, the less the specific energy consumption for that facility. The requirements of GMP have been in a direction away from this, i.e., toward greater separation of operations. Furthermore, a new drug application (NDA) must be filed with FDA for approval everytime there is a change in a process. This is even required if one moves a product from one part of a plant to another. Such approval, if granted, requires about nine months time. Consequently, there is a reluctance by the firms to move anything around. This mitigates against continued energy reductions.

New plants and new process units may be more energy intensive and have poorer energy efficiency than existing ones because of GMP. For example, a new warehouse for storing certain types of products must be air conditioned.

Uncertainties regarding environmental regulations will further restrict potential energy savings. More stringent regulations regarding discharges into waterways (such as a zero discharge goal by 1983 under the 1972 Water Pollution Control Act) will tend to increase the amount of energy consumed. It is conceivable that such regulations could deter the change of incineration to biodegradable wastes back to incineration again. This would constitute a reversal in energy savings already achieved by some firms. A need for faster depreciation on new equipment may be conducive to greater availability of capital for energy savings.

Through January 1, 1980, many firms will be under capital restraints imposed by these various federal regulations. It is estimated that one-third of available capital must be spent to meet these regulations. Another third must be expended just to maintain current plant facilities in order to stay in business. This leaves the remaining third to accomodate new ventures and to achieve continued energy reductions. This approximate distribution of probable use of capital funds is based on information obtained in discussions with industry representatives.

The drug industry's specific energy consumption is thus influenced by its productivity--the amount of product produced in given facilities, since a large portion of total energy use is for space conditioning, Hence specific energy consumption forecasts are dependent upon sales growth forecasts.

GOAL YEAR (1980) ENERGY USE TARGET

Energy consumption in the drug industry is small relative to other manufacturing industries in the chemical industry. In 1971 and 1974, the four-digit SIC industries which make up the drug industry consumed 16.7 and 20.5 billion kilowatt hours of energy, respectively, for heat and power. These consumption levels represented approximately 2.3% of total energy consumption of the chemical and allied products industry for each of the two years.

Estimates of energy consumption have been made by Chem Systems. These estimates differ from those derived from Bureau of Census data, and estimates prepared by the Pharmaceutical Manufacturers Association (PMA). They are based on extensive discussions with representatives of the drug industry and from energy data compiled by Versar Inc., in its study of the drug industry prepared for the FEA and Department of Commerce in 1974.

Energy usage by the drug industry is expected to rise from 88.8×10^{12} Btu in 1975 to the range of 99.8 to 106.2×10^{12} Btu by January 1, 1980 (Table 129).

TABLE 129: ENERGY SAVINGS/INCREASE OF U.S. DRUG INDUSTRY

Year	MM Btu	10^3 Btu/Dollar of Sales	Improvement From 1972 in 10^3 Btu/$
1972	89,320,800	16.62	–
1975	88,835,800	13.62	18.1%
1980 (Jan. 1)	99,827,200	12.80-13.62	18.1 to 23.0%
	106,222,400		

Estimated Energy Improvement Goal 20%

Source: Reference (5)

These data show an absolute increase in usage of 11.8 to 18.9% in the 1972-80 period and a 12.5 to 19.6% gain in the 1975-80 period. These increases reflect the fact that the industry is expected to grow, and energy consumption will therefore increase because of greater production.

Specific energy consumption per constant 1972 dollar of sales is expected to remain at least as good as for 1975, i.e., 13.62 x 10^3 Btu per dollar of sales. This is balancing the facts that most easy, already feasible reductions have been made, and that the industry is subject to the pressure of increased energy usage because of FDA regulations.

An additional 5% increase in the goal is possible (12.80 x 10^3 Btu per dollar of sales) if the industry can maintain its momentum, its record of greater productivity. Its high R&D expenditures and past performance strongly support such a thesis.

However, it is felt that an energy conservation goal of 20% is the maximum practical, considering the uncertainties in productivity improvement, capital availability, and full compliance with federal government regulations.

These projections are based on an anticipated sales growth of 10% annually in the 1975-80 period (variable dollars). Such performance is predicated on domestic sales growth achieved in recent years, coupled with industry consensus of their continuation.

The sales price index of 139 projected for 1980 is an extrapolation of the linear variation shown by this factor since 1973. These data are summarized in Table 130. If sales growth is lower, industry productivity will be lower, and the energy improvement would be lower.

The drug industry has also engaged in substantial waste solvent recovery in contrast to its disposal in earlier years. In their view it constitutes an energy savings for which little credit has been received, since they purchase less solvents from other chemical industry groups. Recognition of this effort is nevertheless warranted.

TABLE 130: U.S. DRUG SALES

Year	Domestic Sales[1], MM $	Index[2]	Corrected Domestic Sales, 1972 $ (MM $)
1972	5,375	100.0	5375
1973	5,913	100.3	5895
1974	6,552	106.7	6141
1975	7,404	113.5	6523
1980 (Jan. 1)	10,840 (3)	139	7799

(1) Annual Survey Report; Pharmaceutical Manufacturers Assoc.

(2) Based on J.M. Firestone, Economics Dept., City University of New York (normalized for 1972).

(3) Estimated 10% compounded annual growth.

Source: Reference (5)

SOAP, DETERGENT
AND TOILETRY INDUSTRY

SIC 284 includes establishments primarily engaged in manufacturing soap and other detergents and in producing glycerin from vegetable and animal fats and oils, specialty cleaning, polishing, and sanitation preparations; and surface active preparations used as emulsifiers, wetting agents, and finishing agents, including sulfonated oil and perfumes, cosmetics and other toilet preparations.

SIC 284 consumes only about 1.8% of total energy consumed by the major group SIC 28, and industry number SIC 2841 (Soap and Detergents, Except Specialty Cleaners) consumes more than 50% of this energy. Battelle has stated (5) in view of the very limited time for their total SIC 28 study for FEA, relatively less effort was placed on this group, and most of this effort on SIC 2841.

For the other industry groups, SIC 2842, 2843 and 2844, judgements were made mainly by inference from general knowledge of similar energy consuming establishments, and by limited cross checks for general validity and order of magnitude of these judgements.

The industry, to date, has not found it practical to relate energy use to physical units of production except in isolated process uses under study for energy conservation schemes. Therefore, in this analysis energy use is related to the value of product shipments in constant 1972 dollars. An explanation of how energy savings goals in such units can be agglomerated with other industry group goals in physical units is covered in the discussion of this total study's methodology.

Each of the 4-digit industry groups is dominated by a small number of large establishments which produce 50 to 60% of the value of shipments. These large establishments (and companies) have well organized energy conservation programs, some of whose internal energy savings goals are probably larger than the goals shown in the following summary for the industries.

It is an observation during both this study and from prior experience that for smaller establishments, even though they may have an equal potential for conservation in specific energy consumption, they do not assign the number or

330

quality of personnel needed. This is due both to their small staff and to their judgement that there is more profit in devoting available time to other production and marketing problems than to energy conservation. This study, therefore, has discounted the smaller establishments' ability to contribute as great energy savings as the larger establishments.

Two factors which lend support to this decision are first that energy is a small portion of production cost, and second that a premise of this study is that no new government incentives for conservation investments will be provided as a new motivating factor.

PROCESS TECHNOLOGY INVOLVED

Below are listed the main processes used in SIC 2841. Detailed descriptions with process flow sheets are available, but do not contain an energy analysis (51). To the extent that such data may have been accumulated by the manufacturers, they have been considered proprietary.

Soap process descriptions:
Soap manufacture by batch kettle
Fatty acid manufacture by fat splitting
Soap from fatty acid neutralization
Glycerin recovery
Soap flakes and powders
Bar soap
Liquid soap

Detergent process descriptions:
Oleum sulfonation/sulfation
Air-SO_3 sulfation/sulfonation
SO_3 solvent and vacuum sulfonation
Sulfamic acid sulfation
Neutralization of sulfuric acid esters
and sulfonic acids
Spray dried detergents
Liquid detergents
Dry detergent blending
Drum dried detergents
Detergent bars and cakes

A scanning of the unit operations involved plus qualitative statements from industry sources on the nature and operation of equipment suggests by inference to similar equipment in other industries that perhaps 25 to 30% of specific energy consumption could be saved by known technology. However there are not any specific studies to support this judgement.

Moreover, the same type inference also suggests that significant capital expenditure would be required. In a directional sense, economic justification may be more difficult than in other process industries because of small equipment size, shorter life span before process changes, and the multitude of products that are manufactured.

An example of a unit operation in which substantial energy savings seem possible, but in which the altered technology is apparently not yet satisfactorily demonstrated, is spray drying to produce synthetic detergents. Temperature of the drying gas is held in the range of 400° to 700°F by use of large amounts of air in excess of that needed for combustion.

If this excess air was replaced by a portion of the effluent gas from the spray drier, energy could be saved. This change, however, could alter the character of the dried product, such as particle size range (the sneeze factor), and solubility characteristics. Any such change which might affect consumer convenience or performance clearly requires careful evaluation in a consumer promotion oriented industry.

MAJOR ENERGY CONSERVATION OPTIONS TO 1980—SIC 2841

The modern soap and detergent industry began in the mid 1930s, when synthetic surface active agents were introduced, and was promulgated by the shortage of natural materials (tallow, coconut oil, etc.) during the war period. Introduction of polyphosphate builders produced synthetics to replace soap for laundering applications during the 1950s.

Environmental requirements began to force reformulations, first toward more biodegradable sulfonate surfactants, followed by the confusing and still indeterminate situation concerning phosphate builders. Equally controversial issues surrounded later use of nitrilotriacetic acid, carbonates, and silicates. Formulations balancing environmental features with safety, cleansing effectiveness, cost and appeal to users are still being sought. The use of citrates, multifunctional compounds, and enzyme containing detergents is still under active research.

The effect of this history on energy utilization is twofold. First, the "improved" formulations are directionally more energy intensive. Second, the physical plants are the same ones, roughly 40 years old, made adaptable for rather rapid changes in relatively small sized units to a constantly changing product mix.

Distribution costs are important in the soap and detergent industry. This leads to large, more efficient plants near metropolitan areas, or limited market areas for small plants. Much toll processing is practiced, as well as packaging under private labels.

The significance is that these larger plants are better organized and seem to have better incentives for energy conservation; also energy use changes due to changing raw materials for changing formulations are accounted for in other industries.

Another practice which effects energy usage results from the industry's historically strong efforts in sales promotion. This has led to a great variety of products, produced in relatively small runs, requiring energy for cleanout to avoid contamination, and energy to handle these waste streams.

While no energy use pattern for either a specific establishment or representative plant has been made available to this study, the inference from generally similar plants (generally confirmed by industry sources) is that: 20 to 40% is used for space conditioning and support services which are independent of production rates, and 60 to 80% is related to production rates. Discussions with several industry sources suggest that 1975 total energy use was 20 to 25% less than 1972. However, production was lower (recession year) on the order of 10 to 30% in various segments of the industry.

The improvement in specific energy consumption (energy per unit of production) appears to have averaged 8 to 10%. Improvements were mostly of the housekeeping type, with some from better scheduling, but little from process changes.

Process Energy Savings: No major changes in process energy use are expected by January 1, 1980, although small improvements for better use of process energy will be possible and economically justifiable in almost every plant. A significant improvement could come from recirculating a portion of the gas in spray driers, if it does not alter the characteristics of the dry product.

An estimated 20 to 40% of energy consumed is not directly related to production rates. Hence, if production rate increases through the same physical plant, as is forecast, specific energy consumption will be lowered.

Housekeeping and Maintenance: This includes attention to energy used for human comfort or process or storage needs of temperature, humidity and ventilation, light levels, use of proper insulation and maintenance of utility distribution systems (steam, air, water, electricity) and equipment. This also includes efficient firing of steam boiler plants. General experience has been that for plants without formal programs before 1972, well directed programs will achieve 15% to 20% savings in such energy use.

In SIC 2841 plants this type saving could amount to about 5% of 1972 use. However, experience also shows that maintaining this saving takes sustained effort and constant search for many new small ideas.

Scheduling: Most SIC 2841 plants are not continuous three-shift-per-day, seven-day-per-week operations. Scheduling so as to shut off unused equipment, and use of minimum start-up periods to reactivate has produced significant energy savings where aggressively pursued.

Every establishment has a different situation which may also differ seasonally and with economic or production cycles. Nevertheless, experience indicates that these savings can equal those from housekeeping and maintenance; furthermore they are easier to control once discovered.

Minor Equipment Additions: The industry does not seem to expect significant major plant additions by January 1, 1980. As minor changes are made to accommodate new products or formulations they are likely to be more energy intensive, although they will be more energy efficient. No significant overall change in energy utilization is expected.

Environmental and Safety Regulations: The full effect of effluent limitation guidelines and new source performance standards on energy consumption for the industry has only been assessed qualitatively, although energy estimates for specific technologies have been made. Industry sources suggest that for the period up to January 1, 1980, the effect will be to increase total energy use in a plant by 5 to 8%. However, the introduction of a higher percentage of low phosphate products does increase energy consumption because it is necessary for product quality to feed a more dilute stream to the spray driers. The additional water per pound of product both reduces production capacity and increases specific energy consumption.

Since energy consumed in spray driers is a large portion of process energy (estimated as high as 66% in some plants), a trend in product mix toward more low phosphate detergents will certainly have a significant effect on energy consumption. At this point, insufficient information has been obtained to calculate a quantitative change for the industry.

Economic Criteria: The large companies treat investment for energy conservation projects basically the same as any other investment. Typically, for the relatively small improvement type investments considered herein, a simple payout period of less than 3 to 5 years is acceptable. In rare cases, 6 years would be considered.

This payout period for energy improvement projects is governed almost as much by expected useful project life as by the usual annual gain criteria. The really useful life of improvement projects is often relatively short because of changing manufacturing requirements rather than a flaw in the technology of the improvement.

In-Plant Power Generation: A number of the larger establishments generate on the order of 50% of their electric power requirements by throttling higher pressure steam across a turbine, the resulting lower pressure exhaust steam being used in the plant for process uses or space conditioning. Low pressure steam use is governed by overall production levels, product-mix, and seasonal weather changes. The amount of electric power generated is directly related to low pressure steam demand. The balance of electric power required is purchased.

Despite the energy saving characteristics of such systems for electric power compared to public utilities that cannot fully utilize the low level heat, economic considerations do not necessarily favor in-plant power generation. There is even a long-range trend toward scrapping older facilities, and purchasing outside power. Each situation is different, but the major considerations are:

   small size (economics of large scale units)
   shorter project life (typically 15 years industrial, 40 years utility)
   high cost of capital and relative availability
   higher cost for fuel
   costs for flexibility for the balance of power needed, in purchase
      contracts with utilities.

Conversion to Different Fuels: As in many industries, SIC 2841 establishments converted old coal fired equipment to natural gas for both lower cost and convenience. As natural gas supplies became curtailed, provision to use petroleum liquids has been made. The lower price for most gas compared to oil still dictates it to be used as much as possible. Except for spray driers which cannot readily use residual fuel oil, any petroleum fuel is generally substitutable for gas without significant change in capacity or thermal efficiency. There is, of course, a capital investment, and generally more maintenance and inconvenience. Conversion back to coal firing in existing plants does not seem likely to any significant extent by January 1, 1980.

MAJOR ENERGY CONSERVATION OPTIONS TO 1980—SIC 2842-43-44

The same factors which apply to energy conservation for industry SIC 2841 should apply to these industries to varying degrees. They differ in the following respects:

SIC 2842 and 2844 industries will use a lesser percentage of energy for processing. Housekeeping and operations scheduling assume greater importance. These industries mostly formulate and package from materials derived from other industries.

SIC 2843 produces surface active and wetting agents, finishing agents, and sulfonated oils. These use a higher percentage of process energy. However, much of the production is batch or semicontinuous and less susceptible to economically justifiable energy improvement projects.

All three of these industries have a higher percentage of small establishments than SIC 2841.

GOAL YEAR (1980) ENERGY USE TARGET

Based upon data from the Census of Manufactures and an econometric input/output model Battelle has estimated the value of product shipments for 1972 and 1979.5. The growth rate for SIC 284 as a group in 1972 constant dollars of 3.5% per annum is consistent with the historical growth rate of 5.5% in current dollars, and in general with the observation of industry expectations for this period.

Using the same techniques for adjustment, 1972 energy use estimates were provided by Battelle. 1972 energy use divided by 1972 product shipment value provides the specific energy consumption for 1972. These data are also shown in Table 131.

TABLE 131: SUMMARY PRODUCT SHIPMENTS AND ENERGY ESTIMATES, SIC 284

SIC No.	Products Shipments Value Million Dollars 1972*	Products Shipments Value Million Dollars 1979.5*	Energy Required (10^{12} Btu) 1972*	Energy Required (10^{12} Btu) . . . . . 1979.5 . . . . . . 1972 Basis	Energy Required (10^{12} Btu) . . . . . 1979.5 . . . . . . 1979.5 Basis	Specific Energy Consumption . . . (Btu/$) . . . 1972	Specific Energy Consumption . . . (Btu/$) . . . 1979.5	Estimated Energy Goal (percent)
2841	2,851.8	3,448.6	29.7	35.9	30.5	10,409	8,840	15
2842	1,735.8	2,494.3	6.10	8.77	7.75	3,515	3,106	12
2843	580.7	769.6	11.8	15.6	14.1	20,342	18,294	10
2844	4,247.0	5,460.9	6.86	8.82	7.49	1,614	1,372	15

*Basic data from Battelle adjusted from data in Census of Manufactures and Annual Survey of Manufacturers.

Source: Reference (5)

It is considered that opportunities exist for which there are proven techniques whose cost can be economically justified to obtain an improvement in SEC of 20% by January 1, 1980. It is further expected that some of the larger companies (and establishments) will achieve this level of improvement. However, in estimating a goal for the industries, the following has also been considered:

The high percentage of small companies (and establishments) whose energy conservation efforts will, in best judgement, be limited by other priorities, smaller staff, and more stringent criteria for use of funds (negative).

The added uses of energy since 1972 for emission controls and changing product mix (negative).

The forecast higher output from existing physical plants (positive).

These estimated improvement goals are shown in Table 131. Specific energy consumption for each industry in 1979.5 is also calculated, together with estimated industry energy usage in 1979.5. In all cases, growth in the industry requires more energy use on an absolute basis than is estimated to be saved if the specific energy consumption improvement goal is attained.

PAINT INDUSTRY

SIC 2851 industry manufactures paints (paste and ready mixed), varnishes, lacquers, enamels, shellac, putties, wood fillers and sealers, paint removers, paint brush cleaners, and allied products. The industry does not include manufacture of carbon black, bone black, lamp black, inorganic or organic pigments, printing inks, or artist paints.

Total energy consumption is less than 0.6% of that for all SIC-28, and therefore has received relatively little of the total effort in the Battelle study for FEA (5) since even a gross error in judgement on the estimated energy improvement goal for it would be insignificant in the estimated target for the total SIC-28.

The industry is unique in that an entrepreneur with a few people, and a minimum of equipment, can produce paint competitively if he has proper formulas. Out of about 1,600 establishments, 40% have less than 10 employees, and over 90% less than 100 employees. Plants are scattered throughout the nation serving local and regional markets. Eight companies account for about 50% of the total market, however, even most of these are collections of small plants.

Even for these, energy costs are almost a trivial factor in business, amounting to less than one quarter of 1% of product shipments value, and less than one half of 1% of value added in manufacturing. There is little economic or business motivation to save energy as long as it is available. Hence, in this study, the potential energy saving has been heavily discounted even for the larger companies in arriving at the estimated goal for January 1, 1980.

PROCESS TECHNOLOGY INVOLVED

Paints can be either oil-base or water-base but there is little difference in the production processes used. The major production difference is in the carrying agent; oil-base paints are dispersed in an oil mixture, while water-base paints are dispersed in water with a biodegradable surfactant used as the dispersing agent.

The next significant difference is in the cleanup procedures. As the water-base paints contain surfactants, it is much easier to clean up the tubs with water. The tubs used to make oil-base paint are generally cleaned with an organic solvent, but cleaning with a strong caustic solution is also a common practice.

Paints are generally made in batches. A small paint plant will make up batches of from 100 to 500 gal while a large plant will manufacture batches of up to 6,000 gal. There are generally too many color formulations to make a continuous process feasible. There are three major steps in the oil-base paint manufacturing processes: (1) mixing and grinding of raw materials, (2) tinting and thinning, and (3) filling operations.

At most plants, the mixing and grinding of raw materials for oil-base paints are accomplished in one production step. For high gloss paints, the pigments and a portion of the binder and vehicle are mixed into a paste of a specified consistency. This paste is fed to a grinder, which disperses the pigments by breaking down particle aggregates rather than by reducing the particle size. Two types of grinders are ordinarily used for this purpose: pebble or steel ball mills, or roll-type mills. Other paints are mixed and dispersed in a mixer using a sawtoothed dispersing blade.

In the next stage of production, the paint is transferred to tinting and thinning tanks, occasionally by means of portable transfer tanks but more commonly by gravity feed or pumping. Here, the remaining binder and liquid, as well as various additives and tinting colors, are incorporated. The paint is then analyzed and the composition is adjusted as necessary to obtain the correct formulation for the type of paint being produced. The finished product is then transferred to a filling operation where it is filtered, packaged and labeled.

In a large plant, these operations are usually mechanized. In a small plant, the operation may entail the use of an overhead crane to lift the tub onto a platform where an employee fills various-sized cans from a spigot on the bottom of the tub while other employees hammer lids on the can and paste on labels.

The paint remaining on the sides of the tubs or tanks may be allowed to drain naturally and the cleavage, as it is called, wasted or the sides may be cleaned with a squeegee during the filling operation until only a small quantity of paint remains. The final cleanup of the tubs generally consists of flushing with an oil-base solvent until clean. The dirty solvent is treated in one of three ways: (1) it is used in the next paint batch as a part of the formulation; (2) it is placed in drums that are sold to a company where it is redistilled and resold; or (3) it is collected in drums with the cleaner solvent being decanted for subsequent tank cleaning and returned to the drums until only sludge remains in the drum. The drum of sludge is then sent to a landfill for disposal.

Cleanup of tanks by use of a strong caustic solution is also practiced. The caustic is used to remove wastes which may have hardened in the tanks and would not be amenable to cleanup with solvent; wastewater from the caustic wash can be (1) collected in holding tanks and treated before discharge; (2) collected in drums and taken to a landfill; (3) discharged directly to a sewer or receiving stream; or (4) reused in the washing operation.

Water-base paints are produced in a slightly different method than oil-base paints. The pigments and extending agents are usually received in proper particle size, and the dispersion of the pigment, surfactant and binder into the vehicle is accomplished with a saw-toothed disperser. In small plants, the paint is thinned and tinted in the same tub, while in larger plants the paint is transferred to special tanks for final thinning and tinting. Once the formulation is correct, the paint is transferred to a filling operation where it is filtered, packaged and labeled in the same manner as for oil-base paints.

Cleanup of the water-base paint tubs is done simply by washing the sides with a garden hose or a more sophisticated washing device. The washwater may be: (1) collected in holding tanks treated before discharge; (2) collected in drums and taken to a landfill; (3) discharged directly to a sewer or receiving stream; (4) reused in the next paint batch; or (5) reused in the washing operation. (Some of the larger manufacturers produce the resins used in paint production, but most simply purchase them from other industries.)

In formulating coatings, a large number of materials must be combined to satisfy many requirements such as color, opacity, gloss, smoothness, durability, plus resistance to wear, environment and use. Proper manufacturing technique, drying time, and method of application must also be considered. The paint industry is probably unequalled in the variety of its finished products. Chemical coatings, in particular, are often custom formulated for a specific application for a single customer.

While the above leads to thousands of formulations, and perhaps hundreds of raw materials, the process technology is still relatively simple and remains basically unchanged.

MAJOR ENERGY CONSERVATION OPTIONS TO 1980

While no firm data were obtained, the quick inspection of several small plants in northern locations suggests that over 50% of energy used is for lighting and space conditioning. As in other industries, one would expect that a 20% savings is technically feasible by well proven methods which have been termed housekeeping.

Processing technology is simple batch operation consisting of mixing and grinding raw materials followed by tinting and thinning to a formula, then filling containers and packaging. A final operation is cleaning equipment between batches, and handling these cleanout products. Energy consumption would appear to be almost independent of whether these are oil based or water based formulations. No significant improvement in energy usage can be anticipated. There has been little change in process methods in the past 50 years. Improvement can come from operations scheduling and practices to minimize losses in cleanout, making the largest batches possible, and shutdown of unused equipment.

These small establishments operate with minimum working capital, and hence tend to shun any investment that does not provide an almost instant payout. It is doubtful that even the larger establishments would invest for payouts over one year unless available energy was simply insufficient to continue operations. There is apparently ample equipment capacity. Few plants ever run on a full three shift basis continuously. There should, therefore, be some improvement in specific energy consumption as the volume of production increases.

Environmental regulations have added some energy usage, mostly to avoid solvents being vented and for better processing of waste solids from cleanout. No reliable estimate of increased usage for these has been obtained, but Battelle believes (5) that it is not large.

The major trend in the coatings industry is toward greater product efficiency, in the sense of better protection or decoration for longer periods at lower cost per unit of coverage. This has placed major emphasis on raw materials evaluation, ability to use thinner coatings, single versus multiple coatings, formulations for longer service life, and techniques to minimize wastage.

Very few changes in manufacturing techniques have occurred except those related to greater worker productivity such as larger batch sizes, higher speed dispersers, and automated materials handling and packaging equipment. Larger plants are moving toward using electronic controls for more uniform quality such as tinting, and some use of computers for scheduling has begun. The use of "stir-in pigments" is increasing, which reduces local grinding and mixing requirements. The net effect of these changes is a gradual increase in process energy which is more or less offset by better use of the non-process energy in the plant per unit of production. However, better scheduling (by computer or human) can lead to significant improvements in specific energy use.

A shift toward use of solvents conforming to restrictions in Rule 66, and toward more water based formulations, with minimum use of solvents (also conforming), is also evident.

All plants are moving toward minimizing contaminants in wastewater discharges in conformance with effluent limitation guidelines, and solvent recovery systems are being installed, this often being profitable since the value of recovered solvent exceeds the cost of recovery. These trends suggest that reduction in specific energy consumption can mainly come from housekeeping, and operations scheduling, which will be offset to some extent by higher energy use from automation and better control devices, plus increased energy use to meet effluent guidelines.

GOAL YEAR (1980) ENERGY USE TARGET

Both census data and industry sources provide production data which accounts for about 75% of the total value of shipments of the industry. Paint is not a spectacular growth industry, but has steadily shown a trend of about 5% annual growth in value of shipments, and about 3% in volume. A forecast was made based on 3.3% per year growth from 1972 to 1979.5. This reflects the most recent trends and conforms very closely to the forecast prepared by Battelle for growth in constant 1972 dollars of the value of product shipments. These estimates are shown in Table 132.

TABLE 132: SUMMARY PRODUCTION AND ENERGY ESTIMATES
SIC-2851

	1972	1979.5
Value of product shipments,* Billion 1972$	3.520	4.498
Paint production, million gallons	898	1.105
Energy required		
1972 basis, 10^{12} Btu	16.93**	21.63
1979.5 basis, 10^{12} Btu	na	19.46
Specific energy consumption 1972 basis		
Btu/constant 1972$	4.810	
Btu/gal paint (approximate)	17,000	
1979.5 basis, Btu/constant 1972$	na	4.328
Estimated energy improvement goal, %	na	10

*Adjusted from census data by Battelle.
**Adjusted from 1971 AMS data by volume and price factors.

Source: Reference (5)

Energy used in 1971 obtained from census data was adjusted to 1972 and then divided by 1972 product shipments to provide a specific energy consumption. The estimate of specific energy consumption for paint based upon discussions with a limited number of plants is about 17,000 Btu/gal, and in actuality this could range from 15,000 to 20,000 Btu/gal depending upon regional location, equipment used, batch production sizes, and clean-out practices. These estimates are also tabulated in Table 132.

The estimate is that 15 to 20% of the energy used in 1972 per unit of output could be readily saved by good practices and modest investment. However, it is equally a judgement that the industry does not see high value in energy conservation investment compared to other uses for its funds. These judgements provide the estimated energy improvement goal of 10%, shown in Table 132.

GUM AND WOOD CHEMICALS INDUSTRY

Gum and wood chemicals are primarily rosin, turpentine and tall oil distillates. They are obtained from pine gum and pine wood. About 1.6×10^9 are produced annually. There is essentially no growth anticipated between 1972 and 1979.5. The primary uses for SIC 2861 chemicals are:

Turpentine,	solvent and/or thinner
Tall oil,	soaps, coatings, and oils
Rosin,	varnish, paper, sizing, soaps, soldering flux.

Because of the small amount of energy used in the processing of gum and wood chemicals only a minor amount of effort was devoted to this analysis. Since gum sources for these chemicals have been declining and are expected to be negligible by 1980, only chemicals from wood distillation were considered.

PROCESS TECHNOLOGY INVOLVED

Wood rosin, turpentine and tall oil fractions are obtained by distillation processes. The spent wood chips are used as fuel. Also the residue of the distillation process, pitch, is burned as a fuel. Based on industry sources, about 8,000 Btu per pound are required to produce rosin and turpentine. About 96% of the energy required is obtained from the spent wood chips. The distillation of tall oil required about 5,000 Btu per pound with 21 to 25% of the energy required provided by the residue pitch.

MAJOR ENERGY CONSERVATION OPTIONS TO 1980

The options for energy conservation are improved housekeeping and better control of the distillation processes. It is unlikely that these can result in savings in excess of 10 to 15%. The industry is not growing, hence, no specific energy reduction can be expected from the construction of new facilities.

GOAL YEAR (1980) ENERGY USE TARGET

It appears reasonable that the overall requirements for energy can be reduced from about 5,740 Btu per pound to 5,390 Btu per pound. The summary of production and energy estimates for SIC (product) 2861 is given in Table 133. This table shows that the total energy requirements can be reduced from 9.12×10^{12} Btu. The above data are summarized in Table 134. The estimated energy efficiency goal for SIC (product) 2861 is 11.3%.

TABLE 133: ESTIMATES OF PRODUCTION AND ENERGY FOR GUM AND WOOD CHEMICALS, 1979.5

Source	Production $(10^6$ lb)	Btu/lb	Total Energy Required (Btu x 10^{12})	Self-Generated Energy (Btu x 10^{12})	Total Purchased Energy (Btu x 10^{12})
1972 Technology					
Gum	Negligible	–	–	–	–
Tall oil	1,200	5,000	6.0	1.26	4.74
Wood	390	8,000	3.12	3.00	0.12
	1,590		9.12	4.26	4.86
1979.5 Technology					
Gum	Negligible	–	–	–	–
Tall oil	1,200	4,750	5.7	1.26	4.44
Wood	390	7,360	2.87	3.00	–0.13
	1,590		8.57	4.26	4.31

TABLE 134: SUMMARY OF PRODUCTION AND ENERGY ESTIMATES, SIC 2861

	All Products SIC 2861 1979.5
Production, 10^9 lb	1.6
Purchased energy required, 10^{12} Btu	
1972 basis	4.86
1979.5 basis	4.31
Specific energy consumption, Btu/lb	
1972 basis	3,057
1979.5 basis	2,711
Energy efficiency goal,* %	11.3

*Calculation of energy efficiency goal:

$$\frac{3,057-2,711}{3,057} = 11.3\%$$

Source: Reference (5)

COAL TAR CHEMICALS, DYES AND PIGMENTS INDUSTRY

Statistical information is readily available about the historical production of the most important (largest volume) SIC 2865 chemicals (Tables 135 and 136). These data have been used as a basis for assuming reasonable growth rates for individual classes of materials in this category as shown in Table 135 and projecting the total production of SIC 2865 chemicals at 96 billion pounds in 1979.5 growing from a base of 61 billion pounds in 1972 at a weighted average rate of 7.6%.

TABLE 135: 1979.5 PRODUCTION ESTIMATES, CHEMICALS BY CLASS, SIC 2865

Chemical Class	Production, 10^6 lb 1972	Production, 10^6 lb 1974	Assumed Average Annual Growth Rate, percent	Projected Production, 1979.5, 10^6 lb
Aromatics (coal tar source)	2,335	2,300	0	2,335
Aromatics and naphthenes (petroleum source)	23,753	26,579	7.0	37,278
Cyclic intermediates	34,967	38,147	8.0	56,049
Dyes	263	275	4.0	335
Pigments	66	70	4.0	85
Total	61,384	67,371	–	96,082
Weighted Average	--	--	6.6	

From: Synthetic Organic Chemicals, United States Production and Sales, 1967-1972, United States Tariff Commission, 1968-1974, Washington D.C.: Synthetic Organic Chemicals, United States Production and Sales, 1973, and Preliminary Statistics, 1974, United States International Trade Commission, 1975 and 1976, Washington, D.C.

Source: Reference (53)

TABLE 136: HISTORICAL GROWTH RATES FOR CHEMICALS BY CLASS, SIC 2865

Chemical Class	Annual Growth Rate, percent			Growth 1974 Over 1967, percent	Seven Year Average Rate, percent
	1972	1973	1974		
Aromatics (coal tar source)	4.3	(11.2)	10.9	(1.0)	0
Aromatics and naphthenes (petroleum source)	10.7	4.3	7.2	61.5	7.09
Cyclic intermediates	16.7	2.6	6.4	83.5	9.06
Dyes	8.0	7.9	(3.2)	33.4	4.20
Pigments	13.0	5.3	(4.9)	30.9	3.92

From: Synthetic Organic Chemicals, United States Production and Sales, 1967-1972, United States Tariff Commission, 1968-1974, Washington, D.C.; Synthetic Organic Chemicals, United States Production and Sales, 1973 and Preliminary Statistics, 1974, United States International Trade Commission, 1975 and 1976, Washington, D.C.

Source: Reference (53)

Of the 50 chemicals sold in the largest volume in the U.S. in 1975 (52) 10 are categorized in SIC 2865:

Benzene*
Toluene
Xylene
Ethylbenzene
Styrene*

Terephthalic acid*
  (and dimethyl terephthalate)
p-Xylene*
Cumene
Cyclohexane*
Phenols*

*Chemicals for which some data on energy usage have been obtained. On the assumption that data on energy use for these 10 products might be used as a basis for extrapolating energy used for making all products in SIC 2865, Battelle sought information from several leading chemical companies and the Manufacturing Chemists Association (MCA). Almost no data on energy usage for these chemical products were available from these sources.

PROCESS TECHNOLOGY INVOLVED

Benzene

No major new process technology is expected to have a major impact on the manufacture of benzene by 1979.5. Most of this material will still be obtained by solvent extraction from catalytic reformate. Peak needs will be supplied by either hydrodealkylation of toluene or transalkylation of toluene (producing jointly mixed xylenes and benzene). Gradually increasing amounts of benzene, toluene, and mixed xylenes will become available as by-products in the manufacture of ethylene owing to the diminishing use of ethane and increasing use of naphtha and gas oils as feedstocks. The net effect of this change of feedstock will be a somewhat lower energy requirement for the manufacture of benzene, toluene, and mixed xylenes. The end-uses for benzene are shown in Table 137. This use pattern is not expected to change materially during this decade.

TABLE 137: MAJOR USES OF SELECTED CHEMICALS, SIC 2865

	Percent
Benzene (1971 data)	
Ethylbenzene	43.3
Phenol (mainly via cumene)	22.8
Cyclohexane	15.5
Miscellaneous nonfuel uses	8.1
Aniline	3.9
Detergent alkylate	3.9
Maleic anhydride	3.8
Terephthalic Acid and Dimethyl Terephthalate (1973 data)	
Polyester fibers	93
Polyester films	7
Styrene (1973 data)	
Styrene homopolymers	59.0
Styrene-butadiene copolymers (mainly latexes)	16.3
SBR elastomers	14.5
Unsaturated polyester resins	6.3
Miscellaneous	3.9
Cyclohexane (1971 data)	
Adipic acid	52
Exports	21
Caprolactam	19
1,6-Hexamethylenediamine	4
Miscellaneous	4
Phenol (1974 data)	
Phenolic resins	49
Bisphenol A	14
Caprolactam	14
Alkylphenols	6
Cresols and anisole	4
Plasticizers	3
Adipic acid	3
Miscellaneous	5
p-Xylene (1974 data)	
Dimethyl terephthalate	65.6
Terephthalic acid	27.8
Export	6.6
Bisphenol A (1973 data)	
Epoxy resins	53.2
Polycarbonate resins	31.3
Other miscellaneous uses	15.5
Cumene (1973 data)	
Acetone and phenol	90
α-Methylstyrene	6.5
Acetophenone	3.5

(continued)

TABLE 137: (continued)

	Percent
Ethylbenzene (1973 data)	
Styrene	97.3
Export	2.7
Phthalic Anhydride (1973 data)	
Plasticizers	49.6
Unsaturated polyester resins	22.5
Alkyl resins	23.4
Miscellaneous	4.4

Source: Reference (53)

Fortunately, data on average usage can be found in articles sponsored by engineering/construction firms, and by U.S. and foreign chemical companies seeking to sell chemical plants and/or process technology. Sources of data were identified for processes for manufacturing the six products asterisked on the foregoing list of 10 chemicals.

In addition, data on energy usage were also obtained for one other product, bisphenol A. The complexity of even this limited number of the many chemicals involved in this SIC product class is illustrated in Table 138 which shows the data for three different processes used to make benzene. Three of the processes convert toluene to benzene and are used only when the demand for benzene is heavy. For this reason, the established energy required for making benzene has been based on that required by the Shell Chemical Company process for extracting benzene from catalytic reformate, although other extraction processes are used from time to time, but to what extent is unknown.

The processes for making benzene illustrate further the difficulty of making a practical estimate of 1972 energy usage for a single chemical. First the estimates in Table 138 are based on engineering data of firms promoting processes and may be representative of the best 1972-1974 technology rather than energy usage in operating plants.

Next, the four processes used for producing benzene vary from 620 Btu per pound to generating energy equivalent to 10,740 Btu per pound. Since the processes also start with different raw materials, and the raw materials require different energy in their preparation, the data in Table 138 alone do not provide comparable energy usage data on the production of benzene.

Terephthalic Acid and Dimethyl Terephthalate

A large increase is expected in terephthalic acid production between 1972 and 1979.5. Much, if not nearly all of the expansion in production will arise from material made by the Amoco Chemical Company process involving direct catalytic oxidation of p-xylene in the liquid phase with air. Amoco is now building a billion-pound-per-year production facility.

Technology is now available to use terephthalic acid directly in the production of high-quality polyester fiber for either industrial or apparel applications. Until

TABLE 138: CONVERSION ENERGY REQUIREMENTS FOR MANUFACTURING SOME SIC 2865 CHEMICALS—1972

Chemical	Process	Conversion Energy per Pound of Product			Total, Btu/lb	
		Steam, lb	Electricity, kwh	Other, Btu	Net(a)	Gross(b)
benzene	Shell sulfolane	--	--	--	620	680
benzene	Detol	0.044(c)	0.0245	(3870)	(3740)	(3560)
benzene	Pyrotol	0.065(c)	0.0236	(10900)	(10740)	(10570)
benzene	Litol	0.387(c)	0.0205	(770)	(236)	(90)
bisphenol A	Union Carbide	5.3(d)	0.23	--	7810	9410
cyclohexane	hydrogenation of benzene -ARCO	nil	nil	nil	nil	nil
dimethyl terephthalate (fiber grade)	Dynamit Nobel from p-xylene	1.2(d)	0.23	1800	4170	5780
terephthalic acid (+ acetic acid)	Eastman Kodak Co.	--	0.018	1475	1540	1660
dimethyl terephthalate (fiber grade)	Eastman Kodak, from terephthalic acid	--	0.1	5910	6250	6950
phenol	cumene	3.28(e) 0.5(c)	0.071	117	5860	6355
styrene	Monsanto	1.35(c) 2.4(d)	0.038	2160	7020	7290
p-xylene	Toray Ind.-UOP	1.04(c)	0.14	--	1670	2650

(a) Electricity equivalence: 3,413 Btu/kwh.
(b) Electricity equivalence: 10,400 Btu/kwh.
(c) Low-pressure steam: 1,150 Btu/lb.
(d) Medium-pressure steam: 1,325 Btu/lb.
(e) High-pressure steam: 1,500 Btu/lb.

Source: Reference (5).

TABLE 139: TOTAL CONVERSION ENERGY REQUIREMENTS FOR MANUFACTURE IN 1972 OF SELECTED CHEMICALS IN STANDARD INDUSTRIAL CLASSIFICATION 2865

Chemical	Process	SEC(1), Btu/lb. Net	Gross	Production, (2) million lb	Total Conversion Energy, billion Btu. Net	Gross
benzene	Shell sulfolane	620	680	8654	5365	5885
bisphenol A	Union Carbide	7810	9410	255.2	1993	2401
cyclohexane	hydrogenation of benzene	nil	nil	2298	nil	nil
dimethyl terephthalate	Dynamit Nobel	4170	5780	2167	9036	12525
phenol	cumene	5860	6355	2052	12025	13040
styrene	Monsanto	7020	7290	5940	41699	43303
p-xylene	Toray Ind.- UOP	1670	2650	2208	3687	5851
Total		--	--	23574	73805	83005
Average Values		3131	3521	--	--	--

Notes: (1) SEC: Specific Energy Consumption from Table 138.
(2) Production statistics from "*Synthetic Organic Chemicals, United States Production and Sales, 1972,*" United States Tariff Commission, Washington, D.C., 1974

Source: Reference (5)

recently, terephthalic acid could be used directly only to make fiber suitable for carpets and industrial purposes. This change will reduce energy consumption in this sector because esterification of terephthalic acid to dimethyl terephthalate will not be necessary. The manufacture of dimethyl terephthalate will be continued in existing facilities.

As shown in Table 137, all but a small amount of the production of both dimethyl terephthalate and terephthalic acid is used in making polyester fibers and films. Steadily increasing amounts of dimethyl terephthalate will be used during this decade to manufacture a recently introduced polyester molding resin (polybutylene terephthalate), but the growth will likely amount to less than 10% of the total amounts of dimethyl terephthalate and terephthalic acid manufactured by 1979.5.

Another example of the complexity of the manufacture of SIC 2865 chemicals is found in the instance of dimethyl terephthalate and terephthalic acid. These chemicals are alternative raw materials for the manufacture of polyethylene terephthalate, familiarly known in industry as polyester.

This polyester is the polymeric material needed to manufacture polyester fiber known best as Dacron, Fortrel, Terylene, and Trevira. Dimethyl terephthalate is made mainly by two technologies. One, which is used by Hercules, depends on a stepwise oxidation of the raw material, p-xylene. The other technology, used by Eastman Kodak Company, depends on first a one-step oxidation of p-xylene to terephthalic acid followed by esterification with methanol. The energy used in processes employing these technologies is shown in Table 138, but energy estimates for the process used by the Amoco Chemical Company to make most of the terephthalic acid manufactured in the U.S. is not available.

The specific energy consumption data in Table 139 were selected (if a selection was required) from the information in Table 138 on the basis that the selected data represented a process that was identical with or similar to that used to manufacture a significant and/or known part of the production of the given material.

The total conversion energy for the production of a chemical was obtained by multiplying the specific energy by the quantity of that chemical produced with the technology (not process) involved. It is this total conversion energy which is shown in the last column of Table 139. The sum of the individual production quantities of these seven chemicals represents 38.4% of Battelle's estimate of the top quantity of SIC 2865 chemicals produced in 1972 (Table 140). Extrapolating from the energy data and production data shown in Table 140, Battelle estimates that in 1972 the total production of 61 billion pounds of SIC 2865 chemicals (Table 135) required (net) about 192×10^{12} Btu or (gross) about 216×10^{12} Btu, depending on whether the electricity equivalent is 3,412 Btu/kwh (net) or 10,400 Btu/kwh (gross). This information is summarized in Table 140.

Ethylbenzene and Styrene

New process technology for the manufacture of ethylbenzene, the immediate precursor of styrene, was demonstrated in 1975 in a 40-million-pound-per-year plant (54). The new technology permits easier recovery of the heat of reaction

TABLE 140: EXTRAPOLATED TOTAL CONVERSION ENERGIES, STANDARD INDUSTRIAL CLASSIFICATION 2865, 1972

Production accounted for, Table 139	- 23,574 million pounds
Total U.S. production, Table 135	- 61,384 million pounds
Total net conversion energy estimate Table 139	- 73,805 billion Btu
Total gross conversion energy estimate, Table 139	- 83,005 billion Btu
Percent of U.S. production SIC 2865 in energy estimate	- 38.4
Extrapolated energy requirement, 1972:	- 192,180 billion Btu
Net (with 3412 Btu per kwh)	- 192,180 billion Btu
Gross (with 10,500 Btu per kwh)	- 216,136 billion Btu

Source: Reference (5)

than is possible with any other known process, and the energy requirement for pollution control with the new process is also lower than that with the older processes. The energy efficiency of the new process is so good that 95% of the new process heat input and heat of reaction can be recovered as useful low- and medium-pressure steam.

In respect to pollution-control problems, the new process has two principal advantages. The catalyst used presents no hazards or waste-disposal problems, whereas the presently used catalysts are difficult to handle and, when spent, are not easily disposed of. The new process produces two by-product streams in small amounts—one, a stream of combustible gas, the other, a viscous liquid. Both can be used as fuel for supplying process heat.

No other major process improvements or new process technologies in the manufacture of styrene are envisioned. Perhaps the improved catalyst could also be used advantageously in manufacturing another large-volume cyclic intermediate, cumene, from propylene and benzene, with similar energy savings. Cumene is manufactured because it affords an excellent route to phenol. Acetone is the principal by-product (Table 137).

The only use for ethylbenzene is for making styrene; the various polymeric materials in which styrene is a component are listed in Table 137 also.

Figure 88 shows the major steps in the ethylbenzene manufacturing process. The process illustrated uses an $AlCl_3$ catalyst to promote alkylation of benzene with ethylene. The major step from an energy consumption viewpoint is the separation of the reaction exit stream into components by distillation. This operation accounts for approximately 75% of the total energy consumption in the production of ethylbenzene.

FIGURE 88:   ETHYLBENZENE* ENERGY CONSUMPTION DIAGRAM

1973 U.S.A. production: 2.94×10^9 kg (6.50×10^9 lb)
1973 process energy consumption (primarily steam):
   430 Mw (13×10^{12} Btu)
1973 total energy consumption (feedstock plus process):
   3,150 Mw (94×10^{12} Btu)

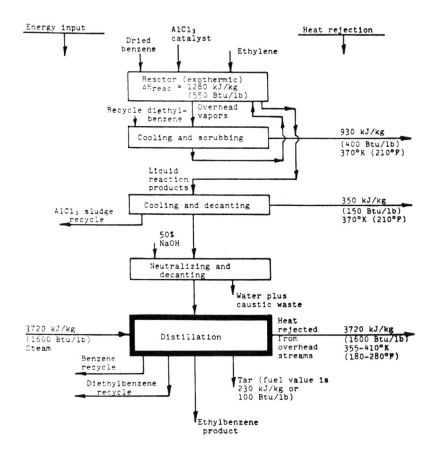

*40% conversion of benzene to products
 95% selectivity to ethylbenzene
 99% conversion of ethylene to products

Source: Reference (3)

Figure 89 shows the distillation operation. Steam provides energy for the three primary columns—the tar removal column, the benzene column, and the ethylbenzene column.

FIGURE 89: ETHYLBENZENE EQUIPMENT DIAGRAM—DISTILLATION

Rejected heat:
  From overhead streams – 3720 kJ/kg
  (1600 Btu/lb) at 355° to 410°K (180° to 280°F)

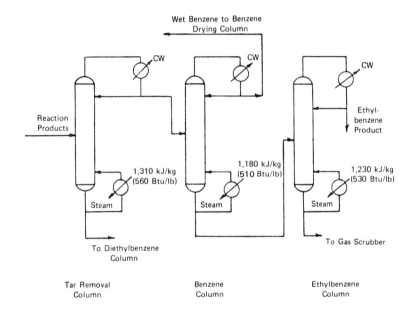

Source: Reference (3)

Figure 90 shows the major steps in the styrene manufacturing process. This process uses a metal oxide catalyst to promote the dehydrogenation of ethylbenzene at high temperature. Major energy consumption operations are the heating of reactants plus steam and the separation of the reactor exit stream into components by distillation. These operations account for more than 90% of the total energy consumption in the manufacture of styrene.

Figure 91 shows the steam superheating, reaction, heat recovery, and desuperheating operations. Steam, natural gas, and waste process gas are used to provide energy for the reaction. The large amount of steam is also provided as a diluent to lower the ethylbenzene partial pressure and thereby allow the reaction conversion to increase.

FIGURE 90: STYRENE* ENERGY CONSUMPTION DIAGRAM

1973 USA production: 2.72 x 10^9 kg (6.01 x 10^9 lb)
1973 energy consumption (primarily natural gas, steam):
    2,600 Mw (78 x 10^{12} Btu)
1973 fuel generation (waste gases, residue): 730 Mw
    (22 x 10^{12} Btu)

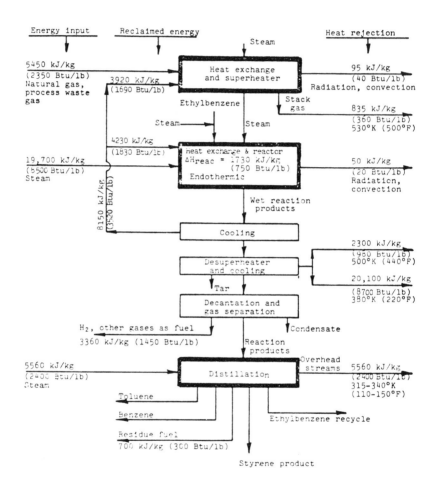

*40% conversion of ethylbenzene to products
 90% selectivity to styrene

Source: Reference (3)

FIGURE 91: STYRENE PROCESS EQUIPMENT DIAGRAM—SUPERHEATER, REACTOR, AND COOLING EQUIPMENT

Rejected heat:
  Radiation, convection – 145 kJ/kg (60 Btu/lb)
  Stack gases – 835 kJ/kg (360 Btu/lb) at 530°K (500°F)
  Hot reaction products – 2300 kJ/kg (980 Btu/lb) at 500°K (440°F)
    and 20,100 kJ/kg (8700 Btu/lb) at 380°K (220°F)

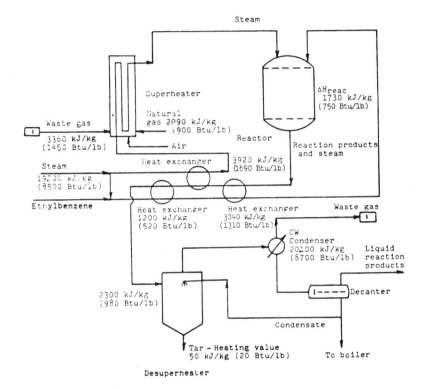

Source: Reference (3)

Figure 92 shows the distillation operation. Steam provides energy for the primary columns—the light ends columns, the ethylbenzene column, and the styrene column.

Figure 93 shows a scheme for utilizing styrene plant waste heat in a water desalination process (55). As shown in the figure, an ethylbenzene feedstock is supplied via a line 2 and admitted to a reactor 4 in admixture with recycle ethylbenzene arriving from column 24 and steam concurrently delivered via a line 7.

The reaction effluent, containing styrene, unreacted ethylbenzene, toluene, benzene, vent gases (such as hydrogen, carbon monoxide, carbon dioxide, ethane, ethylene, etc.), water and other by-products is heat exchanged in a heat exchanger 8 with the ethylbenzene to recover high temperature level heat and then

FIGURE 92: EQUIPMENT DIAGRAM—STYRENE DISTILLATION

Rejected heat:
  From overhead streams – 5560 kJ/kg
  (2400 Btu/lb) at 315° to 340°K (110° to 150°F)

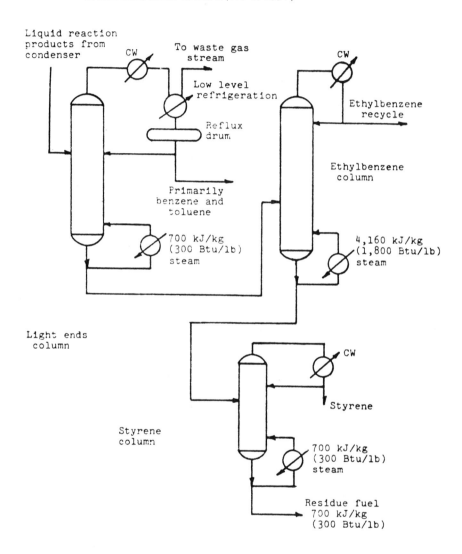

Source: Reference (3)

FIGURE 93: STYRENE PLANT WASTE HEAT UTILIZATION IN A WATER DESALINATION PROCESS

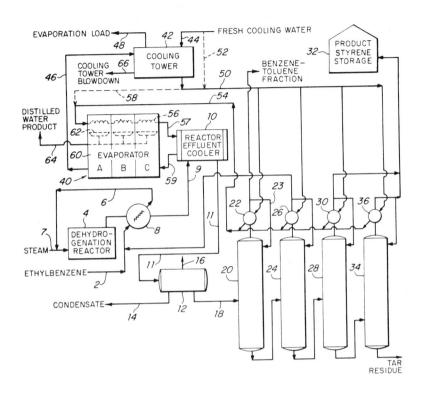

Source: U.S. Patent 3,691,020

is delivered via a line **9** to an effluent cooling heat exchanger **10** where it is cooled further by exchange of heat with cooling water so that steam and certain of its organic constituents are condensed at a relatively low temperature level. From the cooler the effluent is passed via a line **11** to a separator **12** where condensate (mostly water) is removed via a line **14** and vent gases are removed via a line **16**.

The styrene-containing liquid product effluent is removed from the separator and passed via a line **18** to a benzene-toluene column **20** which forms part of a multistage fractional distillation unit. The benzene-toluene column is generally operated in such a manner that benzene and toluene are recovered as an over-head fraction while a styrene-rich fraction containing ethylbenzene is recovered as a bottoms product.

The overhead fraction is condensed in a reflux condenser **22**, with a portion being refluxed via a line **23** while the remainder may be passed to a fractionat-ing column (not shown) to recover separate benzene and toluene concentrates or may be reused to form additional ethylbenzene. The styrene-rich bottoms fraction is passed to a second fractionating column 24 which is operated so as to produce a substantially pure ethylbenzene overhead fraction and a second styrene-rich bottoms fraction. The overhead fraction is condensed in a reflux condenser 26, with a portion being refluxed to column 24 and the remainder being recycled to the reactor via line 6.

The bottoms fraction from column 24 is passed to a finishing column 28 to produce a substantially pure styrene overhead fraction and a bottoms fraction containing some styrene, some ethylbenzene and styrene polymers, and other heavier residual compounds. The overhead fraction is condensed in a reflux condenser 30, with some of it being refluxed to column 28 and the remainder being passed to a product storage tank 32. The bottoms from finishing column 28 is passed to a residual finishing column 34 which is operated so as to recover a substantially pure styrene overhead fraction and a bottoms consisting of tar residues and polymer by-products.

The bottoms from column 34 is withdrawn from the system for disposal or further treatment. The overhead from the column is condensed in a condenser 36, with some of the condensate being refluxed and the rest being passed to storage tank 32.

It is to be noted that the distillation unit comprising columns 20, 24, 28 and 34 is a conventional arrangement for recovering styrene, unreacted ethylbenzene, toluene and benzene from the reaction effluent and thus, need not be described in greater detail.

Recovery of low level waste heat from the effluent from the reactor involves provision of a flash evaporator 40. The flash evaporator preferably consists of a plurality of stages, e.g., three stages A, B, and C as shown, and may be of the type having either horizontal or vertical heating tubes. The number of stages depends upon how much distilled water is to be made. The evaporator is operated at subatmospheric pressure and requires no auxiliary heater elements since enough recoverable heat is available in the reactor effluent to permit vaporization of the water to be distilled.

The evaporator is incorporated within the cooling water loop of the reaction effluent processing equipment together with a cooling tower 42, so that the heat used in the evaporator may be finally rejected to the atmosphere. The cooling tower may be of any convenient design and may include air circulating fans that are electrically or wind-driven.

As shown fresh feed water to be distilled is introduced to the tower via a line 44 while water discharging from the last stage of the evaporator is recycled to the tower via a line 46. Loss of water from the tower by evaporation is indicated schematically by line 48. Cooled water from the tower is pumped via a line 50 to cooler and condensers in the reaction effluent processing equipment, e.g., condensers 22, 26, 30 and 36.

It is to be noted that other cooling heat exchangers in addition to the condensers 22, 26, 30 and 36 may be associated with columns 20, 24, 28 and 34 or with other equipment (not shown) that may be included in the distillation unit, and that such additional heat exchangers may be cooled with water from the tower. Optionally, some or all of the fresh feed water normally introduced to the tower may be passed directly to coolers and condensers in the reaction effluent processing equipment as indicated by line **52**. After passing through users such as condensers **22, 26, 30** and **36**, the cooling water is directed via a line **54** through the heating tubes **56** of the several stages of flash evaporator **40**.

As an optional measure, some or all of the cooling water from tower **42** and/or some or all of the makeup fresh feed water may be passed directly to the evaporator heating tubes as indicated by line **58**. The cooling water passes serially through the heating tubes of evaporator stages **A, B,** and **C** and then via lines **57** and **59** through one side of reactor effluent cooler **10** back to the evaporator where it passes in turn through the flash chambers **60** of successive evaporator stages countercurrent to its direction of flow through the evaporator stages via the evaporator heating tubes.

In passing through the cooler, the cooled water removes reactor effluent heat by cooling and condensing the effluent. This heat effectively raises the temperature of the cooling water so that it can be flashed down in the flash chamber of the evaporator.

The flashed vapor is condensed by exchange of heat with the cooling water flowing in the heating tubes **56** to form a pure distilled water product that is collected in collecting trays **62** and recovered via a line **64**. Residual cooling water is discharged from the last stage of the evaporator and directed via line **46** to the cooling tower where remaining heat absorbed in the effluent cooler **10** and the other condensers and coolers of the reactor effluent processing equipment is dissipated and rejected to the atmosphere.

Thereafter the cooled water is recirculated via line **50** (and/or **58**) for reuse in the manner above described. Sufficient makeup feed water is introduced via line **44** to compensate for (a) evaporation losses in the cooling tower, (b) cooling tower blowdown, and (c) the water recovered from the evaporator as pure distillate product. A blowdown is taken from tower **42** via a line **66**. This blowdown is adjusted so as to maintain a desired dissolved solids content in the circulating cooling water.

p-Xylene

Purified p-xylene is most commonly separated from mixed xylenes by a low-temperature crystallization process which has a relatively high specific energy conversion due to the need for refrigeration. An adsorption process, in commercial use since 1971, likely has a much smaller specific energy consumption. The adsorption process seems to be very promising—eight plants were in operation world-wide at the end of 1975 with a combined capacity of 1,440 million pounds.

At least one of these plants is in the U.S. (Exxon Chemical Company U.S., Baytown, Texas). Eleven more plants in various parts of the world were at different stages of design or construction at the end of 1975. Although the adsorption process cannot displace crystallization processes in existing plants, it seems probable that this newer process will be widely used in furnishing new production capacity, with consequent savings in energy. The only significant uses for p-xylene are in making terephthalic acid and dimethyl terephthalate (Table 137).

Other Products

No other new processes or new technologies have been identified by Battelle for bisphenol A, cyclohexane or phenol. Small but steady improvements in the manufacture of phthalic anhydride by the catalytic vapor phase oxidation of o-xylene are anticipated, but this will be a relatively small contribution to energy conservation in the SIC 2865 category. The uses for bisphenol A and phthalic anhydride are listed in Table 137.

Figure 94 shows the major steps in the cumene manufacturing process as the initial step in the cumene to phenol process. In the process phosphoric acid on

FIGURE 94: CUMENE* ENERGY CONSUMPTION DIAGRAM

1973 USA production: 1.21×10^9 kg (2.67×10^9 lb)
1973 process energy consumption (primarily natural gas**):
    270 Mw (8.1×10^{12} Btu)
1973 total energy consumption (feedstock plus process):
    1270 Mw (38×10^{12} Btu)

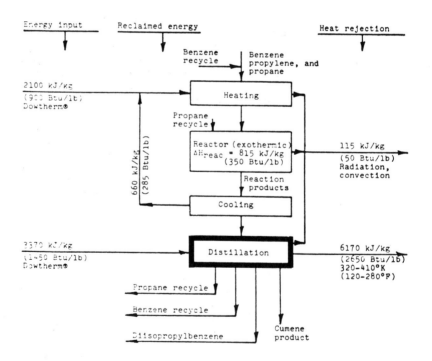

*100% propylene conversion   **Natural gas is used to heat Dowtherm
  92% selectivity to cumene
  14% benzene conversion
  97% selectivity to cumene

Reference (3)

alumina catalyzes the alkylation of benzene with propylene. The major energy consumption step is the separation of reactor effluent components by distillation. This operation accounts for approximately 90% of the total energy consumption in the process.

Figure 95 shows the distillation scheme used to separate the reactor exit stream components. Dowtherm is used to provide energy for the three distillation columns—the propane column, the benzene column, and the cumene column.

FIGURE 95: EQUIPMENT DIAGRAM—CUMENE DISTILLATION

Rejected heat:
  Overhead distillation column streams—
    5350 kJ/kg (2300 Btu/lb) at $320°$-$410°$K ($120°$-$280°$F)

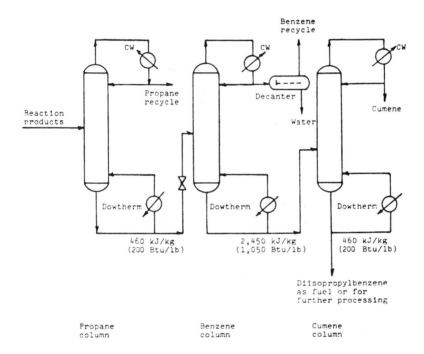

Source: Reference (3)

Figure 96 shows the major steps in the phenol/acetone manufacturing process. Cumene is oxidized to cumene hydroperoxide which is then split into phenol and acetone. Major energy consumption operations are air compression and separation of the reactor exit stream by distillation. These operations account for approximately 70% of the energy consumption in the manufacture of phenol/acetone.

FIGURE 96: PHENOL/ACETONE* ENERGY CONSUMPTION DIAGRAM

1973 USA production: phenol 1.02×10^9 kg (2.25×10^9 lb)
acetone 0.905×10^9 kg (1.99×10^9 lb)
1973 energy consumption (primarily steam): 530 Mw (16×10^{12} Btu)
1973 fuel generation (tar): 60 Mw (1.8×10^{12} Btu)

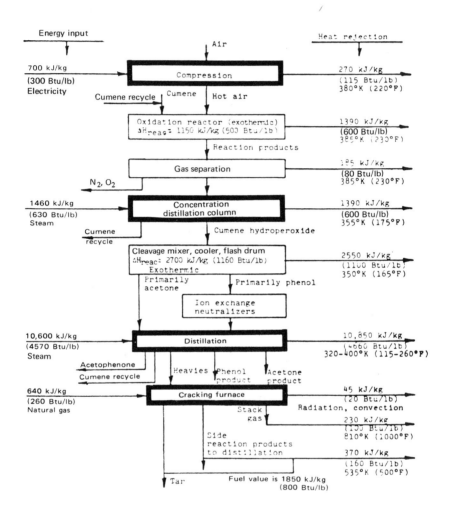

*25% conversion of cumene.
92% selectivity to phenol and acetone.
Energy values are in terms of energy per unit weight of
phenol produced.

Source: Reference (3)

Figure 97 shows the compression of air which is used to oxidize cumene. A two stage compression scheme with intercooling is employed.

Figure 98 shows the separation of unreacted cumene from cumene hydroperoxide by distillation. Steam is used to provide energy for this operation.

Figure 99 shows the separation of acetone from other reactor effluent components by distillation. Steam is the energy source for the three distillation columns—the crude acetone column, the light ends column, and the refined acetone column.

Figure 100 shows the separation of phenol from other reactor effluent components by distillation. Steam is the energy source for the four distillation columns—the heavy ends column, the cumene column, the dehydrogenation column, and the phenol column.

Figure 101 shows the cracking of the bottoms from the heavy ends distillation column and from the acetophenone distillation column. Natural gas is used to provide heat for cracking of the bottoms material.

FIGURE 97: EQUIPMENT DIAGRAM—AIR COMPRESSION FOR CUMENE OXIDATION

Rejected heat:
Hot compressed air — 270 kJ/kg (115 Btu/lb) at $380°K$ ($220°F$)

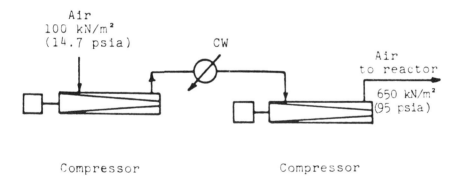

Air
100 kN/m²
(14.7 psia)   CW

Air
to reactor

650 kN/m²
(95 psia)

Compressor   Compressor

Note: Energy values are in terms of energy per unit weight of phenol produced.

Source: Reference (3)

FIGURE 98: EQUIPMENT DIAGRAM—CUMENE HYDROPEROXIDE CONCENTRATION DISTILLATION COLUMN

Rejected heat:
Overhead stream — 1390 kJ/kg (600 Btu/lb) at 355°K (175°F)

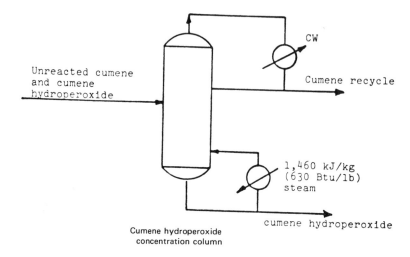

Cumene hydroperoxide
concentration column

Note: Energy values are in terms of energy per unit weight of phenol produced.

FIGURE 99: EQUIPMENT DIAGRAM—ACETONE DISTILLATION COLUMNS

Rejected heat:
Overhead streams — 1350 kJ/kg (580 Btu/lb) at 320° to 335°K (115° to 140°F)

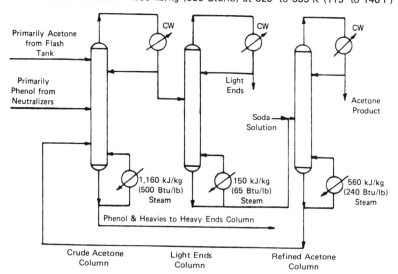

Note: Energy values are in terms of energy per unit weight of phenol produced.

Source: Reference (3)

FIGURE 100: EQUIPMENT DIAGRAM—PHENOL DISTILLATION COLUMN

Rejected heat:
Overhead streams — 9500 kJ/kg (4080 Btu/lb) at 320° to 400°K (115° to 260°F)

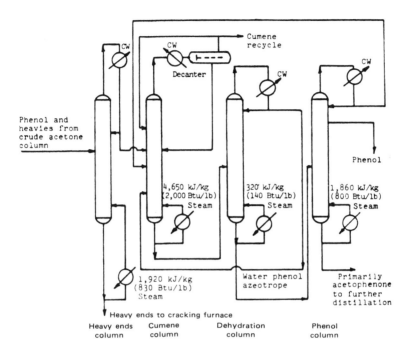

Note: Energy values are in terms of energy per unit weight of phenol produced.

FIGURE 101: EQUIPMENT DIAGRAM—ACETOPHENONE CRACKING FURNACE

Rejected heat:
Radiation, convection — 45 kJ/kg (20 Btu/lb)
Hot stack gases — 230 kJ/kg (100 Btu/lb) at 810°K (1000°F)
Hot process streams — 370 kJ/kg (160 Btu/lb) at 535°K (500°F)

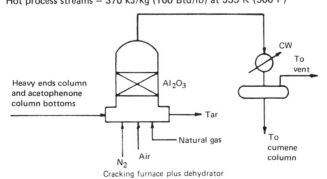

Note: Energy values are in terms of energy per unit weight of phenol produced.

Source: Reference (3)

MAJOR ENERGY CONSERVATION OPTIONS TO 1980

Interviews with energy coordinators at several leading chemical companies by Battelle in the course of their study for FEA (5) revealed a strong consensus regarding the amounts of energy which might be saved by January 1, 1980 as compared with the amounts used in 1972, and the ways in which the savings could be achieved. A unanimity of opinion was that the increased cost of energy provided ample reason for any company to give a great deal of attention to the opportunities for saving energy. In a typical large chemical company with a profit of $5 per $100 of sales, a savings of $1 in energy costs is equivalent to an increase of $20 in sales.

Virtually all chemical companies have formal energy conservation programs. Program structure varies from company to company, ranging from highly centralized to decentralized, depending on management philosophy. The best programs are directed strongly at educating and involving even the lowest level workers in the plant. There was more or less general agreement that energy conservation programs would reduce energy between 1972 and 1979.5 as follows:

> About 5 to 6% of the total savings would come from improved house-
>   keeping (better maintenance, employee awareness, turning off
>   energy consumption not needed, etc.)
> About 6 to 8% of the total savings would come from engineering im-
>   provements with a significant capital investment being required
>   in many instances (See following discussion.)
> Additional savings would have to be generated by major process re-
>   structuring, or introducing a basically new technology
> Compliance with Environmental Protection Agency standards leads
>   to a 1 to 2% increase in energy expended in 1975 over 1972.

A wide variety of low-investment actions can be used in processing cyclic organic chemicals. Some of the conservation options are indicated in Table 141. The effectiveness of the boiler house and steam options can be readily calculated and generalized. The options involving processes however cannot be generalized. The practical energy savings are a function of the product being processed and the other adjacent production facilities.

TABLE 141: OPTIONS FOR ENERGY CONSERVATION

> Boiler House and Steam System
>   Increased condensate return
>   Replace improperly sized steam traps and maintain
>   Install continuous boiler combustion analysis equipment for
>   boiler control
>   Upgrade level of insulation in plant
>   Reuse hot water streams previously sewered
>   Reclaim and use or sell waste oil
>   Revise start-up, operational, and shutdown procedures to con-
>   serve steam, water, or electricity
>   Adjust process temperatures for most efficient energy use con-
>   sistent with safety, production requirements, and quality
>   Use low-pressure steam whenever possible

(continued)

TABLE 141: (continued)

        Improve heat recovery from boiler blowdown
        Install automatic or remote-control valves for shutting off steam
          when not needed, e.g., vacuum jets, strippers, concentrators
        Install economizers on boilers
    Energy Recovery from Processes
        Increase use of off gases with usable energy content
        Cascade heat to other processes
        Install heat exchanger for heating and cooling instead of using
          steam or cooling water
    Process Control and Scheduling
        Modify dryer controls
        Control system for distillation columns
        Reset temperatures in control systems
        Use minicomputer for process control
        Modify solvent-recovery systems

Source: Reference (5)

The use of natural gas as fuel has been quite extensive in making SIC 2865 products. The industry has recognized the necessity of substituting liquid fuels where possible, and in most companies, preparations for this substitution are at various stages of completion. At this time, costs per Btu show a small advantage in using a liquid fuel, but certain disadvantages still make natural gas economically more attractive, principally because of its clean-burning characteristics. Because of the economic penalties of switching from natural gas to liquid fuels, no company will switch as long as low-cost natural gas is available. No company interviewed by Battelle in the course of their study for FEA (5) has taken any steps to use solid fuels.

Estimates have been made by the Dow Chemical Co. in a study for EPA (3) of possible energy conservation approaches in specific processes. Table 142 shows such approaches for ethylbenzene manufacture.

TABLE 142: ETHYLBENZENE ENERGY CONSERVATION APPROACHES

Causes of Energy Losses	Approximate Magnitude of Losses	Energy Conservation Approaches
Rejected heat		
Cooling of reactor exit streams	1,280 kJ/kg (550 Btu/lb)	Design modification (waste heat recovery)
Cooling of overhead streams from distillation columns	3,720 kJ/kg (1,600 Btu/lb)	Design modification (waste heat recovery)
Overall process		
Failure to use tar as fuel	230 kJ/kg (100 Btu/lb)	Waste utilization
Low conversion of benzene	1,850 kJ/kg (800 Btu/lb)	Research and development
High reflux ratios in distillation columns	700 kJ/kg (300 Btu/lb)	Design modification (more plates)

Source: Reference (3)

Table 143 shows possible energy conservation approaches for styrene manufacture, Table 144 for cumene manufacture and Table 145 for the phenol/acetone process.

TABLE 143: STYRENE ENERGY CONSERVATION APPROACHES

Causes of Energy Losses	Approximate Magnitude of Losses	Energy Conservation Approaches
Superheater, reactor and cooling operations		
Heat in stack gases	835 kJ/kg (360 Btu/lb)	Design modification (waste heat recovery)
Radiation, convection	145 kJ/kg (60 Btu/lb)	Insulation Maintenance
Heat discarded from process stream	22,400 kJ/kg (9,680 Btu/lb)	Design modification (waste heat recovery)
Rejected heat in distillation operation	18,500 kJ/kg (2,400 Btu/lb)	Research and development (waste heat recovery)
Overall process		
Low ethylbenzene conversion	18,500 kJ/kg (8,000 Btu/lb)	Research and development
Fuel value of tar	50 kJ/kg (20 Btu/lb)	Waste utilization
High reflux ratios in distillation columns	930 kJ/kg (400 Btu/lb)	Design modification (more plates)

TABLE 144: CUMENE ENERGY CONSERVATION APPROACHES

Causes of Energy Losses	Approximate Magnitude of Losses	Energy Conservation Approaches
Rejected heat		
Radiation, convection	115 kJ/kg (50 Btu/lb)	Insulation Maintenance
Overhead streams from distillation columns	6,170 kJ/kg (2,650 Btu/lb)	Design modification (waste heat recovery)
Overall process		
Low benzene conversion	3,500 kJ/kg (1,500 Btu/lb)	Research and development
High reflux ratio in distillation columns	700 kJ/kg (300 Btu/lb)	Design modification (more plates)

Source: Reference (3)

TABLE 145: PHENOL/ACETONE ENERGY CONSERVATION APPROACHES

Causes of Energy Losses	Approximate Magnitude of Losses	Energy Conservation Approaches
Rejected heat		
Hot compressed air	270 kJ/kg (115 Btu/lb)	Design modification (waste heat recovery)
Overhead streams from distillation columns	12,130 kJ/kg (5,260 Btu/lb)	Design modification (waste heat recovery)
Other hot process streams	4,500 kJ/kg (1,940 Btu/lb)	Design modification (waste heat recovery)
Hot stack gases*	230 kJ/kg (100 Btu/lb)	Design modification (waste heat recovery)
Radiation, convection**	45 kJ/kg (20 Btu/lb)	Insulation Maintenance
Overall process		
Low cumene conversion to cumene hydroperoxide	1,460 kJ/kg (630 Btu/lb)	Research and development
Fuel value of tar	1,850 kJ/kg (800 Btu/lb)	Waste utilization
High reflux ratios in distillation columns	2,040 kJ/kg (880 Btu/lb)	Design modification (more plates)
Nonisothermal compression of air*	95 kJ/kg (40 Btu/lb)	–
Nonisentropic compression of air**	95 kJ/kg (40 Btu/lb	Maintenance

Note: Energy values are in terms of energy per unit weight of phenol produced. Overall process losses * and ** are electrical. The fuel value of these losses would be approximately three times the values listed.

Source: Reference (3)

GOAL YEAR (1980) ENERGY USE TARGET

Statistical information is readily available about historical production of the largest volume SIC 2865 chemicals. These data have been used as a basis for assuming reasonable growth rates for individual classes of materials in SIC 2865. The total production of SIC 2865 chemicals is projected to be 96 billion pounds in 1979.5 growing from a base of 61 billion pounds in 1972 at a weighted average rate of 7.6%.

In the period devoted to the Battelle study for FEA (5) and with the data available it was not possible to identify on a product-by-product basis the energy conservation options and their resultant energy savings. For the purposes of an initial report (5), there was no choice but to rely on some aggregate estimate of energy that might be consumed between 1972 and 1979.5. One method, which appears to be realistic (without much evidence in support to the contrary) is that employed by R.E. Doerr of Monsanto. Mr. Doerr contends that this method applies to most organic chemicals. It is based on the following premises:

> Plants designed and built prior to the energy crisis, will show a specified energy consumption 15% lower in 1979 than in 1972.
>
> New plants, designed and built after the energy crisis, will have a specific energy consumption 30% lower in 1979 than that for the same product in 1972.
>
> The savings of 30% in new plants should be discounted 15% to allow for inefficiencies in plants starting up in 1979.
>
> The ratio of production to plant capacities will be the same in 1979 as it was in 1972.

In subsequent efforts it is expected that additional data can be developed that will confirm or reject the estimates of energy conservation potential that follow.

Production in 1979.5 (Table 135) is projected at 96,000 million pounds. Total conversion energy in 1979.5, using 1972 energy consumption data, would be 301×10^{12} Btu (net). Total conversion energy needed for the manufacture of products in SIC 2865, calculated by Doerr's method, is estimated to be 242×10^{12} Btu (net), with a savings of 19.6% in SEC over 1972. The 1979.5 specific energy consumption would be 2520 Btu/lb. The results of these calculations are summarized in Table 146. Expressed in another way, production in 1979.5 will be 56% higher than in 1972, but energy consumption will be only 26% higher.

TABLE 146: PROJECTED CONVERSION ENERGY STATISTICS, SIC 2865, FOR YEAR 1979.5

Projected total production, Table 135	96,082 million pounds
Projected new production in new facilities in 1979.5 (1979.5 projected production minus 1974 production)	28,711 million pounds
Projected 1979.5 production in existing facilities	67,371 million pounds
Average SEC*, for 1972, Table 139	3131 Btu per pound
1972 SEC* less 15% savings	2661 Btu per pound
1972 SEC* less 30% savings	2192 Btu per pound
1979.5 SEC*	2520 Btu per pound
Total conversion energy, 1979.5 using 1972 SEC*	301×10^{12} Btu
Total conversion energy needed for production in existing facilities	179×10^{12} Btu
Total conversion energy needed for production in new facilities	63×10^{12} Btu
Total 1979.5 conversion energy, SIC 2865	242×10^{12} Btu
Savings over 1972, percent	19.6 percent

*Specific energy consumption, Btu/lb.

Source: Reference (5)

ALIPHATIC ORGANIC CHEMICALS INDUSTRY

This category (SIC 2869) includes many hundreds, and perhaps even thousands, of individual chemical species. All but a hundred or so of these materials are made in relatively small quantities. Like the cyclic organic chemicals any of the miscellaneous organic chemicals considered in SIC (product) class 2869 is likely to be manufactured by more than one process each starting with different raw materials and each having different energy requirements. It is advantageous to use one process and raw material in one geographic location and another process and raw material in another.

Some large volume chemicals are made in large dedicated plants, but most of them are made in large organic complexes. Each company has its own pattern of related products and markets. The processes in these complexes are inter-related and only recently has it been important to plant managers to attempt to ascertain the energy used in specific groups of processes. Data are not available, even at the plant level, on the energy used in producing specific chemicals. In some cases, even when specific energy consumption data are claimed to be plant practice, they are arbitrary allocations among products. Valid data on energy consumption per pound of product for 1972 are not available according to Battelle in their report to FEA (5).

Of the 50 chemicals sold in the largest volume in the United States in 1975 (52) the following 18 materials are categorized in SIC 2869:

Ethylene*	Acetic acid
Propylene*	Acetone*
Ethylene dichloride	Isopropyl alcohol*
Methanol*	Propylene oxide
Formaldehyde*	Acetic anhydride
Vinyl chloride	Adipic acid
Ethylene oxide*	Ethanol*
Ethylene glycol	Acrylonitrile
Butadiene*	Vinyl acetate

*Chemicals for which some data on energy usage have been obtained.

371

On the assumption that process energy data for these 18 products could be extrapolated to approximate the total energy used to make SIC 2865 materials, data were requested from the Manufacturing Chemists Association (MCA) and several leading chemical manufacturers that make these and related materials. Almost no process energy data were available from these sources. A few chemical companies professed not to know specific energy consumption statistics for their products. Even where data have been assembled by companies, they are considered highly proprietary, and in the hands of a rival organization could damage their competitive position.

The 1972 production of these selected organic chemicals required an estimated 230 x 10^{12} Btu, which represents a SEC of 5,346 Btu/lb. When extrapolated to the total production of chemicals in SIC (product) class 2869, the estimated total energy consumption in 1972 was 670 x 10^{12} Btu.

PROCESS TECHNOLOGY INVOLVED

Ethylene

Information related to the calculation of process energy for ethylene manufactured by cracking naphtha is shown in Table 147. Ethylene is an especially important material in this study because it is produced in huge quantities and it entails a high SEC as well.

TABLE 147: APPORTIONMENT OF ENERGY AMONG PRODUCTS FROM STEAM CRACKING OF NAPHTHA*

Product**	Amount, lb	Fuel, M Btu/lb	Electricity, kwh	Total Energy, Btu/lb Net	Gross
Ethylene	1,000	6,035	0.0079	6,060	6,120
Propylene	495	2,987	0.0039	3,000	3,030
Butadiene	136	821	0.0011	825	832
Butylenes/butane	125	754	0.0010	757	764
Benzene/toluene	300	1,811	0.0024	1,820	1,840
C_8 aromatics	52	314	0.0004	315	318
Hydrogen	46	278	0.0004	279	282
Total	2,154	13,000	0.0171	13,056	13,186

*Based on data provided by the Lummus Co., Sources and Production Economics of Chemical Products, 1973-1974, McGraw-Hill, Inc. New York, 1974, p 145.
**In addition to these listed chemical products, the steam cracking of naphtha (and other petroleum liquids) yields pyrolysis gasoline, pyrolysis fuel oil, and a pyrolysis gas (mainly methane) with a total heating value exceeding the fuel requirement by a significant amount.

Source: Reference (5)

Figure 102 shows major steps in an ethylene manufacturing process that uses ethane as the feedstock. Major steps from an energy consumption viewpoint are pyrolysis of the feed, compression of the gases from the pyrolysis furnace, and liquefaction of the gases before distillation.

FIGURE 102: ETHYLENE ENERGY CONSUMPTION DIAGRAM*

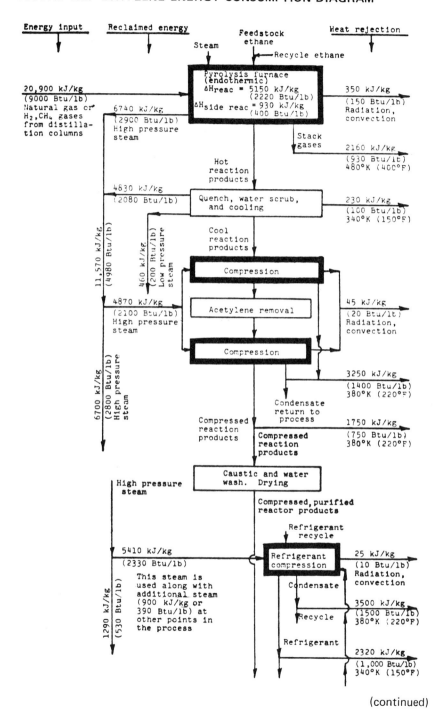

(continued)

FIGURE 102: (continued)

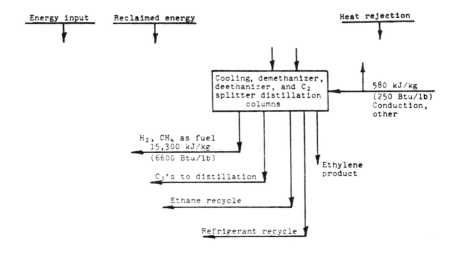

*65% conversion of ethane in furnace with an 80% yield of ethylene

1973 U.S.A. production: 10.1×10^9 kg (22.4×10^9 lb)
1973 process energy consumption (primarily natural gas plus
H_2, CH_4 gases generated in the process): 9,300 Mw
(280×10^{12} Btu)
1973 total energy consumption (feedstock plus process
requirements) 23,000 Mw (700×10^{12} Btu)

Source: Reference (3)

Both the compression and liquefaction by refrigeration are necessary to allow separation by distillation of the components in the furnace exit stream. The pyrolysis, compression, and refrigeration operations account for approximately 90% of the total energy consumption in the manufacture of ethylene.

Figure 103 shows the furnace and associated waste heat recovery equipment in a modern ethylene plant. Waste heat is recovered from furnace stack gases and from hot process gases leaving the furnace.

Figure 104 shows the compression of process gases from the pyrolysis furnace. Intercooling and gas-liquid separation are necessary between each stage of compression. Acetylene hydrogenation is shown between the third and fourth compression stages.

Figures 105 and 106 show a refrigeration system for a modern ethylene plant. Figure 105 shows the compression of propylene and ethylene refrigerants. Various levels of compression and expansion lead to a number of temperature levels that are needed in the distillation columns. The system is referred to as a cascade system because some of the propylene refrigerant is used to cool the ethylene refrigerant which is then used to cool several process streams.

FIGURE 103: EQUIPMENT DIAGRAM—PYROLYSIS FURNACE AND WASTE HEAT RECOVERY EQUIPMENT

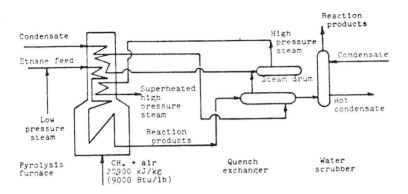

Furnace energy balance

Energy input to furnace	20,900 kJ/kg (9,000 Btu/lb)
Heats of reaction	
Ethylene	5,150 kJ/kg (2,220 Btu/lb)
Side reactions	930 kJ/kg (400 Btu/lb)
Energy change from feed to product	5,570 kJ/kg (2,400 Btu/lb)
Energy transferred to steam	6,740 kJ/kg (2,900 Btu/lb)
Energy lost	2,510 kJ/kg (1,080 Btu/lb)
Waste heat recovery	
Quench exchanger	4,830 kJ/kg (2,080 Btu/lb)
Water scrubber	460 kJ/kg (200 Btu/lb)

Rejected heat: Radiation, convection—350 kJ/kg (150 Btu/lb)
Stack gases—2,160 kJ/kg (930 Btu/lb) at 480°K (400°F)

FIGURE 104: EQUIPMENT DIAGRAM—ETHYLENE COMPRESSORS

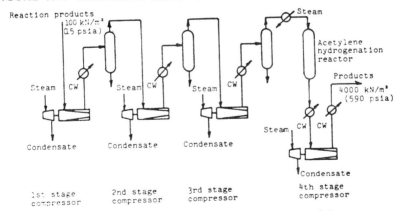

Steam input to each turbine = 1,220 kJ/kg (525 Btu/lb)

Rejected heat: Radiation, convection—45 kJ/kg (20 Btu/lb)
Condensate (vapor)—3,100 kJ/kg (1,350 Btu/lb) at 380 °K (220°F)
Hot compressed reaction products—1,850 kJ/kg (800 Btu/lb) at
380°K (220°F)

Source: Reference (3)

FIGURE 105: EQUIPMENT DIAGRAM—PROPYLENE AND ETHYLENE REFRIGERATION SYSTEMS

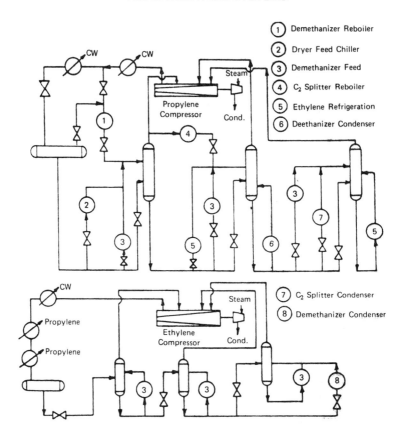

Rejected heat: Radiation, convection—25 kJ/kg (10 Btu/lb)
Condensate (vapor)—3,500 kJ/kg (1,500 Btu/lb) at 380°K (220°F)
Hot refrigerant streams—2,320 kJ/kg (1,000 Btu/lb) at -340°K
(-150°F)

Source: Reference (3)

An energy saving feature of the scheme in Figure 105 is the use of the coldness in the bottom streams from the demethanizer and C_2 splitter distillation columns to cool a portion of the propylene refrigerant. Figure 106 shows the refrigeration of process gases by cold process streams leaving distillation columns, by propylene and ethylene refrigerants, and by the gases in the demethanizer overhead stream. The demethanizer overhead stream is usable as a coolant because it has been supercooled by passing through a turbo-expander. Figure 106 also shows the efficient practice of multifeeding the demethanizer column. In effect the feed stream has been partially separated before it enters the demethanizer column.

FIGURE 106: ETHYLENE PROCESS REFRIGERATION

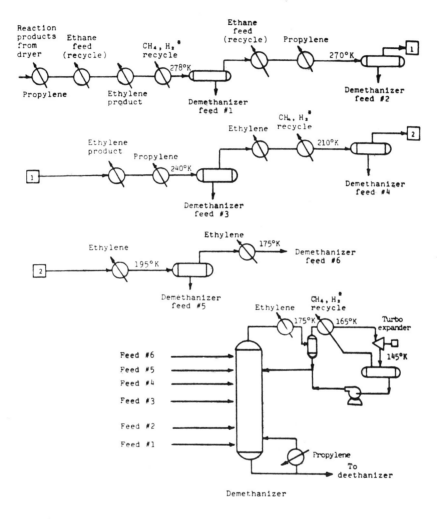

Demethanizer

*Burned in pyrolysis furnace

Source: Reference (3)

Methanol

The I.C.I. low-pressure methanol process offers a number of opportunities for energy savings as compared to established processes as noted by M.J. Pettman and G.C. Humphreys (56). It is stated that 10% or more energy savings are possible by adding relatively inexpensive equipment. Such savings result from less stringent distillation conditions and recovery of heat from the reformer synthesis gas and methanol synthesis sections. Additionally, energy may be recovered by production of electricity using an expansion turbine driven by the gas purge from the synthesis section. Details of heat recovery from the methanol synthesis reactor have been given by D.D. Mehta (57).

Ethylene Glycol

Improved ethylene oxide processes and new direct ethylene glycol processes offer opportunities for energy savings as reviewed by A.M. Brownstein (58).

Other Products

The flowsheets that are available from which energy requirements can be calculated are those prepared by firms seeking to sell chemical plants and/or process technology. As such they represent 1972 to 1975 technology, not average 1972 plant practice.

Organic chemicals production has been growing rapidly for the past 20 years. Technologies have been changing as rapidly as production has been growing. Plants built 15 to 20 years ago to meet market requirements with newly developed technology may be still operating, while new plants with substantially increased capacity have been built with improved or different technology. The relative volumes and technologies vary from company to company. And, of course, data on these are proprietary.

To date no way has been found to estimate the average energy consumption in 1972 for even the large volume organic chemicals. Using an analogy from the automotive industry, data are available representing a few of the 1972 models. Estimates have been on the changes in technology and energy usage some companies are proposing to use in some 1979 models. There has been found to date no general and reliable method for estimating the energy use in pre-1972 models operating in 1972.

The overall data on energy consumption available from census surveys were considered, but they were collected on an establishment basis. They are related to the mix of products manufactured in all SIC 2869 establishments and not to the total production of SIC 2869 products. Previous studies were reviewed to see if any one had devised a way to establish the average technology and energy usage for specific organic chemicals in a particular year. Project Independence Reports, the series of reports on Energy Policy supported by the Ford Foundation, and other reports were reviewed. Analyses in such reports either ignored or did not recognize this basic problem.

Recognizing their potential inaccuracies, it has been necessary in this study to use data published in articles sponsored by engineering/construction firms, as well as by United States and foreign chemical companies seeking to sell chemical plants and/or process technology. Useful data on specific energy consumption were obtained on processes for the manufacture of the 9 products marked with an asterisk in the list of selected chemicals at the beginning of this chapter. In addition to these 9 chemicals, data were also obtained for the following smaller volume products: acetaldehyde, acrylate esters and methyl methacrylate, alpha-olefins, butadiene, sec.-butyl alcohol, caprolactam, chlorofluoromethanes, maleic anhydride and methylamines. The data from which the specific energy consumption estimates were calculated are shown in Table 148.

TABLE 148: CONVERSION ENERGY REQUIREMENTS FOR MANUFACTURING SOME SIC 2869 CHEMICALS—1972

Chemical	Process	Conversion Energy per Pound of Product			Total, Btu	
		Steam, lb.	Electricity, kwh.	Other, Btu	Net(d)	Gross(e)
acetaldehyde	Wacker ethylene	1.2(a)	0.023	--	1500	1680
acetone	cumene-phenol	3.28(b)	0.071	117	5280	5780
acrylic acid	propylene oxidn. Toyo	1.0(a)	0.5	--	2900	6400
acrylic acid	propylene oxidn. (Nippon Shokubai)	3.0(a)	0.25	--	4300	6050
acrylate esters	esterif. acrylic acid	5.5(a)	0.02	--	6400	6530
acrylate esters	esterif. + Toyo oxidn.	6.4(a)	0.47	--	9000	12300
acrylate esters	esterif. + Nip. Shok. oxidn.	8.2(a)	0.245	--	10270	12000
α-olefins	Mitsui ethylene olig.	7.0	0.070	--	8290	8780
butadiene	naphtha cracking	--	0.001	604	607	614
butadiene	separation only	3.23(b)	0.036	36	5000	5260
butadiene	comb. of above	3.23(b)	0.037	960	5620	5890
sec.-butyl alcohol	n-butene hydration	4.32(a)	0.082	--	5250	5820
caprolactam	DSM low sulfate	2.61(a)	0.068	332	8430	8910
chlorofluoro-methanes	Rhone-Poulenc	0.61(c), 1.32(b); 1(a)	0.1	--	1490	2190
chloromethanes	--	4(a)	0.16	900	5150	6260
ethanol	Veba-Chemie	3.5(b)	0.04	--	6290	6570
ethylene	ethane cracking	--	--	--	11700	12000
ethylene	naphtha cracking	--	0.0079	6035	6060	6120
ethylene oxide	--	(1.05)(c)	0.214	--	(660)	(834)
formaldehyde	--	(0.41)(c)	0.043	--	(400)	(96)
formaldehyde	C.E. Lummus	-(1.8)(a)	0.10	--	(1730)	(1030)
isopropyl alcohol	Deutsche Texaco	6.18(a), 0.59(b)	0.058	--	8190	8600
maleic anhydride	Veba-Bayer benzene oxidn.	(6.0)(b)	0.40	--	(7635)	(4840)
methanol	ICI low pressure (Kellogg)	(0.1355)(a)	0.0024	2570	2420	2440
methylamines	Leonard	13(a)	0.09	--	15300	15900

(a) Low-pressure steam, 1,150 Btu/lb
(b) High-pressure steam, 1,500 Btu/lb
(c) Medium-pressure steam, 1,325 Btu/lb

(d) Electricity equivalence at 3,413 Btu/kwh
(e) Electricity equivalence at 10,400 Btu/kwh

Source: Reference (5)

TABLE 149: TOTAL CONVERSION ENERGY REQUIREMENTS FOR MANUFACTURE OF SELECTED STANDARD INDUSTRIAL CLASSIFICATION 2869 CHEMICALS IN 1972

Chemical	Process	SEC, Btu/lb. Net	Gross	Production, million lb.	Total Conversion Energy, billion Btu Net	Gross
acetaldehyde	Wacker ethylene	1500	1680	1447.6	2171	2432
acetone	cumene-phenol	5280	5780	1086.0	5734	6277
acrylic acid	Nippon Shokubai propylene oxidn.	4300	6050	123.2	530	745
methyl methacrylate and acrylate esters		8600	9500	1046	8996	9937
α-olefines	Mitsui ethylene oligomerization	8290	8780	210	1741	1844
butadiene	naphtha cracking extraction(3)	5610	5620	803	4505	4512
butadiene	butane-butylenes	14600	15000	2723	39756	40845
sec.-butyl alcohol	hydration n-butenes	5250	5820	520	2730	3026
caprolactam	DSM low sulfate	8430	8910	640	5395	5702
chlorofluoromethanes	Rhone-Poulenc	1490	2190	920	1371	2015
chloromethanes	CH$_3$OH chlorination	5150	6260	820	4223	5133
ethanol	Veba-Chemie	6290	6570	1850.7	11641	12159
ethylene	ethane, etc., cracking	11700	12000	8341.0	97590	100092
ethylene	naphtha cracking	6060	6120	4170	25270	25520
ethylene oxide	ethylene oxidn.	(660)	(834)	3961.7	(2615)	(3304)
formaldehyde	methanol oxidn.	(1730)	(1030)	5651.8	(9778)	(5821)
isopropyl alcohol	Deutsche Texaco propylene hydration	8190	8600	1790.0	14660	15394
maleic anhydride	Veba-Bayer benzene oxidn.	(7635)	(4840)	274.4	(2095)	(1328)
methanol	ICI low pressure, Kellogg Leonard	2420	2440	6471.6	15661	15790
methylamines		15300	15900	157.8	2414	2509
Total		---	---	43007.8	229900	243479
Average Values		5346	5661	---	---	---

Source: Reference (5)

The data on specific energy consumption in Table 149 were selected from the information in Table 148 on the basis that the data selected represent a process identical with or similar to that used to manufacture a significant and/or known part of the production of a given material. The total conversion energy for the production of a chemical was obtained by multiplying the specific energy consumption in Btu/lb by the quantity of that chemical produced by the technology (not process) involved. This total conversion energy is shown in the last column of Table 149.

The sum of the individual production quantities of these chemicals (43 billion pounds) represents 34.3% of Battelle's estimate of the total quantity (125 billion pounds) of SIC 2869 chemicals produced in 1972. Extrapolating from the energy and production data shown in Table 149, Battelle estimates that in 1972 the total production of 125.4 billion pounds of SIC 2869 chemicals required (net) about 230×10^{12} Btu or (gross) about 243.5×10^{12} Btu, depending on whether the electricity equivalent is 3,412 Btu/kwh (net) or 10,400 Btu/kwh (gross).

MAJOR ENERGY CONSERVATION OPTIONS TO 1980

Interviews with energy coordinators at several leading chemical companies revealed a strong consensus regarding the amounts of energy which might be saved by January 1, 1980, as compared with the amounts used in 1972, and the ways in which the savings could be achieved. A unanimity of opinion was that the increased cost of energy provided ample reason for any company to give a great deal of attention to the opportunities for saving energy. In a typical large chemical company with a profit margin of $5/$100 of sales, a savings of $1 in energy costs is equivalent to an increase of $20 in sales.

Virtually all chemical companies have formal energy conservation programs. Program structure varies from one company to another, ranging from the highly centralized to the decentralized, depending on management philosophy. The best programs are directed strongly at educating and involving workers in the plant at all levels. The opportunities for energy conservation in all organic chemicals manufacture are similar, at least in principle. Ongoing conservation programs have shown that energy reduction can be achieved as follows:

> About 5 to 6% of the total savings would come from improved housekeeping (better maintenance, employee awareness, turning off energy consumption not needed, etc.).

> About 6 to 8% of the total savings would come from engineering improvements with a significant capital investment being required in many instances.

> The remaining savings would be generated by major process improvements and restructuring, or introduction of a basically new technology.

> Compliance with Environmental Protection Agency standards leads to a 1 to 2% increase in energy in 1979.5 over 1972.

The options for conserving energy in organic chemicals manufacture are the same as those for coal tar chemicals and were shown earlier in Table 141. This is the same list of options that apply to cyclic organic chemicals as well. In general the unit processes involved are the same for both classes of chemicals.

There is no basis however for making generalized estimates of the energy savings that might be achieved by these options except those noted above. Each plant with its pattern of integration with other facilities has its own unique set of options to identify and evaluate.

The industry has recognized the necessity of substituting liquid fuels wherever possible, and in most companies, preparations for this substitution are under way. On the basis of costs per Btu, a small advantage accrues in using a liquid fuel, but certain disadvantages still make natural gas economically the more attractive fuel, principally because of its clean-burning characteristics. Because of the economic penalties of switching from natural gas to liquid fuels, no company will make the change as long as low-cost natural gas is available. No company interviewed has taken any steps to use solid fuels, although some are studying the possibilities of doing so in the long term.

Table 150 shows the causes of energy losses in the pyrolysis operation, compression operation, refrigeration operation, and overall process for ethylene production. It also shows the approximate magnitude of the losses and some possible energy conservation approaches.

TABLE 150: ETHYLENE ENERGY CONSERVATION APPROACHES

	Causes of Energy Losses	Approximate Magnitude of Losses	Energy Conservation Approaches
(1)	Furnace losses		
	(a) Hot stack gases	2,160 kJ/kg (930 Btu/lb)	Design modification (waste heat recovery)
	(b) Radiant and convection heat losses	350 kJ/kg (150 Btu/lb)	Insulation Maintenance
(2)	Heat rejection in compression operation		
	(a) Unavailability of latent heat in steam to drive the turbine	3,250 kJ/kg (1,400 Btu/lb)	Process integration (find use for the low pressure steam exiting the turbine)
	(b) Loss of heat imparted to compressed gases	1,750 kJ/kg (750 Btu/lb)	Design modification (waste heat recovery)
	(c) Radiation, convection	45 kJ/kg (20 Btu/lb)	Insulation
(3)	Heat rejection in refrigeration operation		
	(a) Unavailability of latent heat in steam to drive the turbine	3,500 kJ/kg (1,500 Btu/lb)	Process integration (find use for the low pressure steam exiting the turbine)
	(b) Loss of heat imparted to compressed refrigerant	2,320 kJ/kg (1,000 Btu/lb)	Design modification (waste heat recovery)
	(c) Radiation, convection	25 kJ/kg (10 Btu/lb)	Insulation
(4)	Overall process		
	(a) Low conversion of ethane to products	8,100 kJ/kg (3,500 Btu/lb)	Research and development

(continued)

TABLE 150: (continued)

Causes of Energy Losses		Approximate Magnitude of Losses	Energy Conservation Approaches
(b)	Low yield of ethyl- ene from ethane	4,600 kJ/kg (2,000 Btu/lb)	Research and development
(c)	Nonisothermal com- pression of process gases and refrigerants	580 kJ/kg (250 Btu/lb)	—
(d)	Nonisentropic com- pression of process gases and refrigerants	700 kJ/kg (300 Btu/lb)	Maintenance
(e)	Excessive temperature differences between hot and cold fluids in the refrigeration oper- ation	460 kJ/kg (200 Btu/lb)	Design modification (use more heat exchange sur- face) Maintenance Insulation
(f)	Nonoptimization of distillation-refrigera- tion-compression scheme	460 kJ/kg (200 Btu/lb)	Design modification

Source: Reference (3)

The Energy Policy Project of the Ford Foundation (14) has evaluated engineering data on the major processes for producing twenty-one of the large volume
basic organic chemicals. Both utility and feedstock energy use were estimated
for existing technology in 1967 and for anticipated technology for 1980. Table
151 lists each chemical, the technologies considered and the estimated change
in utility energy consumption per unit of output between 1967 and 1980. The
projected technologies and energy utilization changes are based on constant
relative fuel and construction costs (i.e., they do not reflect current high fuel
prices).

TABLE 151: ESTIMATE OF THE IMPACT OF TECHNOLOGICAL CHANGES
IN ORGANIC CHEMICAL MANUFACTURING

Chemical	1967 Process(es)	1980 Process(es)	Percent Change in Utility Energy Consumption
Acetaldehyde	Ethyl alcohol Ethylene	Ethylene	-50%
Acetic acid	Acetaldehyde	Butane Carbon monoxide plus methyl alcohol	-38%
Acetic anhydride	Acetic acid	Acetic acid	-14%
Acetone	Isopropyl alcohol Cumene	Isopropyl alcohol Cumene	-41%
Acrylonitrile	Acetylene Propylene	Propylene	-98%

(continued)

TABLE 151: (continued)

Chemical	1967 Process(es)	1980 Process(es)	Percent Change in Utility Energy Consumption
Adipic acid	Cyclohexane	Cyclohexane	-27%
Carbon disulfide	Methane	Methane	-33%
Ethyl chloride	Ethyl alcohol Ethylene Ethane	Ethylene	-29%
Ethyl alcohol	Ethylene (sulf) Ethylene (DR)	Ethylene	-9%
Ethylene glycol	Ethylene oxide	Ethylene oxide (carbonation)	-76%
Ethylene oxide	Ethylene (Cl) Ethylene (air) Ethylene (O$_2$)	Ethylene (air) Ethylene (O$_2$)	-178%*
Formaldehyde	Methyl alcohol	Methyl alcohol	-150%*
Hexamethylene diamine	Adipic acid Butadiene Acrylonitrile	Butadiene	-44%
Isopropyl alcohol	Propylene (sulf)	Propylene (sulf) Propylene (DR)	-25%
Methyl alcohol	Natural gas, HP	Natural gas, LP	-7%
Perchloroethylene	Acetylene Propane/propylene Ethylene dichloride	Ethylene dichloride	-17%
Trichloroethylene	Acetylene Ethylene dichloride	Ethylene	+250%
Vinyl chloride monomer	Acetylene Ethylene	Ethylene Ethane	-24%
Propylene oxide	Propylene (Cl)	Propylene (oxirane)	+233%
Vinyl acetate monomer	Acetylene/acetic acid Ethylene/acetic acid	Ethylene/acetic acid	+65%
Ethylene	Ethane-propane	Naphtha	-134%*

*Ethylene oxide, formaldehyde, and ethylene production are net energy producers

Source: Reference (10)

For most of the organic chemicals substantial reductions in utility energy use are anticipated. Only three (vinyl acetate, propylene oxide, and trichloroethylene) are expected to increase their utility energy requirements. When these twenty-one chemicals are weighted according to 1967 production and the level of production predicted for 1980, an average utility energy use decrease of 66% between 1967 and 1980 is observed.

This reduction in consumption is not representative of the entire organic chemical industry as it is greatly influenced by the net production of energy in ethylene production. With ethylene excluded the average decrease in utility energy use is 37% rather than 66%. Over the same period of time average feed-stock consumption increased slightly.

GOAL YEAR (1980) ENERGY USE TARGET

Statistical information is readily available on the historical production of the several classes of SIC 2869 chemicals (Table 152). These data have been used as a basis for assuming reasonable growth rates for individual classes of materials in this category and projecting the total production of SIC 2869 chemicals in 1979 in 211 billion pounds (Table 153). The growth is thus at an annual 7.3% rate from the 1972 base of 125 billion pounds. There is no readily available independent way to estimate the aggregate savings in fuel and purchased energy that can be achieved between 1972 and 1979.5.

TABLE 152: HISTORICAL GROWTH RATES FOR CHEMICALS BY CLASS, SIC 2869

Chemical Class	Annual Growth Rate, % 1972	1973	1974	Growth 1974 over 1967, %	Seven Year Average Rate, %
Aliphatic hydrocarbons	5.78	5.46	1.95	78.4	8.62
Rubber processing chemicals	13.9	11.1	n.a.	99.5*	12.20**
Plasticizers	16.6	9.7	1.0	62.8	7.21
Miscellaneous chemicals	17.7	9.4	1.9	286.9	21.32
Flavor and perfume chemicals	14.6	6.4	n.a.	4.9*	0.80**

*Growth rate for 1973 over 1967
**Six year average growth rate

Taken from: *Synthetic Organic Chemicals, United States Production and Sales, 1967-1972*, United States Tariff Commission, 1968-1974, Washington, D.C.; *Synthetic Organic Chemicals, United States Production and Sales, 1973, and Preliminary Statistics, 1974*, United States International Trade Commission, 1975 and 1976, Washington, D.C.

Source: Reference (5)

It was therefore necessary to use some approximation method based on per-centages. One such method that appears to be reasonable is that proposed by R.E. Doerr of Monsanto. The basic premises proposed by Mr. Doerr are:

Plants designed and built prior to the energy crisis can be modified to achieve a specific consumption 15% lower in 1979.5 than in 1972.

New plants, designed and built after the energy crisis will have a specific energy consumption 30% lower in 1979.5 than that for the same product in 1972.

The savings of 30% in new plants should be discounted 15% to allow for inefficiencies in plants starting up in 1979.

The ratio of production to plant capacities will be the same in 1979.5 as it was in 1972.

TABLE 153: 1979.5 PRODUCTION ESTIMATES, CHEMICALS BY CLASS, SIC 2869

Chemical Class	Production, million pounds		Assumed Average Annual Growth Rate, 1974-1979, percent	Projected Production, 1979.5, million pounds
	1972	1974		
Flavor and perfume chemicals, cyclic and acylic	110.5	117(a)	1	123
Rubber processing chemicals, cyclic and acylic	361.0	401(a)	2	443
Plasticizers, cyclic and acylic	1708.3	1891.7	4	2302
C_2 hydrocarbons	19075.0	24216.8	6	32406
C_3 hydrocarbons	8000.0	10474.9	8	15391
C_4 hydrocarbons	5150.0	5526.8	4	6724
C_5 hydrocarbons	691.7	1066.1	6	1427
All other aliphatic hydrocarbons	2216.5	7541.6	6	10092
Miscellaneous chemicals, acyclic	88064.9	97037.2	8	142580
Total	125377.9	148273.1		211488
Weighted Average	--	--	7.3	--

(a) 1973 statistic

Taken from: *Synthetic Organic Chemicals, United States Production and Sales, 1972,* United States Tariff Commission, 1974, Washington, D.C.; *Synthetic Organic Chemicals, United States Production and Sales, 1974 Preliminary Statistics,* United States International Trade Commission, 1976, Washington, D.C.

Source: Reference (5)

Lacking a better basis, energy usage and improved efficiencies were calculated as follows. Production in 1979 (Table 153) is projected at 211,000 million pounds. Total conversion energy in 1979.5, using the specific energy assigned for 1972 would be 1,131 x 10^{12} Btu (net). Total conversion energy needed for the manufacture of products in SIC 2869, is calculated to be 910 x 10^{12} Btu (net) with savings of 19.6% in SEC over 1972. The 1979 SEC would be 4,300 Btu/lb. Expressed in another way, production in 1979.5 will be 69% higher than in 1972, but the total conversion energy (net) will be only 35% greater.

As shown in Table 154, the 211.5 x 10^9 lb production in 1979.5 would require 1,131 x 10^{12} Btu (net) on a 1972 energy usage basis. After the possible energy conversion options are applied, the 1979.5 production will require 910 x 10^{12} Btu total energy, which translates to an SEC of 4,300 Btu/lb. The energy efficiency goal for SIC 2869, thus, is 19.6%.

TABLE 154: SUMMARY OF PRODUCTION AND ENERGY ESTIMATES, SIC 2869

	All Products, SIC 2869 1979.5
Production, 10^9 lb	211.5
Total Energy Required, 10^{12} Btu	
1972 Basis	1131
1979.5 Basis	910
Specific Energy Consumption, Btu/lb	
1972 Basis	5346
1979.5 Basis	4300
Energy Efficiency Goal,[a] percent	19.6

(a) Calculation of energy efficiency goal:

$$\frac{5346 - 4300}{5346} = 19.6 \text{ percent.}$$

Source: Reference (5)

NITROGEN FERTILIZER INDUSTRY

SIC 2873 represents chemical product industries of considerable importance because it is:

(a)   One of the leading chemical areas in regard to tonnage produced of
      finished products.

(b)   One of the major users of energy in the form of natural gas consump-
      tion for use as feedstock and fuel in the manufacture of ammonia.

(c)   The overwhelming major source of critically needed nitrogen for U.S.
      agriculture is ammonia by itself, and in derived products, nitric acid
      (SIC 28731) and urea (SIC 28732), also in phosphatic fertilizers, such
      as monammonium phosphate (MAP), diammonium phosphate (DAP),
      in SIC 28742 and 28743, and fertilizers, mixing only (SIC 2875).

Industry products included in the SIC 2873 segment are classified as follows:

SIC Number	Product
28731	Synthetic ammonia, nitric acid, ammonium compounds
28732	Urea
28733	Fertilizer, materials of organic origin, including sewage sludge and natural fertilizer materials

Information is lacking on SIC 28733. These products are manufactured in small,
widely dispersed facilities throughout the country. Neither the Census of Manufactures nor available trade literature covers adequately the production of such
materials. Further, this industry segment is not a significant energy consumer,
especially when compared to the other elements in SIC 2783.

PROCESS TECHNOLOGY INVOLVED

Ammonia

The major process for ammonia manufacture in the U.S. for some time has been
based on gas reforming of natural gas. This will remain the preferred choice for

388

new plants in the period up to year 1980. Alternate options for gas reforming via naphtha or partial oxidation of fuel oil will most likely not be adopted in the near future up to year 1980. Beyond 1980, it is most probable that many new plants will be based on syn gas produced by coal gasification.

From Natural Gas and Naphtha: The process is called gas reforming. A summary process diagram is provided in Figure 107 for natural gas and Figure 108 for naphtha. The natural gas or vaporized hydrocarbon (naphtha) is admixed with steam and converted to hydrogen and carbon monoxide over nickel catalyst, inside of externally fired tubes in refractory furnace modules. This operation is designated primary reforming. The process side (exit basis) conditions are in the range of 1500°F at 460 psia. The reactions are parallel to the representation for methane, as follows: $CH_4 + H_2O = 3H_2 + CO$.

The resultant gases still carry up to approximately 11% CH_4 by volume, and as yet are deficient in the nitrogen component of the ammonia synthesis requirement. Atmospheric air, in equivalence according to the nitrogen requirement, is therefore compressed to fit the 460 psia process level and admixed (after preheating) into the gases. The oxygen content of the incoming air burns out in reaction with a portion of the $H_2/CO/CH_4$ content of the gases and provides a temperature rise of approximately 500°F, permitting an adiabatic pass over a second portion of nickel catalyst to final temperatures of approximately 1800°F. This operation is termed secondary reforming and is accomplished over a packed-bed arrangement. The residual methane is as low as the 0.3% range.

The secondary-reform gases are next (after partial cooling, with heat recovery) processed for CO conversion for additional hydrogen production. This is accomplished via water-gas shift over iron oxide catalyst (typically promoted with chromium oxide in first-stage beds operating at 450° to 500°F). The reaction may be represented as follows: $CO + H_2O = H_2 + CO_2$.

The resulting gases (after cooling, with heat recovery) are scrubbed for CO_2 removal, typically with chemical-absorbent processes such as Benfield. The gases are then treated for last-trace (e.g., as low as 0.3% CO from two-stage systems) CO removal. The favored CO removal system in modern plants is methanation, utilizing nickel catalyst (actually the reverse of gas reforming itself). An alternative process that was popular in older plants utilized copper liquor scrubbing. The respective methods (in particularly simplified form for the copper liquor) may be represented as follows:

$$Methanation:\ CO + 3H_2 \rightleftharpoons CH_4 + H_2O$$
$$Cu\ Liquor:\quad CO + 2CuO \rightleftharpoons CO_2 + Cu_2O$$

The final syn gas obtained is in requisite 3/1 H_2/N_2 balance, plus up to approximately 1% CH_4 and 0.3% Ar (the latter as entered in the secondary reform air). Two to three stages of compression are applied to bring pressure drops through CO conversion, CO_2 removal, last-trace CO removal) to 2,000 to 5,000 psia discharge for ammonia synthesis conversion. The choice in each case of actual pressure level for ammonia synthesis is determined mainly according to size of plant; units from 600 to 1,000 T/D NH_3 often utilizing 2,000 to 3,000 psia staged in two casings; and larger capacity units having the option to the higher pressures, with staging then in three casings.

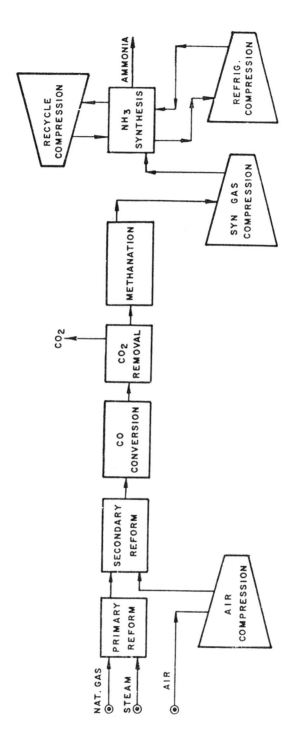

FIGURE 107: AMMONIA MANUFACTURE/PROCESS DIAGRAM—GAS REFORMING/NATURAL GAS

Source: Reference (5)

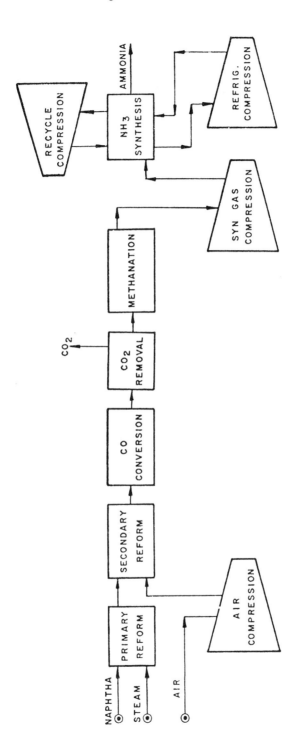

FIGURE 108: AMMONIA MANUFACTURE/PROCESS DIAGRAM—GAS REFORMING/NAPHTHA

Source: Reference (5)

The ammonia synthesis is accomplished over promoted iron oxide catalyst (in the reduced state under actual operation) in packed bed arrangement. Conversion is not complete in single pass, and multiple pass via recycle is utilized. The recycle or synthesis loop includes the catalyst in multiple beds (typical design); refrigerated cooling and condensing for removal of product ammonia; and auxiliary equipment including heat-exchange, product separation, etc. The recycle ratios (of recycle gas to makeup gas) employed in modern plants are in ranges typically from 3/1 to 6/1 by volume. The pressure drops through the loops are in the order of 10% of the synthesis pressure. The resulting flow and head requirements for the recycle gas represent substantial compression load.

Figure 109 is a block flow diagram for ammonia synthesis from natural gas showing the energy values for the various unit operations in the boxes in units that represent millions (10^6) of Btu. Feed and products are underlined with energy values assigned. No fuels or electrical energy consumption is outlined below.

FIGURE 109: PRODUCTION OF AMMONIA

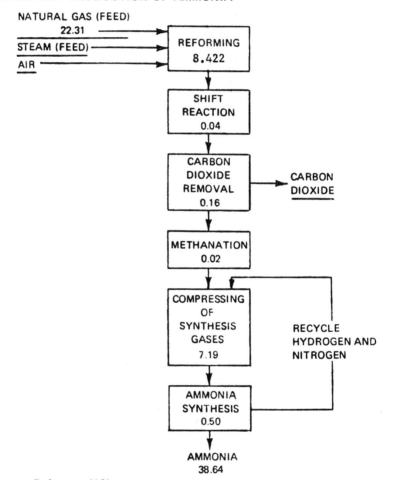

Source: Reference (10)

From Heavy Oils: The process is called partial oxidation. A summary process diagram is provided in Figure 110. The heavy hydrocarbon liquid feedstock, together with preheated oxygen and steam, is atomized-injected into a high-temperature refractory-lined burner chamber. This operation is the actual partial oxidation step. The process conditions are in the range of 2400° to 3000°F at 600 to 1,000 psia. The oxygen-based and the steam-based reactions may be represented respectively as follows:

$$2C_nH_m + nO_2 \rightleftharpoons 2nCO + mH_2$$
$$2C_nH_m + 2nH_2O \rightleftharpoons 2nCO + (2n+m)H_2$$

The actual conversion proportions assumed by the oxygen, versus that for the steam, are according to the usual factors of material balance, energy balance, and process equilibrium.

The resultant gases are obtained directly with methane content at the 0.3% range, and from such aspect therefore directly suited to next stage processing through CO conversion, without requirement for a secondary reforming operation, as described previously. The partial oxidation gases, however, do contain a proportion of unreacted carbon (in quantity approximately 1 to 3% of total carbon of original oil feedstock), with a solids removal stage (usually water scrubbing) thereby necessitated to effect separation of the carbon. The partial oxidation gases are therefore processed through an appropriate processing sequence, including cooling (with heat recovery), carbon-removal scrubbing, and requisite reheating prior to admission to CO conversion.

The sulfur of the oil feedstocks appears in the partial oxidation gases nearly all as H_2S, but with some minor proportions as COS and CS_2. These are removed either before or after the CO conversion, depending upon the type of CO conversion catalyst and desired nature of operation. Any of a number of suitable chemical-absorbent or physical-absorbent systems (e.g., chemicals such as Alakazid or Benfield; or physical such as Selexol or Rectisol) may be used for the removal of the sulfur compounds (higher pressure ranges favoring the physical, versus lower pressure ranges favoring the chemical).

The gases, after CO conversion, are processed for CO_2 removal. Options parallel to those described for the H_2S removal are applicable. The two removal requirements may, in fact, be integrated in appropriate stage-scrubbing or stage-regeneration to permit recovery of separate and sufficiently rich H_2S, as desirable for minimal cost downstream conversion to elemental sulfur.

The gases obtained are now pure other than for last-trace CO removal. At the same time, they are lacking in the nitrogen component required for the final ammonia synthesis (oxygen having been used in the gas generation, and air-admitting secondary reform not employed). A favored system for this variant therefore is liquid nitrogen wash, effecting the requisite nitrogen addition as well as the CO removal; and utilizing as makeup nitrogen source the otherwise entirely waste nitrogen from the process-related oxygen-producing air separation plant. An alternative system is the parallel of the methanation process described in conjunction with gas reforming, with requisite nitrogen then added downstream to the treated gas. The final gas is compressed to synthesis level and processed for ammonia production. The compression and synthesis are parallel to the equivalent operations described earlier for plants based on gas reforming.

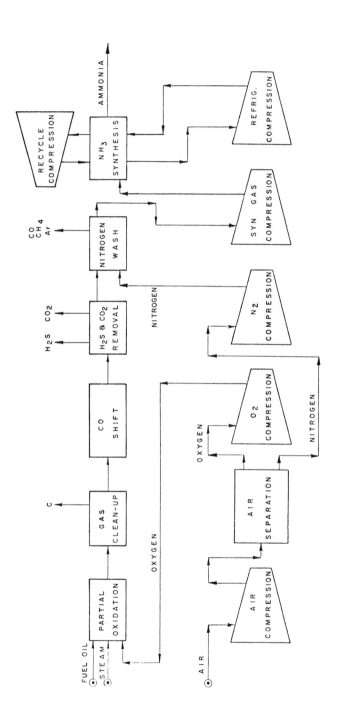

FIGURE 110: AMMONIA MANUFACTURE/PROCESS DIAGRAM–PARTIAL OXIDATION/FUEL OIL

Source: Reference (5)

The syn gas pressure will be in the 600 to 1,200 psia range, however, instead of the 380 psia level when obtained from gas reforming. The syn gas compression from partial oxidation, therefore, would be typically one stage less than the parallel equipment working from gas reforming.

From Coal: The process is called steam-oxygen gasification. A summary process diagram is provided in Figure 111. The gasification reaction may be accomplished parallel to that described previously for partial oxidation of heavier petroleum fractions.

Commercial installations to date, however, of this version of steam-oxygen gasification have been solely at, or not far from, atmospheric pressure, i.e., the Koppers-Totzek process, rather than at substantially elevated pressure. Newer generations of such plants, on the other hand, will undoubtedly operate at elevated pressure, e.g., as projected on the process diagram.

An alternative method for effecting the steam-oxygen gasification can be according to the Lurgi AG type of elevated pressure gas generator. This is essentially a pressurized shaft kiln. Lump coal is fed at the top; the bottom is steam-oxygen blown.

The product gas carries tars and condensible liquids, removable on cooling and scrubbing. The cleaned gases carry methane which is separated during the course of the gas processing, and may be converted separately via the primary reforming described previously, for supplementing the total synthesis gas. (Alternatively the methane may be disposed to fuel utilization, or supplementing of SNG production if in an adjacent area).

A third method for the gasification from coal is according to the Winkler process of Davy Powergas, Inc. The Winkler process gasifies its coal from fluid bed reaction, utilizing coal ground to minus three-eighths inch (in between the Koppers type of processes utilizing 70% minus 200 mesh in entrained suspension, and the Lurgi gasifier utilizing ½ to 1½ inch in shaft kiln type of arrangement).

For any of the options for gasification of solid fuels, the raw gas requirements for ammonia manufacture may be taken from larger base-load SNG-oriented coal gasification plants.

The raw gas in such cases would be drawn at the optimum point for subsequent downstream conversions to ammonia manufacture and, in any case, prior to the methanation operations of the mother SNG plant.

This piggy-back type of arrangement for source of raw gas would undoubtedly find favor in areas of coincident coal-based SNG manufacture and suitable ammonia markets or logistics (providing the problems of association between public-agency regulated SNG manufacture and private-industry independent ammonia manufacture could, in fact, be resolved).

The gasification reactions for steam-oxygen gasification from solid fuels may be represented identically to those shown previously for partial oxidation. Downstream processing for purification, compression, synthesis follows the requirements outlined previously, and according to the various considerations of sulfur content, pressure levels, etc.

FIGURE 111: AMMONIA MANUFACTURE/PROCESS DIAGRAM—STEAM-OXYGEN/COAL OR LIGNITE

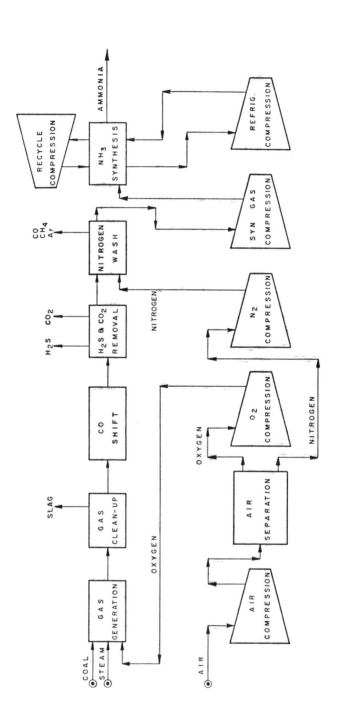

Source: Reference (5)

Nitric Acid

Single Pressure Plant: A simplified schematic representation of conventional U.S. practice is attached for reference as Figure 112. The essence of this conventional practice is air compression to 8 to 10 atmospheres followed by burning with vaporized ammonia over a platinum gauze catalyst.

The resulting nitrogen oxide gases are cooled and subsequently joined by secondary air from the compressor for absorption into water to form the required nitric acid product.

The air supply to the burner is required at temperatures of 450° to 500°F. This is accomplished in most plants by control of intercooling during compression. In some plants, full intercooling is utilized followed by air preheating against the hot gases leaving the burner. The newer and larger plants generally utilize the controlled intercooling method.

The burner operation accomplishes conversion in high yield of the fixed nitrogen content of the feed ammonia into the form of nitrogen oxides. The subsequent cooling operation is generally effected in a sequence of units to permit optimum heat recovery, a typical line consisting of waste heat boiler, tail gas heater, feed water heater, weak acid condenser.

The cooled nitrogen oxide bearing gases, after separation from the condensed weak acid, are fed to the bottom of an absorption tower. The weak acid is fed to the same tower, but at a point of equivalent acid concentration. Makeup water is fed to the tower top. The product acid from the tower is bleached by blowing with hot secondary air, the resulting air plus stripped gases joining the main stream gases prior to their feed to absorption. The bleached final acid is generally obtained at 58% HNO_3 content.

The absorption is cooled and therefore delivers its tail gases at temperatures in the range of 100°F, depending upon time of year and site location. The general application is to reheat the tail gases in exchange with hot burner gases prior to using them for power recovery in an expander turbine. The reheating sequence may include one or more stages of exchangers and, in addition, in newer plants, often incorporates fuel injection and catalytic combustion for superheating and fume abatement. The reheated gases at the expander throttle may therefore be available at temperatures of 350° to 450°F for low temperature application, 750° to 950°F in the case of higher temperature heat exchange and 1250°F as typical in the case of catalytic combustion.

The pressure of the gases at the expander throttle may be in the range of 20% lower than the air from the compressor discharge. The mass flow of gases to the expander is generally in the range of 80% of the original compressed air, reflecting oxygen removal equivalent to the ammonia burning and the conversions to nitrogen oxides and subsequently to nitric acid.

A common feature in most nitric acid plants is to utilize the still pressurized absorber tail gases for aid in driving the air compressor.

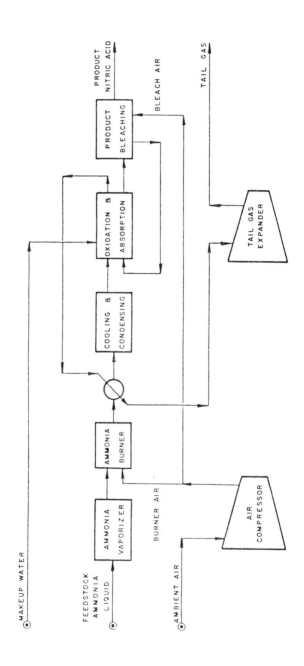

FIGURE 112: NITRIC ACID MANUFACTURE/PROCESS DIAGRAM—SINGLE-PRESSURE PLANTS

Source: Reference (5)

Split Pressure Plant: A simplified schematic representation of the Chemico approach is shown in Figure 113. The process is identical to that for the single pressure operation described previously, up to the completion of the cooling and condensation steps following ammonia burning, and the heat exchange to tail gas reheat.

The cooled NO_x gases in the split pressure process are now boosted additionally in pressure prior to the oxidation and absorption stage. The bleaching operation remains at air compressor discharge pressure, the bleach exit air joining the cooled main-stream NO_x gases, for handling through the booster compressor.

Reheat and power recovery from tail gas is as referenced for the single pressure plants, but working from the greater pressure level from the higher-pressure absorption/oxidation operation.

Urea

A simplified schematic representation of conventional U.S. practice is shown in Figure 114. The essence of this system is carbon dioxide compression from atmospheric to 2,000 to 5,000 psig, followed by reaction at such pressures with ammonia to form urea plus water. The synthesis equation may be represented as follows: $CO_2 + 2NH_3 \rightleftharpoons (NH_2)_2CO + H_2O$.

For plants in the 2,000 to 3,000 psig pressure ranges centrifugal compressors may be used for capacities as low as 500 ST/D. For plants in the 4,000 to 5,000 psig pressure ranges such machines can be used for capacities as low as 1,000 ST/D. Combinations of centrifugal compressors in the lower stages, followed by reciprocating compressor to the final stages, may also be used.

The compressed carbon dioxide and the companion reactant liquid ammonia are supplied to the synthesis reactor maintained at 350° to 400°F. Recycled incompletely reacted material is fed to the reactor as well. The reactor consists essentially of an empty pressure vessel with volume appropriate to required reaction time and fabricated of corrosion-resistant materials such as special grades of stainless steels and, in one instance, zirconium.

The reaction products consist of a complex urea in water solution together with incompletely reacted material such as carbamate (a compound of ammonia and carbon dioxide that has not yet gone through the water release accompanying full conversion to urea) and excess recycle ammonia. The products are separated by decomposition and stripping into recycled fractions and a simple urea in water solution. The resulting simple urea in water solution is treated by evaporation combined with crystallization and/or prilling or flaking to yield a final merchant urea product.

A process for urea manufacture described by P.F. Kaupas and D.F. Bress (59) is one in which contaminated condensate containing waste heat from the process is passed in heat exchange relation with a stream of contaminated air and entrained fertilizer solids. In this manner it is possible to remove substantially pure water vapor and air together with the fertilizer-containing solution which can be used in the manufacture of the fertilizer.

FIGURE 113: NITRIC ACID MANUFACTURE/PROCESS DIAGRAM–SPLIT-PRESSURE PLANTS

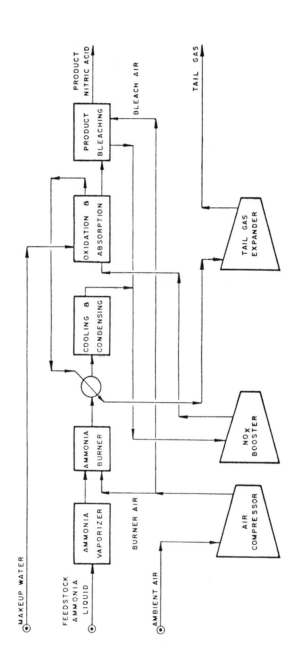

Source: Reference (5)

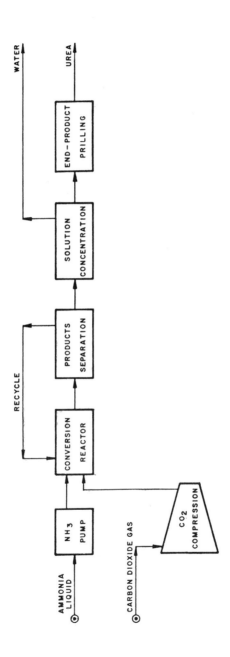

FIGURE 114: UREA MANUFACTURE/PROCESS DIAGRAM

Reference (5)

MAJOR ENERGY CONSERVATION OPTIONS TO 1980

Production of ammonia used over 98% of the total energy, nitric acid about 1% and urea about 0.49%. Clearly, improvements in the use of energy for manufacturing ammonia are of fundamental importance.

During the past 15 years design of ammonia plants has progressed rapidly by operating at lower pressure, using more efficient compressors, recovering waste heat, a more efficient steam balance, and more efficient refrigeration circuit. Older plants can recover heat from the stack gas of the reformer furnaces; also recovery of hydrogen from purge gas by cryogenic techniques can reduce both feedstock and energy requirements by about 3%. These are all becoming more economically attractive as fuel costs rise.

Plans for converting to oil firing instead of gas firing of furnaces are being evaluated. The effect on specific energy consumption is expected to be quite small. Nitric acid plants can use a combination of a hot gas expander and auxiliary steam turbine operating on steam generated from waste heat to jointly drive the main air compressor, partially replacing electric motor drives. This is probably economical only in plants over about 400 tons/day of capacity.

In urea plants latest technology uses less energy for recycle streams. Part of this technology can be retrofitted to older plants. All plants have active programs of improved housekeeping, maintenance, closer process control, and closer energy balancing.

Based on information developed thus far the estimate is that the above factors would permit a reduction in specific energy consumption by over 20% by January 1, 1980. The most important increases in energy use over 1972 are to meet emission regulation for the liquid effluents from urea plants, and tail gases from nitric acid plants. Another factor, difficult to quantify, is delay in corporate decisions relating to ammonia plant due to not being able to estimate the long range supply of natural gas feedstock, and to consideration of whether to shut down older plants, rather than revamp, and build new ones.

GOAL YEAR (1980) ENERGY USE TARGET

Production estimates for 1972 and 1979.5 are shown in Tables 155 and 156. Note that total production, from all SIC industries producing nitrogenous fertilizers is approximately 45% greater than that produced from establishments in SIC 2873 only. Growth is estimated at 4.5% compounded annually to January 1, 1980, or an increase of 39.1%.

Data from TVA, The Fertilizer Institute, and Census of Manufactures have been analyzed over the period 1971 to 1975 to arrive at the specific energy consumption (SEC) of 14,504 Btu/lb, and total energy consumption of 364.9 x 10^{12} Btu for production from all SIC groups for 1972. These are also shown in Table 155.

SEC is on the average considered to be essentially constant irrespective of the establishments classification where manufactured.

TABLE 155: PRODUCTION AND ENERGY ESTIMATES—SIC 2873

	All Products SIC 2873 1972	1979.5
(A) Production, 10^9 pounds as N		
Total, all SIC	25.16	35.00
Made in SIC 2873	17.38	24.18
(B) Total energy required, 10^{12} Btu		
1972 basis, total all SIC	364.9	507.6
1979.5 basis, total all SIC	na	431.4
(C) Specific energy consumption, Btu/lb		
1972 basis	14,504	14,504
1979.5 basis	na	12,328
(D) Estimated energy efficiency goal, %		15

Notes: Production, 10^9 pounds N, means contained nitrogen.
       Compound average growth rate for period CY 1972 through
        January 1, 1980 is projected at 4.5% per year for 7.5 years.

Growth factor for 7.5 years = $1.045^{7.5} = 1.391$.

Source: Reference (5)

TABLE 156: CAPACITY AND PRODUCTION ESTIMATES, SIC 2873 CHEMICALS

Product	No. Plants 1972	1980	Total Capacity Million Pounds N 1972	1980	Estimated Production Million Pounds N 1972	1980
Ammonia	86	110	28,068	38,271	24,915	34,700
Nitric acid	72	88	3,956	4,971	3,560	4,950
Urea	41	53	4,009	7,119	2,376	3,300
Ammonium nitrate	65	68	5,198	5,698	4,045	5,630
Ammonium sulfate	67	70	1,360	1,432	1,040	1,450

Notes: Nitric acid data for usage in fertilizer industry is lacking. Estimate made from
       fertilizer end products.
     Weight data are in million pounds of contained nitrogen N.
     The initial product of manufacture is ammonia, which in turn serves as the
       intermediate in manufacture of the other SIC 2873 products. The overall
       capacity and production totals therefore are not directly additive.

Source: Reference (5)

The application of specific energy conservation steps has been estimated to have
the impact shown in Table 157. Note that the overall 20% reduction is decreased by a 5% increase in operations energy resulting from compliance with
EPA and OSHA standards. Therefore, the estimated overall energy conservation
reductions for SIC 2873 by January 1, 1980 may reasonably be expected to
amount to 15%. Considering all factors, the energy improvement goal has thus
been estimated at 15% reduction in specific energy consumption by January 1,
1980. Specific energy consumption and total energy consumption in 1979.5 have

been calculated and are also shown in Table 155. Due to the forecast growth in production, total energy use is estimated to increase about 39% despite the decrease in specific energy consumption.

These estimates also assume that 1979.5 average operating rate and seasonal changes in relation to capacity will be similar to 1972. No details of energy use versus operating rate are available, but it is estimated that a 10% decrease in production rate will decrease energy use less than 5% in these plants.

TABLE 157: IMPACT OF ENERGY CONSERVATION STEPS

	Energy Conservation Option	Energy Reduction, %
(1)	Improved plant housekeeping and maintenance	5
(2)	Improved plant operations in-house techniques	10
(3)	Conversion of ammonia plant reformer furnaces from natural gas firing to fuel oil (HHV basis)	1
(4)	Install cryogenic unit on ammonia plant tail-gas to recycle hydrogen to synthesis	1
(5)	Improve operation and control equipment on waste heat recovery system of ammonia plant to increase by-product steam generation and utilization within plant	2
(6)	Install in nitric acid plants a combination of steam turbine operating off waste heat recovery steam generation to jointly drive the air compressor and replace the electric motor drive	0.2
(7)	Revamp MAP, DAP, and urea solution operations	0.8
	Total Reduction	20.0

Source: Reference (5)

SOME PROJECTIONS BEYOND 1980 TO 1990

As part of the Project Independence Blueprint (10) it was estimated that between 1974 and 1980, U.S. ammonia demand would increase by about 10 billion pounds. Of this it was assumed that only 25% would be produced in the U.S. and the remaining 7.5 billion pounds would be imported. It was further assumed that all new capacity required after 1980 would be met by increases in domestic production based, in part, on non-natural-gas fuels and feedstock including naphtha, fuel oil, and gasified coal.

Plants in existence in 1974 were assumed to retire at a rate of 1.5% per year and be replaced by new domestic capacity. Replacement and new capacity between now and 1980 was estimated to be 20% more fuel efficient than present plants and all new plants built after 1980 were assumed to be 30% more efficient than present plants.

All new capacity was assumed to be of the low pressure type. Projections of energy consumption are shown in Table 158.

TABLE 158: ENERGY CONSUMPTION PROJECTIONS FOR AMMONIA PRODUCTION THROUGH 1990*

Consumption Factors	Fuel (MBtu/lb)	Electric (kwh/lb)
Pre-1974 plants	9.95	0.078
New 1974-1980	7.96	0.027
New 1980-1990	6.97	0.027

Total Consumption—Fuels (10^{12} Btu)

	1967	1971	1975	1977	1980	1985	1990
Existing	243	279	302	294	280	257	234
New	0	0	10	21	51	126	214
Total	243	279	312	315	331	383	448

Total Consumption—Electric (10^9 kwh)

	1967	1971	1975	1977	1980	1985	1990
Existing	1.9	2.2	2.4	2.3	2.2	2.0	1.8
New			0.0	0.1	0.2	0.5	0.8
Total	1.9	2.2	2.4	2.4	2.4	2.5	2.6

*Feedstock not included, does not include conservation at existing facilities.

Source: Reference (10)

PHOSPHATE FERTILIZER INDUSTRY

SIC 2874 products are based on the primary step of acidulation of phosphate rock by sulfuric acid, and include the following:

SIC No	Product
28741	Phosphoric acid
28742	Superphosphate (normal and triple ammonium phosphates, other phosphate fertilizer
28743	Mixed fertilizers (made in same plant wherein phosphate fertilizer materials are produced)

Phosphoric acid is manufactured by acidulation of finely ground phosphate rock with sulfuric acid, recovery of the raw production acid, and concentration to desired P_2O_5 level by evaporation. Normal superphosphate is prepared by acidulating pulverized phosphate rock with sulfuric acid and allowing the reaction products mix to assume a solid form and cure by storage. Normal superphosphate is declining in agricultural consumption in favor of triple superphosphate.

Triple superphosphate is made by acidulating pulverized phosphate rock with phosphoric acid, followed by drying and granulation. Diammonium phosphate (DAP) is made by partial neutralization of wet process phosphoric acid with anhydrous ammonia. Mixed fertilizers are mixtures guaranteeing N, P_2O_5, and K_2O contents in various combinations of these components. The K_2O component is supplied by purchased potash mineral.

PROCESS TECHNOLOGY INVOLVED

Phosphoric Acid

Wet process phosphoric acid is a term now used for the product made by digestion of phosphate rock with sulfuric acid. Although any strong inorganic acid may be used, no others presently are of commercial interest for both economic and technical reasons. The major principal reaction taking place in the manufacture of wet process phosphoric acid probably is best represented by the following:

$$Ca_{10}(PO_4)_6F_2 + 10H_2SO_4 + 20H_2O \longrightarrow 10CaSO_4 \cdot 2H_2O + 6H_3PO_4 + 2HF$$

This is carried out in the diagrammed system over a period of approximately eight hours. The reaction itself is completed to a large extent in a matter of minutes, but more time is needed to allow for the proper formation of gypsum crystals. Sulfuric acid (93% H_2SO_4) and finely ground phosphate rock are continuously added to a slurry consisting of reactants, products, and enough recycling weak phosphoric acid to maintain sufficient fluidity. The slurry is continously being drawn off and filtered.

The gypsum filter cake is reslurried with water and discarded in some cases. In others, it may be purified and sold or used as a by-product. Weak phosphoric acid washings are returned to the digestors. Product phosphoric acid (about 32% P_2O_5 equivalent) is drawn off and concentrated to the desired strength.

For most uses, raw production acid (28 to 30% P_2O_5) must be concentrated before it is used. Most "DAP acid" (for the production of diammonium phosphate) is in a 40 to 45% P_2O_5 range (typically a blend of raw production acid and evaporator acid). Acid used to produce triple superphosphate may be 40 to 45% P_2O_5, depending on the type of product desired (granular or ROP) and individual plant preferences. Most plants concentrating production acid to the 50 to 54% P_2O_5 level (evaporator acid) use forced circulation vacuum evaporators. During this concentration step, as much as 40% of the fluorine originally in the rock may be released, and some type of recovery (economic or waste) is typically employed in order to comply with pollution control regulations.

A block flow diagram for phosphoric acid manufacture is given in Figure 115. The energy values for the various unit operations are given in the boxes in units that represent millions (10^6) of Btu. Feed and product are underlined with energy values assigned. No fuels or electrical energy consumption is itemized on the flowsheet.

Triple Superphosphate (TSP)

Triple superphosphate (also called treble, double, or concentrated superphosphate) is an impure monocalcium phosphate made by phosphoric acid with phosphate rock:

$$Ca_{10}(PO_4)_6F_2 + 14H_3PO_4 \longrightarrow 10Ca(H_2PO_4)_2 + 2HF$$

Although it may be made with any phosphoric acid, the major portion of this material is now made with wet process phosphoric acid. The P_2O_5 equivalent of the product averages 46 to 47%, but may range from 44 to 52% depending upon the purity of the acid and rock, and the efficiency of the manufacturing process. Triple superphosphate plants are, as a rule, much larger than normal superphosphate plants. For economic reasons they are usually located near

FIGURE 115: PRODUCTION OF PHOSPHORIC ACID (PER TON OF P_2O_5 AS 40% OF P_2O_5 ACID)

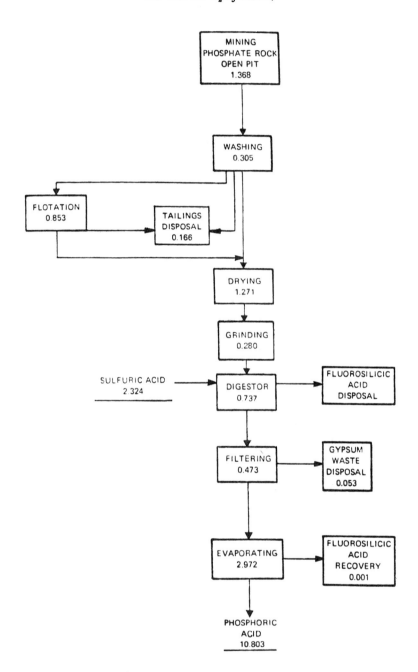

Source: Reference (12)

phosphate rock deposits or near port facilities or other receiving points if the phosphate rock must be imported. Being more concentrated than the raw materials from which it is made, the greatest shipping economies can be realized if triple superphosphate is manufactured close to the source of phosphate rock, which is the least concentrated raw material. Moreover, wet process phosphoric acid, unlike sulfuric acid, is not generally available on the open market, and requires large and continuously operating plants in order to minimize investment and operating costs per ton of output.

Presently, there are three principal methods of manufacturing triple superphosphate. One of these produces a pulverized product which, being soft and porous, is particularly suited to the manufacture of ammoniated fertilizers. The second produces a granulated product in a multistep process which is well suited for direct application as a phosphate fertilizer, or for inclusion in mixed bulk blends made by dry blending solid raw materials. The third method combines the features of quick drying and granulation in a single step.

Pulverized Triple Superphosphate: This process uses phosphoric acid concentrated to 65 to 75% H_3PO_4 (47 to 54% P_2O_5 equivalent). Mixing of phosphoric acid and phosphate rock, formerly a problem because of a quickly developing plastic consistency, now is readily accomplished in a cone mixer developed by TVA. Small, simple, but effective, this device employs tangetial streams of acid under pressure to accomplish quick mixing. The slurry produced is discharged to a belt conveyor where it hardens within less than a minute. It is then disintegrated mechanically before storage on the curing pile. While this product normally is used for manufacturing ammoniated fertilizers, it may be granulated after the curing period is completed.

Because of the lower capital investment and reduced labor and maintenance costs, as compared with granular triple superphosphate, the production cost of a ton of the pulverized material might be as much as $3.00 less. However, if an extensive gas scrubbing system for the storage area is required to meet local air pollution regulations, this cost differential might be considerably narrowed.

Granular Triple Superphosphate: A slurry method produces a granular triple superphosphate directly. This method employs phosphoric acid of 32 to 42% P_2O_5 equivalent (45 to 58% H_3PO_4). In a series of reaction tanks, the rock and acid slurry is steam heated to 80° to 100°C until the reaction is completed. Quick drying is facilitated by blending the slurry into a bed of recycling crushed oversize fines and product. The partially dried granules then are fed to a rotary drier. A multilayered, dense product with excellent handling qualities is thus built up.

Single-Step Granular Triple Superphosphate: This method combines the features of quick drying and granulation in a single step. In this process, developed recently by TVA, acidulation and granulation are accomplished simultaneously in a rotary drum mixer. Wet process phosphoric acid and steam are fed through distributors under a rolling bed of material in the drum. The product has a short curing time, can readily be ammoniated, and has good storage properties. The process has the added advantage of low equipment and processing costs. It also may be adapted to existing continuous ammoniators.

Ammonium Phosphates (MAP and DAP)

There are various processes and process variations in use for manufacturing ammonium phosphates. All are essentially the same in principle. A typical flow diagram is illustrated in Figure 116 for DAP.

FIGURE 116: TVA FLOW DIAGRAM FOR DIAMMONIUM PHOSPHATE MANUFACTURE

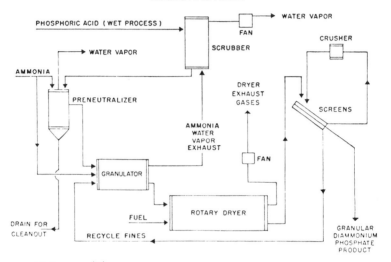

Source: Reference (5)

In general, phosphoric acid, sulfuric acid and anhydrous ammonia are reacted to produce the desired grade of ammonium phosphate. Potash salts are added, if desired, and the product is granulated, dried, cooled, screened and stored. All processes for the manufacture of diammonium phosphate (grades 16-48-0 and 18-46-0) from wet process phosphoric acid are essentially the same in principle. Wet process phosphoric acid of about 40 to 42% P_2O_5 equivalent is partially neutralized (to about 89%) by anhydrous ammonia (liquid or gaseous), the pH being adjusted to about 5.6. This required about 1.3 mols of ammonia per mol of phosphoric acid.

Sulfuric acid often is metered in along with phosphoric acid. This provides a simple and convenient means of controlling the grades of the fertilizer. The temperature is maintained at about 240°F by adjusting the acid dilution so that water evaporation balances the heat of reaction.

In the Dorr-Oliver process, the reaction is completed in the fluid state by using additional reactors. The slurry overflows to a turn-shaft paddle mixer containing recycling dry product (about 8:1 ratio). Dry salts, such as potassium chloride, also are added at this point if desired. The moist granules formed in the mixer are then dried, cooled, screened and stored. Ammoniation in the reactors can only proceed to the point at which about two-thirds of the product is diammonium phosphate. To complete the conversion, the semisolid mass in the blunger is further ammoniated using liquid ammonia. A process developed by the Tennessee Valley Authority appears to be the most popular in recently built plants. This process features efficient use of reaction heat, low recycle (6-1 to 4-1) and ease of control. After the initial neutralization step, the slurry and recycling material are fed into a revolving drum where final ammoniation and granulation take place simultaneously. The moist granules then are dried, cooled, screened and stored as with other processes.

After half of the water introduced into the process is evaporated in the pre-neutralizer, an additional 40% is driven off in the ammoniator-granulator. Thus, less drying capacity is required than in the Dorr-Oliver slurry process. Power requirements also are less because of the reduced recycle rate, and the granules are more uniform. The flow sheet of the TVA process is illustrated in Figure 116.

MAJOR ENERGY CONSERVATION OPTIONS TO 1980

The two major areas in phosphatic fertilizers manufacture which offer significant energy savings are:

1. Improved plant housekeeping and maintenance
2. Improved plant operations in-house techniques

The processes are well-known and long-established. There is very little likelihood of new process breakthroughs by January, 1980 which would feature significant basic energy reduction. The quality of phosphate rock mineral is declining. To process poorer grades will require from 3 to 7% more energy. Only a few producers are likely to face this problem and then in a small way, by January 1, 1980. But, it will be important to most producers in the following years.

Higher energy costs have motivated programs throughout the industry toward better housekeeping, maintenance, and process scheduling and optimization. Between 1972 and 1975 an improvement of about 10% has generally been obtained. Experience suggests this improvement can continue.

Specific energy consumption in the industry can be reduced by 30% by January 1, 1980 relative to 1972. However, some of these savings will be offset by increased energy requirements to meet OSHA and EPA standards. While OSHA standards themselves should not affect this industry segment, EPA standards require fluorine emissions compliance in phosphoric acid production. The industry is installing air pollution controls. At most, this may increase energy consumption in specific plant situations not more than 1%. Therefore, a net specific energy improvement of 20% is considered feasible.

Energy improvements will result from the use of wet rock and wet grinding in place of the current use of dry rock and dry grinding. In installations where the process change is feasible, fuel requirements will be reduced since the drying equipment will not be required. Additional savings will call for more efficient plant systems, including improved housekeeping, maintenance, process control and energy balancing.

GOAL YEAR (1980) ENERGY USE TARGET

Production estimates for 1972 and 1979.5 are shown in Tables 159 and 160. Total production from all SIC industries producing phosphatic fertilizer is about 8% higher than that produced in establishments in SIC-2874 only. Growth has been estimated at 4.5% per year, or an increase of 39% between 1972 and January 1, 1980.

Data from TVA, The Fertilizer Institute, and the Census of Manufactures have been analyzed for the period from 1971 to 1975 to estimate the specific energy consumption for 1972 of 7,457 Btu per lb, and total energy consumption of 81.8×10^{12} Btu for production from all SIC groups in SIC 2874. These are also shown in Table 159.

TABLE 159: PRODUCTION AND ENERGY ESTIMATES, SIC 2874

	All Products SIC 2874	
	1972	1979.5
Production, 10^9 pounds P_2O_5		
Total, all SIC	10.97	15.26
Made in 2874	10.14	14.10
Total energy required, 10^{12} Btu		
1972 basis, total all SIC	81.80	113.76
1979.5 basis, total all SIC	na	91.00
Specific energy consumption, Btu/lb		
1972 basis	7,457	7,457
1979.5 basis	na	5,966
Estimated energy improvement goal, %		20

Note: Compound average growth rate for period 1972 through January 1,
1980 projected at 4.5% per year for 7.5 years.
Growth factor = $1.045^{7.5}$ = 1.391

TABLE 160: CAPACITY AND PRODUCTION ESTIMATES, SIC 2874 CHEMICALS

	No. Plants		Total Capacity Million Pounds P_2O_5		Estimated Production Million Pounds P_2O_5	
Product	1972	1979.5	1972	1979.5	1972	1979.5
Phosphoric acid*	30	38	12,714	19,340	11,188	19,450
Normal superphosphate	60	32	–	–	1,354	800
Triple Superphosphate	14	18	3,934	6,196	3,318	5,600
Ammonium phosphates	39	43	7,264	10,546	5,154	9,700
All other 2874	na	na	na	na	1,140	1,500

*The phosphoric acid is mainly wet process. Its use is largely as intermediate in manufacture of other SIC 2874 products. The overall P_2O_5 capacity and production totals therefore are not directly additive.

Source: Reference (5)

Considering all factors, the energy improvement goal has been estimated at 20% reduction in specific energy consumption by January 1, 1980. Using this goal, specific energy and total energy consumption for 1979.5 have been calculated as shown in Table 159. Due to the forecast growth in production, total energy use in 1979.5 will be higher than 1972 by about 11% despite the estimated decrease in specific energy consumption. These estimates have also assumed that operating rate in relation to capacity for the industry will be essentially the same in 1979.5 as it was in 1972.

SOME PROJECTIONS BEYOND 1980 TO 1990

Phosphoric acid is primarily used in the manufacture of fertilizer and detergent. It is estimated that between 1973 and 1980 fertilizer production will grow at a rate of 3.7% per year and soap products will grow at a rate of 4.0% per year. During the period 1980 to 1985, fertilizers will slow to 3.0% per year and soap products will continue at 4.0% per year.

If high grade and low grade acid manufacturing are assumed to grow at the same rates as soap and fertilizer production, respectively, and the 1980 to 1985 growth rates are extended to 1990, the projected growth will be that shown in Table 161.

TABLE 161: PROJECTED U.S. PHOSPHORIC ACID PRODUCTION

	1967	1971	1975	1977	1980	1985	1990
Low grade acid	8.0	9.8	11.4	12.6	13.9	16.9	20.6
High grade acid	2.1	2.1	2.3	2.5	2.7	3.3	4.0
Total	10.1	11.9	13.7	15.1	16.6	20.2	24.6

Source: Reference (10)

The projection of energy consumption assumes that all wet and electric furnace plants operating in 1967 consumed energy as follows. Wet process: steam, 4,750 lb at 1,340 Btu per lb = 6.4×10^6 Btu and electricity, 120 kwh at 10,400 Btu per kwh = 1.2×10^6 Btu for a total of 7.6×10^6 Btu. Electric furnace process: electricity, 5,500 kwh at 10,400 Btu per kwh = 57.2×10^6 Btu.

New wet process plants coming on line between 1967 and 1975 are assumed to consume, on the average, 1,300 pounds of steam and 105 kwh of electricity per ton of P_2O_5. After 1977 all new wet process plants are assumed to have consumption characteristics similar to the Kellogg-Lopker process with steam ejectors. New plants utilizing solvent extraction upgrading are assumed to consume 30% more steam and electricity. No information has been obtained on energy conservation advances in electric furnace processing and no process specific consumption reduction is assumed.

Using the fuel and power consumption factors cited above, a projection of total energy consumption was made on the basis of the demand shown in Table 161. Two key assumptions were made concerning the mix of new and old capacity and wet and furnace processing. Firstly, it has been assumed that capacity on line in 1974 would be retired at the rate of 1.5% per year through 1990. Secondly, it has been assumed that by 1980 solvent extraction upgrading of wet process acid would be competitive in the detergent market and would capture all new capacity in the high grade acid market between 1980 and 1990. Based on these assumptions the projections in Table 162 were developed.

TABLE 162: PROJECTED ENERGY CONSUMPTION IN PHOSPHORIC ACID PRODUCTION

	1967	1971	1975	1977	1980	1985	1990
	. . . . .Fuels (10^{12} Btu)						
Pre-1974 wet process	25.60	27.2	26.6	26.3	25.1	23.1	20.9
New wet process	–	–	0.8	1.7	3.2	6.5	10.5
Electric furnace	–	–	–	–	–	–	–
New wet process with extraction	–	–	–	–	–	–	–
Total	25.6	27.2	27.4	28.0	28.3	29.6	31.4
	. . . . .Electric (10^9 kwh)						
Pre-1974 wet process	0.48	0.57	0.61	0.59	0.56	0.51	0.46
New wet process	–	–	0.03	0.12	0.16	0.24	0.33
Electric furnace	5.78	5.91	6.33	6.88	7.43	7.00	6.57
New wet process with extraction	–	–	–	–	–	0.04	0.09
Total	6.26	6.48	6.97	7.59	8.15	7.79	7.45

Source: Reference (10)

FERTILIZER MIXING INDUSTRY

Industry products included in this segment (SIC 2875) are classified as follows:

SIC No.	Product
28752	Complete mixtures (grades guaranteeing N, P_2O_5 and K_2O). Shipped in dry and liquid forms.
	Incomplete mixtures, including dry and liquid forms
	Grades guaranteeing N and P_2O_5 only, including ammoniated superphosphates.
	Grades guaranteeing P_2O_5 and K_2O only.
	Grades guaranteeing N and K_2O only.
	Grades guaranteeing N, P_2O_5 or K_2O only.

The key distinction between SIC 2875 and product classification SIC 28743 (mixed fertilizers) is that SIC 2875 covers "establishments primarily engaged in mixing fertilizers from purchased materials," that is, made in plants which are not manufacturing these fertilizer materials.

The SIC 2875 industry is by its nature a collection of small plants regionally located to be conveniently close to their agricultural customers, and to cater to their specific requirements by custom-blending either the dry or liquid fertilizer mixes needed for specific soil and crop requirements.

In 1972 there were over 8,000 plants, and this number has been growing at about 3% per year. This trend is expected to continue both because of transportation economies and the close regional association required between farmer and fertilizer mixers to attempt to optimize crop yields and fertilizer costs. Typical product formulations are shown in Table 163.

415

TABLE 163: MAJOR USES—FERTILIZERS, MIXING ONLY

Material	Pound/ton of product.							
	12-12-12		14-14-14		5-20-20		16-20-0	
anhydrous ammonia	440	-	-	-	-	55	125	-
ammoniating solution	-	649	445	446	568	146	-	187
ammonium sulfate	340	-	136	293	191	-	-	400
diammonium phosphate	-	-	166	-	149	-	902	286
normal superphosphate	537	729	775	761	470	344	-	530
triple superphosphate	291	207	-	-	-	717	-	204
phosphoric acid (75% H_3PO_4)	-	-	-	168	195	-	412	85
potassium chloride	388	400	408	400	466	646	-	-
sulfuric acid (100% basis)	116	138	139	62	120	140	667	118

The products in this industry segment provide specifically measured amounts of plant nutrient foods in specific ratios which are applicable to specific plants or crops at specific growing times. The blends or mixes have been compounded to answer definitive agronomic needs of farmers; the plant operations may be categorized as "custom made."

Source: Reference (5)

PROCESS TECHNOLOGY INVOLVED

Mixed fertilizers are essentially prepared in one of three ways, namely, granulation, liquid mix or bulk blending. In each process, there are various steps of manufacture and the manufacturing procedure will vary somewhat with the equipment used and the plant nutrient food ratios required.

Granulation

Granulation practice is based upon (1) use of a relatively large volume of liquid phase, (2) use of soluble salts in conjunction with high temperature in order to develop the proper volume of liquid phase with a minimum amount of water and (3) careful adjustment of the proportion between the liquid and solid phases. No water is added; instead, raw-material solutions are left more dilute than they otherwise would be. The process can be divided into four major phases.

Raw Materials Feeding: Solid materials may be premixed but often are fed separately; recycled fine material, an important feed material, normally is fed separately. Liquid materials also may be fed either separately or premixed. For processes involving reaction between liquid materials, a prereactor often is used.

Granulation Proper: Liquid and solid materials are mixed in a granulation vessel, which may be one of several types. The granulation occurs at elevated temperature, either because chemical reaction takes place during granulation or because the incoming liquid has been heated in a previous reaction step.

Screening and Drying: Material from the granulator is passed over screens to remove particles too large or too small. The fines are recycled to the granulator and, in some processes, serve the important function of adjusting the ratio between liquid and solid phases. Oversize particles are crushed and either rescreened or recycled. Onsize product is dried, the degree of drying depending upon the caking tendency of the product; sometimes little or no drying is required because the heat of reaction removes most of the moisture. For some formulations, particularly those high in soluble-salt content, material from the granulator is dried before screening.

Cooling: Storage properties are improved if the product is cooled before storage.

Although this sequence of steps is generally followed, there is wide variation in the way each is carried out in regard to both type of equipment and manipulation of process variables. No system for classifying these variations has proved entirely satisfactory. Perhaps a summary of the various classification methods that have been attempted is the best way to introduce the technology.

Superphosphate-Based Fertilizer vs Ammonium Phosphate- or Nitric Phosphate-Based Fertilizer: Granulation of fertilizer based upon superphosphate differs from granulation of ammonium phosphate or nitric phosphate in that much more solid material is fed to the granulator in superphosphate granulation. Also, the main reaction is ammoniation of monocalcium phosphate rather than of phosphoric acid. However, the classification is not a good one because the two types approach each other when phosphoric acid is used along with superphosphate, as it often is, and when sulfuric acid is used in both superphosphate and ammonium phosphate granulation.

Slurry vs Nonslurry Operation: In some processes, particularly for ammonium phosphate and nitric phosphate, all the liquids are reacted in a tank, or tanks, and the resulting slurry is fed to the granulator along with recycled fine solids (plus potash if required). The term slurry rather than solution is used because the water content is kept as low as possible, seldom sufficient to keep all soluble salts in solution. In contrast to the slurry technique, a so-called solids process is used in most plants that use superphosphate in the formulation. The simplest version is granulation of a superphosphate-potash mix with ammoniating solution. No slurry is involved and solids make up the major part of the formulation; in contrast, there may be no solids at all in a slurry process formulation. Again, the classification is not clear-cut because in many plants production of super-phosphate-based granular fertilizer involves the use of a preneutralized slurry.

Layering vs Agglomeration: Two fairly clear-cut processes have emerged in gran-ulation of ammonium phosphate and nitric phosphate. In the older one, some-times called "layering" a slurry of reacted liquids is applied to the surface of a relatively large amount of recycled solids. The objective is to coat-out the slurry on the granule surfaces rather than to stick particles together. The resulting building up of the granules, layer by layer, gives a dense structure, facilitates dry-ing and gives an attractive spherical product. The main disadvantage is the large amount of recycle required; since the amount of liquid that can be applied at each pass without incurring agglomeration is limited, many passes must be made.

In contrast, the process termed "agglomeration" involves restriction of the recy-cling of solids as much as possible. The proper amount of recycle is used to ad-just the solids/liquid ratio; if the recycle rate is too high, there is not enough liquid to stick the particles together as agglomerates of adequate size; if the re-cycle rate is too low, the excess liquid causes overagglomeration and formulation of agglomerates that are too large.

The difficulty with this classification is that, in practical operation, layering and agglomeration often occur together, probably in the majority of granulation plants; large particles may be coated in the layering process, and fine ones may stick together or stick to the surface of larger ones. And even with proper ad-justment of the liquid to solid ratio in an agglomeration process, larger particles may take on a coating rather than join with other particles to form agglomerates. Moreover, the plant operator may adjust his process to give mixed operation, especially to reduce recycle in the layering process.

Ammoniation in Granulator: Possibly the most clear-cut distinction is whether or not any ammonia is introduced into the granulator. This is done in practi-cally all plants that use superphosphate, but practice differs in ammonium phos-phate and nitric phosphate production. The process called "ammoniation-granu-lation" has several advantages in manufacture of these products, as will be dis-cussed later. Although it has been associated mainly with the agglomeration technique, a layering method can and has been used in conjunction with ammo-niation-granulation, and both slurry and nonslurry operations are applicable. In most such plants, part of the ammonia is introduced in a preneutralization step and the remainder into the granulator.

The major problem in the manufacturing of granular fertilizer based upon super-phosphate is formulation. For any given nutrient ratio, numerous combinations of raw materials are feasible; each of them is a compromise between advantages and disadvantages. The general objective is to get low raw-material cost, high plant throughput, low process loss and low pollution of air and streams. Product objectives, some of which depend upon formation, are high nutrient concentration, smoothness and sphericity of granules, uniformity in particle size and good storage properties.

Liquid Mix

The basic operation in making liquid mixed fertilizers is the preparation of an ammonium phosphate solution; calcium phosphates cannot be used to supply phosphate because of their low solubility. Degree of ammoniation is important in making the ammonium phosphate solution because of its effect on solubility; a mixture of mono and diammonium phosphate is much more soluble than either compound alone. The degree of ammoniation generally used gives a solution with an $N:P_2O_5$ weight ratio of 1:3 ($NH_3:H_3PO_4$ mol ratio of 1.68). This is not the optimum ratio for solubility (mol ratio of 1.53) but the 1:3 weight ratio is convenient when the ammonium phosphate solution is sold as such without mixing it with supplemental materials. The resulting standard solution has a grade of 8-24-0.

Most liquid mixed fertilizers contain other compounds in addition to ammonium phosphate to supply either potash or additional nitrogen. The resulting solubility system becomes quite complex, containing as many as six components. Hence, although solubility relationships are important in solid fertilizer production, the advent of liquid mixes made solubility the major consideration in production. If the salt concentration is too high, crystallization (commonly called salting out) may occur during storage or transport and the resulting crystals may clog transfer or application equipment. Most producers use enough water in the formulation to prevent crystallization until the temperature falls below 32°F, but under special conditions of climate or time of year the product may be formulated for lower or high salting out temperature. The concentration is kept as high as possible to reduce shipping and handling cost.

Production of liquid mixed fertilizer is a relatively simple operation, amounting to little more than mixing liquids together and dissolving potash. About the only additional operation is cooling of the solution during neutralization. Plants are divided generally into two categories termed hot-mix and cold-mix. In a typical hot-mix plant, part or all of the water is introduced into the reactor tank and the other liquids, including ammonia and acid, then are fed in continuously through meters. Potash is batch weighed and fed in as convenient, or is sometimes weighed in the reactor by mounting the reactor on a scale. Solution normally is circulated through a cooler to remove heat of reaction.

Cold-mix plants are extremely simple, consisting usually only of storage vessels and a mix-weigh tank. Phosphate is supplied as neutral base solution (8-24-0, 10-34-0, or 11-37-0). The low investment cost, lowest in the industry for a multinutrient fertilizer plant, has made this method quite popular.

Low concentration, in comparison with solid fertilizers, continues to be one of the major drawbacks to liquid mixed fertilizer. Introduction of polyphosphate has helped in this respect, but has not been very beneficial for the popular high-potash mixes. Suspension fertilizers (generally called slurries) solve this problem by carrying part of the potash, as well as other salts in some formulations, in suspension as finely divided crystals. A suspending agent (usually attapulgite clay) is used to thicken the suspension, reduce settling and inhibit crystal growth; normally only 1 to 3% of suspending agent is required. The suspension method allows the production of grades as high as 15-15-15, 10-30-10- and 20-20-0, as compared with 7-7-7, 6-18-6 and 13-13-0 for typical clear liquid formulations. When made properly, the suspensions settle very slowly and are easily redispersed by stirring or recirculation. The crystals do not give trouble in application if spray nozzles of adequate size are used.

In addition to higher analysis, suspensions can reduce formulation cost if superphosphate or nitric phosphate is used to supply phosphate. Suspensions generally have not grown as fast as clear liquids, however, mainly because they present more problems in production and application.

Liquid mixed fertilizers have outstanding advantages over solids by virtue of lower plant investment, better homogeneity and less labor in handling and application. They formerly had the advantage also of better service to the farmer, by eliminating bagging and providing a local custom-mixing service; bulk blends now offer the same advantage. Formulation cost for liquids generally is lower for nitrogen and higher for phosphate, but in the majority of situations overall raw material cost for bulk blends is lower because of the relatively low cost of the diammonium phosphate used.

Bulk Blending

Bulk blending is a dry-mixing operation and, therefore, appears little different from the old dry-mixing practice. The main distinctions that set it apart are the following:

(a) Dry-mix plants were located mainly in the older fertilizer-using areas of the country, whereas bulk blending is concentrated in newer areas, such as the Central States.

(b) Dry mixing lost ground because the trend to higher analysis made it difficult to make mixes with good physical condition. Availability of granular materials changed this, but granular fertilizers were more popular in the newer fertilizer areas.

Dry-mix producers bagged most of their product and sold through dealers, whereas bulk blenders serve both as producers and dealers, providing a local service not generally provided in the older dry-mix practice.

From this it is evident that bulk blends are partly a departure and partly the moving of an old practice to a new location. When blending first attained prominence in the Midwest, the main competition was from homogeneous granular fertilizers based upon ammoniated superphosphate, sold mainly in bags to dealers who generally sold the product along with other merchandise with limited advice on how to use it.

The bulk blender was able to capitalize on several drawbacks to this system, as (1) bulk blending shortens the marketing channel by combining the mixer and dealer functions; (2) handling and distribution cost less for bulk material than for bagged product; (3) bulk blending reduces handling costs by eliminating the transfer from producer to dealer; (4) shipping distance of materials such as potash is shortened because the material goes directly from primary producer to the mixer-dealer rather than detouring to a granulation plant; (5) a custom application service can be offered; and (6) the bulk blender, through his close contact with the farmer, can work with agricultural advisors in guiding the farmer's use of fertilizer. Assistance with soil testing and sampling is an important part of such a service.

The main drawback to bulk blending is the higher cost of raw materials that results from the need for granular solid materials in blends, whereas the producer of granular mixed fertilizer can use less expensive solutions and nongranular solids. Blending transfers the profit in granulation from the mixed fertilizer producer to the primary producer who makes the granular materials that the bulk blender uses. Segregation of nutrients during handling and application is another disadvantage, but the rapid growth of blending indicates that the drawbacks are outweighed by the advantages.

The materials commonly used in bulk blending are ammonium nitrate, ammonium sulfate, triple superphosphate, diammonium phosphate, and potassium chloride. Other materials sometimes used are urea, ammonium phosphate nitrate (30-10-0), ammonium phosphate (11-48-0), ammonium phosphate sulfate (16-20-0), and normal superphosphate.

The materials should be closely sized, dry enough to prevent caking in storage, and sufficiently strong to prevent fragmentation in handling. Early granular materials fell considerably short of this criterion but primary manufacturers have improved their products considerably. Particle size is becoming well standardized at 6-16 mesh with only small percentages above or below these limits, but it would be desirable to have even closer sizing. Size distribution within this range varies widely; this is also undesirable. Some prilled materials contain as much as 90% particles smaller than 10 mesh, while granular products sometimes contain 85% or more particles coarser than 10 mesh. The shape of particles varies from spherical to irregular and specific gravity from 1.27 to 2.15.

Type of mixers and layout of storage, conveying and mixing facilities vary widely, so much so that probably no two plants are alike. Since the plants are small and quite often built on a very limited budget, they tend to be homemade.

Mixing is usually of the batch type, with materials introduced one at a time from a weigh hopper. Usual capacity is 1 to 2 tons per batch and mixing time is 2 to 3 minutes. As much as 15 tons per hour can be mixed even in a one-ton mixer. In some cases the entire cycle of weighing, mixing and discharging is done automatically.

Mixers are mainly of the rotating drum type, but various other types including ribbon mixers, mixing screws, gravity mixing towers and a volumetric metering device are used. The volumetric metering device is a continuous type in which materials feed by gravity through adjustable gates onto a common belt. The materials mix as they flow into the receiving hopper and in the following cut-screw conveyor.

MAJOR ENERGY CONSERVATION OPTIONS TO 1980

There are no significant energy saving improvements expected in the relatively simple blending and mixing processes in use by January 1, 1980. Furthermore, there is no trend in the custom blending nature and wide variety of products made, or change in seasonal cycles which suggests a change in specific energy consumption.

Energy savings must, therefore, come from more general techniques. This group of small business operators appears to be highly conscious of the energy shortage, and high energy costs. They have apparently made significant progress to date in reducing energy use through housekeeping and maintenance; better in-plant operation practice including complete shutdown of idle equipment, batch scheduling and less off-specification production; improved seasonal scheduling and production for inventory. The overall effect of the implementation of these three options will result in a net energy consumption reduction of 25%.

Only a very small increase in energy use is anticipated from tighter controls on plant air emissions, waste control and liquid effluents. It is apparent from energy consumption trends in SIC 2875 since 1972 that the industry is adopting conservation measures because the equivalent Btu per pound of product is decreasing noticeably, as reported by the U.S. Department of Commerce. It is considered, however, that the extent of favorable trends will decrease somewhat, even to the point of offsetting some of the earlier gains, in reflection of newer additions of more stringent pollution control measures.

Important to the realization of savings will be actual operating rates in 1980. The assumptions for this examination are that they will parallel those for 1972. It must be emphasized, however, that plant idling periods or part-load operation or startup/shutdown activity are energy wasteful. These considerations are of especial significance to the mixed fertilizer industry, in the light of its very high seasonality aspect. No data have been obtained, however, from which a quantitative estimate of change in energy use with change in operating pattern can be made.

GOAL YEAR (1980) ENERGY USE TARGET

Production during 1972 was 42.6 billion pounds, of which less than 5% was made outside the SIC 2875 establishments. From analysis of data from the Fertilizer Institute, TVA and Census of Manufactures from 1971 through 1975, the specific energy consumption for 1972 has been estimated as 156.8 Btu per pound and the total energy used as 6.68 trillion (10^{12}) Btu.

Considering the foregoing factors, the reduction in specific energy consumption between 1972 and January 1, 1980 is estimated at 25%. This is shown in Table 164.

Since growth in production of mixed fertilizers is forecast the same as for the rest of the fertilizer industry at 4.5% per year through this period, total energy consumed in the year 1979.5 is estimated to increase approximately 15% over 1972 despite the improvement in specific energy consumption. These calculated data are also shown in Table 164.

Energy consumption for SIC 2875 varies widely with the season and will continue to do so. An underlying assumption is that plant operation patterns in relation to total capacity will be similar in 1979.5 to 1972. However, no data have been accumulated either on ultimate plant capacity nor on percent operation from which an estimate of change on energy consumption can be made.

TABLE 164: PRODUCTION AND ENERGY ESTIMATES — SIC 2875

	All Products SIC 2875	
	1972	1979.5
Production, 10^9 pounds		
Total, all SICs	42.60	59.26
Made in 2875	41.24	57.37
Total energy required, 10^{12} Btu		
1972 basis, total, all SICs	6.68	9.29
1979.5 basis, total, all SICs	na	6.97
Specific energy consumption, Btu/lb		
1972 basis	156.8	156.8
1979.5 basis	na	117.6
Energy improvement goal, %	–	25

Note: Compound average growth rate of period 1972 through
January 1, 1980 projected at 4.5% per annum for 7.5 years.

Growth factor = $1.045^{7.5} = 1.391$

Source: Reference (5)

AGRICULTURAL CHEMICALS INDUSTRY

Included in SIC 2879 are the following 5-digit classifications and respective product groups:

SIC	Products
28791	Insecticides
28792	Herbicides
28793	Agricultural Chemicals, nec
28794	Household Preparations

SIC 2879 products are the nonfertilizer side of the agricultural chemicals industry. The overall products spectrum, consisting mainly of herbicides, insecticides, fungicides, and rodenticides, could alternatively be described as "Pesticides and Related Products," which is the designation in the publication "*Synthetic Organic Chemicals,*" of the U.S. International Trade Commission.

PROCESS TECHNOLOGY INVOLVED

The chemical processes which may be used in the production of some of the major pesticides are outlined below. The various products are covered under five main groups: Insecticides, Fumigants, Fungicides, Herbicides and Rodenticides, each group being further subdivided on the basis of chemical structures of the compounds.

Insecticides

Based upon their chemical nature the insecticides can be divided into the following classes.

Chlorinated Hydrocarbons: These insecticides are highly chlorinated compounds. The percentage of chlorine in the molecules of chlorinated cyclic hydrocarbons varies between 50 to 73. Some of the most important insecticides belong to this class of compound.

A listing of these chemicals follow. Asterisked compounds indicate those now banned by the EPA.

DDT*
BHC
Lindane
Chlordane*
Heptachlor*
Aldrin*
Endrin
Toxaphene
Thiodane

Dichloro Diphenyl Trichloroethane (DDT) — DDT is produced when chloral or chloral hydrate is condensed with chlorobenzene in the presence of 20% oleum at 20°C. Chlorobenzene is prepared by passing dry chlorine through benzene in an iron vessel at 40°C using ferric chloride as a catalyst. Chloral is produced by chlorinating ethanol to form the hemi-acetal of trichloroacetaldehyde from which chloral is liberated by treatment with concentrated sulfuric acid.

Hexachlorocyclohexane (BHC) — When benzene is chlorinated in the liquid phase in the presence of light, a mixture of five isomers of 1,2,3,4,5,6-hexachloro-cyclohexane is produced; only the gamma isomer is active. BHC technical contains 10 to 13% of the gamma isomer.

Gamma-Hexachlorocyclohexane (Lindane) — Lindane is a refined product containing at least 99% of the gamma isomer. It is prepared by treating the crude product with methyl alcohol or acetic acid in which alpha and beta isomers are nearly soluble and then fractionally crystallizing the alcohol soluble fraction from chloroform.

Octachloro Dihydrodicyclopentadiene (Chlordane) — Chlordane, heptachlor, aldrin, endrin, toxaphene and thiodane are also known as cyclodiene derivatives as they are prepared from the basic material hexachlorocyclopentadiene. Chlordane is formed by the reaction of cyclopentadiene and hexachlorocyclopentadiene (from chlorination of pentane or from chlorination of cyclopentadiene in alkaline hypochlorite solution) at 75° to 85°C. The hexachlorocyclopentadiene is subsequently chlorinated in carbon tetrachloride at 50°C.

Heptachlorotetrahydromethanoindene (Heptachlor) — Chlorination of the Diels Alder adduct of hexachlorocyclopentadiene and cyclopentadiene in benzene in the presence of fuller's earth, in the dark yields heptachlor. The chlorination temperature may be varied between 40° and 110°C.

Hexachlorohexahydrodimethanonaphthalene (Aldrin) — Hexachlorocyclopentadiene·and bicyclo[2.2.1] 2,5 heptadiene are refluxed at 85° to 90°C for 16 to 18 hours. Bicyclo[2.2.1] 2,5 heptadiene is produced by the Diels Alder reaction between cyclopentadiene and acetylene. Dieldrin is the epoxide of aldrin, formed by the action of peracetic acid.

Hexachloroepoxyhexahydrodimethanonaphthalene (Endrin) — Endrin is prepared by reacting hexachlorocyclopentadiene with vinyl chloride to form 1,2,3,4,5,7,7-heptachlorobicyclo[2.2.1] heptane which is slowly reacted with cyclopentadiene and heated to give Isoderin. Isoderin is converted to endrin by the action of peracetic acid.

Octachlorocamphene (Toxaphene) — Toxaphene is made from camphor by re-action with chlorine which first adds rapidly to the double bond and then un-dergoes substitution with more difficulty.

The reaction is carried out at 85° to 90°C and being exothermic, some cooling is required. It is carried out in the liquid phase using 5 parts of carbon tetra-chloride per part of camphene feed, and takes 15 to 30 hours to reach comple-tion. The reaction is catalyzed by ultraviolet light and is performed at atmos-pheric pressure. Carbon tetrachloride is distilled off after excess chlorine and hydrogen chloride have been blown out, and the residue is allowed to solidify.

Hexachlorohexahydromethanobenzodioxathiepin Oxide (Thiodane) — The manufacture of thiodane which is a relatively new member of this group involves the preparation of HET diol made by the condensation of cis-2-butene-1,4-diol with hexachlorocyclopentadiene. The HET diol is then treated with thionyl-chloride whereby hydrogen chloride is evolved and the cyclic sulfite ring is formed.

Organophosphates: Organophosphate is the general term for insecticides contain-ing a high percentage of phosphorus (21 to 39%). These are poisonous, highly corrosive and often explosive compounds. For the manufacture of organophos-phates, glass-lined vessels are usually employed. Some examples of these in-secticides are as follows:

> Diazinon
> Malathion
> Parathion
> Rogor or Dimethoate
> DDVP

O',O-diethyl phosphorochlorodithioate is the key intermediate for this group of compounds and is generally made by the interaction of ethanol with thiophos-phoryl chloride.

Diazinon — Diazinon is prepared by reacting the above intermediate with the condensation product of ethyl acetoacetate and isobutyramide.

Dicarbethoxyethyl Dimethyl Dithiophosphate (Malathion) — The feed materials for the manufacture of malathion consist of O,O-dimethyl phosphorodithioic acid and diethyl maleate or fumarate.

An antipolymerization material such as hydroquinone may be added to the reaction mixture to inhibit the polymerization of the maleate or fumarate com-pound under the reaction conditions: O,O-dimethyl phosphorodithioic acid is obtained by the interaction of phosphorus pentasulfide with methyl alcohol, and maleic and fumaric acids are obtained by the oxidation of butenes.

The reaction is carried out at a temperature between 20° and 150°C at atmos-pheric pressure using an aliphatic tertiary amine catalyst. Suitable solvents in-clude low molecular weight aliphatic monohydric alcohols, ketones, etc., and a reaction time of 16 to 24 hours is necessary. When the reaction is complete the mixture is taken up in benzene. It is then washed with 10% sodium carbonate solution and water and the organic layer is dried over anhydrous sodium sulfate, filtered and concentrated under vacuum to obtain the final product.

Diethyl Nitrophenyl Thiophosphate (Parathion) — Phosphorus trichloride is combined with sulfur by heating at 130°C. The thiophosphoryl chloride formed is combined with sodium ethoxide. Sodium p-nitrophenate in chlorobenzene is then combined with diethoxy thiophosphoryl chloride by heating at 130°C.

Dimethyl Dithiophosphoryl N-Methyl Acetamide (Dimethoate or Rogor) — Dimethyl dithiophosphoryl acetic acid is added to aqueous methylamine at 0°C. The mixture is stirred for 2 hours after which it is neutralized with 10% sulfuric acid, centrifuged and washed. Dimethyl dithiophosphoryl acetic acid is manufactured from methyl alcohol, phosphorus pentasulfide and acetic acid.

Dimethyl Dichloro Vinyl Phosphate (DDVP) — Also called Vapona, DDVP is manufactured from chloral and trimethyl phosphite which are usually employed in equimolar quantities although mol ratios ranging from 1:10 to 1:1 may be used. Chloral is obtained by the chlorination of acetaldehyde which in turn is made from ethylene. The other reactant, trimethyl phosphite, is produced by reacting phosphorus trichloride with methanol or by the methanolysis of triphenyl phosphite.

The reaction between chloral hydrate and trimethyl phosphite to produce Vapona is exothermic. Temperatures of 10° to 150°C at atmospheric pressure may be used with suitable inert solvents such as benzene, toluene, ether, dioxane or hexane. No catalyst is required and the reaction takes 10 minutes to 2 hours to reach completion. The reaction product can be separated by distillation or by extraction with selective solvents.

Carbamates — The important compound in this group is Sevin, which is widely used because of its wide spectrum, good residual activity and low levels of mammalian toxicity. It is manufactured by the interaction of sodium 1-naphthoxide with a toluene solution of phosgene and reacting the resultant compound with aqueous methylamine.

Fumigants

One of the most commonly used fumigants is EDCT, a mixture containing 3 parts of ethylene dichloride and 1 part of carbon tetrachloride to remove the fire hazard. Two other compounds are used as fumigants. These are:

> Aluminum phosphide
> Calcium cyanide

Ethylene Dichloride: Ethylene dichloride is commercially manufactured by the direct addition of chlorine to ethylene. The ethylene, 98% pure and free of acetylene, should be present in excess in order to minimize substitution reactions. The ratios of ethylene to chlorine may vary from 1:1 to 1.3:1.

The liquid phase addition of chlorine to ethylene may be carried out at temperatures from 80° to 120°C, usually at the boiling point of ethylene dichloride (83.7°C) which is then volatilized from the reaction mixture as it is formed.

The addition reaction which is highly exothermic is usually carried out at pressures ranging from 1 to 10 psig in the presence of a catalyst like ferric chloride. The reaction time in the liquid phase process is a matter of minutes.

Aluminum Phosphide: Aluminum phosphide is made by passing phosphorus vapor in a stream of hydrogen over hot aluminum.

Calcium Cyanide: Calcium cyanamide is reacted with carbon in the presence of common salt at a temperature above 1000°C attained through the use of an electric furnace. Calcium cyanamide, which is the basic raw material for the production of calcium cyanide is produced by passing nitrogen through finely ground calcium carbide, which is heated to 1000°C. Calcium cyanide acts as a fumigant by liberating hydrogen cyanide by the action of moisture on calcium cyanide.

Fungicides

The most important groups of fungicides are those listed and described below.

Dithiocarbamates: The earlier compounds comprised iron and zinc salts of dimethyldithiocarbamic acid, but the use of these was marred by their rather high mammalian toxicity. Recently these have been replaced by compounds of the ethylene bisdithiocarbamate group. The earliest compound Nabam was prepared from ethylenediamine, carbon disulfide and caustic soda. This is now replaced by the more stable salts:

> Zineb
> Maneb
> Thiram

Zinc Ethylene Bisdithiocarbamate (Zineb) — The ammonium salt of ethylene bisdithiocarbamate is reacted with zinc oxide. The particle size of zinc oxide used should be in the range of 1 to 20 microns. The reaction which is carried out in an aqueous medium is aided by the use of a dispersing agent such as lignin sulfonate to disperse the metal oxide and to prevent the formation of a deposit when a stable crystallization product is formed.

Manganous Ethylene Bisdithiocarbamate (Maneb) — Maneb is made by the interaction of ethylenediamine and carbon disulfide in the presence of a free base; the salt formed is reacted with a manganous salt at room temperature.

Tetraalkyl Thiuram Disulfide (Thiram) — An aqueous solution of sodium dialkyl dithiocarbamate is treated with air or nitrogen containing about 1% chlorine. The reaction temperature is maintained at 30° to 50°C; thiuram disulfide is filtered off and dried below 60°C.

The synthesis of alkylamines from alcohol is carried out in the vapor phase. Methyl alcohol and ammonia are passed over a dehydrating catalyst (alumina) at 380° to 450°C under a pressure of 50 atmospheres. The resultant mixture of primary, secondary and tertiary amines is separated by azeotropic distillation.

Organomercurials: The second classification of compounds used as fungicides are mercury based compounds. These compounds include the following:

Ceresan
Ethyl mercury bromide

Methoxy Ethyl Mercury Silicate (Ceresan) — In the first step of the process for making Ceresan, aniline is condensed with p-toluene sulfonyl chloride giving p-toluene sulfonanilide as an intermediate product. This intermediate is reacted with ethyl mercury acetate at a temperature of $-20°$ to $-30°C$ when the micro-crystalline organometallic product precipitates out.

Ethyl Mercury Bromide — Ethyl alcohol, potassium bromide and sulfuric acid are refluxed in an atmosphere of nitrogen; magnesium and mercury chloride are added and the product is filtered off.

Captan: Captan is the other important fungicide. It is prepared by reacting butadiene and maleic anhydride to form tetrahydrophthalic anhydride and ammoniating the resultant product to form tetrahydrophthalimide. The imide is dissolved in caustic soda and reacted with perchloromethyl mercaptan at $6°$ to $14°C$ to yield captan. Perchloromethyl mercaptan is itself prepared from carbon disulfide and chlorine in the presence of catalytic amounts of iodine.

Herbicides

Commercially important herbicides include those described below.

Phenoxyacetic Acids: Listed among the phenoxyacetic acids are the following:

2,4-D
2,3,5-T
MCPA

2,4-Dichlorophenoxyacetic Acid (2,4-D) — The manufacture of 2,4-D involves the preparation of monochloroacetic acid and 2,4-dichlorophenol. Monochloroacetic acid is produced by passing chlorine through glacial acetic acid at $100°C$ in the presence of phosphorus trichloride catalyst. The acid is liberated by the addition of hydrogen chloride. 2,4-Dichlorophenol is obtained by chlorination of phenol below $90°C$.

2,4,5-Trichlorophenoxyacetic Acid (2,4,5-T) — Since phenol cannot be chlorinated in the 2,4,5-positions the first step in preparing 2,4,5-T involves hydrolysis of 1,2,4,5-tetrachlorobenzene with sodium hydroxide. The hydrolyzed product is then reacted with monochloroacetic acid to yield 2,4,5-T.

4-Chloro-2-Methyl Phenoxyacetic Acid (MCPA) — Chlorine and methyl phenoxyacetic acid are reacted together to form MCPA.

Bipyridyls: Bipyridyls are compounds which are pyridine derivatives.

1,1'-Ethylene-2,2'-Dipyridinium Dibromide (Diquat) — Diquat is prepared by heating an excess of ethylene dibromide with 2,2'-dipyridyl at elevated pressure and at a temperature above the boiling point of the diester. The heating is carried out in a sealed tube at $170°C$ for 8 hours. The product is washed with acetone and dried at $100°C$.

Halogenated Aliphatic Acids: This group includes very potent grass killers like

    Trichloroacetic acid
    Dalapon

Trichloroacetic Acid (TCA) — TCA is made by the reaction of acetic acid and chlorine in the presence of a metal halide catalyst. Acetic acid is warmed to 30°C before introduction of chlorine; the temperature is then raised to 190° to 220°C for half an hour.

Dalapon — Replacing one chlorine by a methyl group in TCA yields dalapon.

Substituted Phenols: The two most important members of this group are pentachlorophenol and 4,6-dinitro,2-sec-butylphenol (Dinoseb). These compounds are also known respectivly as

    PCP
    DNBP

Pentachlorophenol — This compound may be made by either of two routes. One involves the direct chlorination of phenol in the presence of aluminum chloride catalyst. The alternative route involves the hydrolysis of hexachlorobenzene. The product of hydrolysis is either partially or completely evaporated and the residue dissolved in water. The solution is filtered and the filtrate acidified to a pH greater than 3.0 to obtain high purity polychlorophenols.

Rodenticides

Some of the important rodenticides are:

    Zinc phosphide
    ANTU
    Warfarin

Zinc Phosphide — Zinc phosphide is prepared by mixing together zinc dust and red phosphorus in a furnace heated to 60°C.

2-Naphthylthiourea (ANTU) — 2-Naphthylamine, obtained by the reduction of nitrated naphthalene is dissolved in glacial acetic acid to which sodium or potassium thiocyanate is added at 104°C. The mixture is heated for 3 to 4 hours. The product is filtered off and washed with both water and alcohol.

3-(2-Acetonylbenzyl)-4-Hydroxy Coumarin (Warfarin) — This compound is obtained by condensing 4-hydroxy coumarin with benzal acetone in water in the presence of a catalytic amount of ammonia. The mixture is refluxed for 2 to 3 hours, cooled to room temperature and filtered.

MAJOR ENERGY CONSERVATION OPTIONS TO 1980

The major areas for energy savings are improved control of plant operations, improved housekeeping and maintenance, improved processing technology, and improved scheduling of plant operations. The main terms within improved control of plant operations include tighter control on excess air for boilers and addition or extension of heat recovery means on stack gases. The main items for house-keeping and maintenance include daytime turning off of lights, steam tracer con-trol, steam trap maintenance and heat exchanger maintenance. This group of im-provements can be accomplished with relatively modest expenditure and reason-able times for installation and implementation.

The opportunities for improved processing technology are of continuous interest to the respective operating companies, but generally mean either wholly new plants or substantial revamps for the upgrading of existing plants. The lead times are long.

The question of improved scheduling is one in which the operating companies are more the acceptors of the facts rather than the initiators. The major im-pact is from the market tone over any given period. High demand means high production; the plants run full-out, round the clock, seven days; energy-wasteful startup and shutdown and idling effects are eliminated. The alternative position of weak demand results in poorer operating and energy efficiency. The season-ality aspect of the pesticides industry, its products sold mostly to agriculture, aggravates the scheduling difficulties.

One negative factor, working against energy savings for the pesticides industry, is the energy required to meet EPA and OSHA regulations. These weigh more heavily on operations dealing with toxic products, as is true with pesticides, than on some other classes of industries. The necessity for incineration of residues, for example, has largely turned out to be an energy balance, apparently being insufficient for full-compliance incineration, and requiring the addition of supplemental purchased fuel.

No specific study of capital investment requirements for energy saving project was made. However, no indication that either capital funds or manpower will be a significant limiting factor to achieving the estimated energy savings goal is evident.

GOAL YEAR (1980) ENERGY USE TARGET

A summary of the production and energy estimates is shown in Table 165. It should be noted that the total production from all industries of the SIC 2879 group of products is approximately 44% greater than the output of the prod-ucts from SIC 2879 establishments only. The Specific Energy Consumption figures of the table are essentially independent of the establishment classification where manufactured. The Total Energy Required data, however, must be related to the correct total output quantity, e.g., whether the total output from all SIC groups or the output from SIC 2879 establishments only.

The compound average growth rate for the overall industry for the period 1972 through 1980 is estimated at 9.5% per year. This is made up of 5.0% per year projected for 1975 through 1980, and 17% per year for the period 1972 through 1974. The growth rate for the past several years has indeed been phenomenal, but it appears to be slowing down markedly. The major conquest of the agri-cultural market in terms of total acreage covered has been essentially accom-plished. Any new gains will represent a mixture of additional products, some additional intensity of application, and some additional acreage remaining to be covered. The data in Table 165, for the total output from all industries, show production growing from 1.12 billion pounds in 1972 to 2.21 billion pounds in 1980.

TABLE 165: CAPACITY AND PRODUCTION ESTIMATES

	Number of Plants		Total Capacity 10^6 lb		Total Capacity 10^6 lb	
	1972	1980	1972	1980	1972	1980
Insecticides, Total	-	-	520	-	420	-
Mfd in SIC 28791 Only	96	-	-	-	-	-
Herbicides, Total	-	-	560	-	450	-
Mfd in SIC 28792 Only	34	-	-	-	-	-
Ag. Chems. NEC, Total	-	-	180	-	148	-
Mfd in SIC 28793 Only	22	-	-	-	-	-
Household Preps, Total	-	-	130	-	100	-
Mfd in SIC 28794 Only	33	-	-	-	-	-
Total All SIC's	-	-	1,390	2,740	1,118	2,210
Mfd in SIC 2879 Only	185	298	964	-	775	-

Notes:

(1) 1972 figures are from reported values in aggregate, and are estimated values for breakdown. 1980 figures are projections.

(2) 1980 projections are made in aggregate only, the breakdown growth rates being non-uniform, herbicides for example having the highest.

Source: Reference (5)

The corresponding figures for output from SIC 2879 establishments only as shown in Table 166, are 0.775 billion pounds in 1972 and 1.53 billion pounds in 1980. The unit energy requirements are about 18,000 Btu/lb on 1972 estimated basis, projected to reduce to about 15,000 Btu/lb by January 1, 1980.

The energy savings to obtain the lower specific energy consumption will come primarily from improvements in plant systems including: housekeeping, maintenance, process control and energy balancing. The maximum savings which can be justified on a technical and economic basis for the period 1972 through January 1, 1980 are projected at 25%, but the goal is estimated at 15% after allowance for the increased energy to meet EPA and OSHA regulations.

Fuel substitutions will be largely from natural gas to distillate oil, primarily in boiler firing applications. The effect on energy usage of this fuel substitution will not be large; oil is marginally poorer on boiler LHV efficiency, but marginally better when evaluated on boiler HHV efficiency.

TABLE 166: PRODUCTION AND ENERGY ESTIMATES, SIC 2879

	All Products SIC 2879	
	1972	1979.5
Production, 10^6 Pounds		
Total, All SIC's	1,118	2,208
Made in 2879	775	1,531
Total Energy Required, 10^{12} BTU		
1972 Basis, All SIC's	20.1	39.7
1979.5 Basis, All SIC's	n.a.	33.8
Specific Energy Consumption, BTU/Pound		
1972 Basis	17,950	19,950
1979.5 Basis	n.a.	15,260
Estimated Energy Efficiency Goal, %		15

Note: Compound average growth rate for period CY 1972 through January 1, 1980 projected at 9.5%/Y for 7.5 years.

Growth Factor = $1.095^{2.5}$ = 1.975

Source: Reference (5)

Because of almost doubling production over this period, total energy required for the year 1979.5 will increase 68% over 1972 in spite of the 15% reduction in specific energy consumption.

These energy estimates assume that plant operating ratio, and seasonality factors will be essentially similar for 1972 and 1979.5. Operating ratio is important to the specific energy consumption in this industry; part load operations, and start-up on shutdowns waste large amounts of energy. It is estimated, for example, that a 10% reduction in plant output will only produce a 5% reduction in energy used.

ADHESIVES AND SEALANTS INDUSTRY

An adhesive or bonding agent is any substance that produces a bond between two or more similar or dissimilar substrates. The term adhesive has become generic and includes more popular terms such as cement, glue, mucilage and paste. The adhesives business is particularly difficult to define since many substances with adhesive properties are used in borderline applications where bonding is secondary to such primary functions as coating. The products covered in the adhesives and sealants industry (SIC 2891) are outlined in Table 167.

The industry includes establishments primarily engaged in manufacturing industrial and household adhesives, glues, calking compounds, sealants, and linoleum, tile and rubber cements from vegetable, animal, or synthetic plastics materials, purchased or produced in the same establishments. Establishments primarily engaged in manufacturing gelatin and sizes are classified in SIC 2899, and vegetable gelatin or agar-agar in SIC 2833.

TABLE 167: ADHESIVES AND SEALANTS

Adhesives	Iron cement, household
Adhesives, plastic	Laminating compound
Calking compounds	Mucilage
Cement (cellulose nitrate base)	Paste, adhesive
Cement, linoleum	Porcelain cement, household
Cement, mending	Rubber cement
Cement, rubber	Sealing compounds for pipe
Epoxy adhesives	threads and joints
Glue, except dental: animal,	Sealing compounds, synthe-
vegetable, fish, casein and	tic rubber and plastic
synthetic resin	Wax, sealing

Source: Reference (5)

PROCESS TECHNOLOGY INVOLVED

The adhesives industry is different from the traditional chemical company in that very little processing is involved. Mainly mixing and blending operations are required in order to manufacture desired product mix.

The manufacturing technology for adhesives is not expected to change over the 1972 to 1979.5 period. The most significant factor affecting industry processing will be a changing product mix to meet consumer demands and particularly new end-use applications.

Over the next five years, there will be further market penetration by hot melt adhesives, hot melt pressure-sensitive adhesives, and hot melt sealants. In addition, the coatings industry has been hard at work developing new polymer technology to be applied as nonpolluting powder coatings, high-solids coatings, and radiation coatings.

MAJOR ENERGY CONSERVATION OPTIONS TO 1980

The changing product mix of the adhesives industry is not expected to significantly affect specific energy consumption over the 1972 to 1979.5 period, since adhesive manufacture is mainly a mixing and blending operation. The incentive for fuel economy in the adhesive industry has been spurred by rising fuel costs. In this regard, the adhesives industry is in a position similar to other industries in the United States.

Offsetting to some extent price-induced conservation effects have been increased energy requirements, resulting from stricter federal regulations concerning working conditions, waste disposal and other aspects of manufacturing. The best estimate of an achievable energy efficiency goal for the adhesives industry is a reduction of 8% in specific energy consumption by January 1, 1980 versus 1972. This goal is attainable by the industry through application of proper energy housekeeping.

GOAL YEAR (1980) ENERGY USE TARGET

The production of adhesives during 1972 was 7.6 billion pounds according to a number of private industry sources, of which 6.0 billion pounds was for forest products and 1.6 billion pounds for non-forest products. The growth rate for adhesives production between 1972 and 1979.5 is estimated to be between 5 to 6% annually. By 1979.5 total adhesives production is expected to be about 11.4 billion pounds. Forest and non-forest products will consume about 8.7 and 2.7 billion pounds respectively.

The specific energy consumption in adhesive manufacture is estimated to have been 1,300 Btu per pound in 1972. About 70% of this energy consumption is independent of production levels and is consumed in such uses as heating tanks, buildings, and warehouses. Regional location of adhesive plants is therefore an important parameter in evaluating specific energy consumption. Product mix is also important, since individual products vary in their energy requirements, and a change in product mix is forecast.

The remaining 30% of the energy required in adhesive manufacture varies directly with production. No new technological changes, which will significantly affect specific energy consumption in processing are expected in this industry by January 1, 1980. It is therefore expected that these variable energy requirements will remain at about 1972 levels through January 1, 1980.

The large proportion of fixed energy requirements requires that special attention be given to production rates, since per unit energy usage will decline as production expands. Of particular importance is the production rates in 1979.5 of those plants which were operating in 1972 compared with required production to meet demand at that time.

If new plants are required to be built to meet demand, industry-wide fixed requirements would be duplicated, thereby increasing specific energy consumption above what it would have otherwise been if demand would have been fully met by plants which were in operation in 1972. New adhesive manufacturing plants will be an important factor in forest adhesive products, for as new forest sites are brought into production adhesives manufacturing plants generally are also built. With regard to non-forest adhesives, it is estimated that about 20% of the 2.7 billion pounds of production in 1979.5 will come from plants built after 1972.

Since new manufacturing plants will be an important factor in both forest and non-forest adhesive products, the industry will be unable to avail itself of all of the potentially large specific energy savings which would have been possible if no new facilities were built and all production came from plants which were operating in 1972.

The adhesives industry, like others examined in the Battelle study (5), is subject to government regulations affecting effluent streams, working conditions and other elements of manufacturing. Based upon discussions with industry, it is estimated that in the neighborhood of 8% of the industry's energy consumption in 1979.5 will be related to satisfying various federal government regulations which affect manufacturing operations.

On balance, the adhesives industry is expected to be able to achieve a specific energy improvement goal of 8% by January 1, 1980 as compared to 1972. Production and energy estimates are summarized in Table 168.

TABLE 168: PRODUCTION AND ENERGY ESTIMATES, SIC 2891

| | All Products SIC 2891 | |
	1972	1979.5
Production, 10^9 pounds	7.6	11.4
Total energy required, 10^{12} Btu		
1972 basis	9.88	14.82
1979.5 basis	na	13.68
Specific energy consumption, Btu/lb		
1972 basis	1,300	1,300
1979.5 basis	na	1,200
Estimated energy improvement goal, %		8

Source: Reference (5)

EXPLOSIVES INDUSTRY

An explosive is a material that can undergo very rapid, self-propagating decomposition, or reaction of its ingredients, with the liberation of heat and large volumes of gas at high pressure. There are two principal types called low explosives and high explosives. Only high explosives are now used in industry; the low explosives, principally black powder which decompose at a lower speed, have not been used in industry since 1971.

Most explosives contain oxygen, usually in the form of nitro compounds, nitrates, or perchlorates. Energy is formed by the production of various oxides by combustion. However, some explosives, such as lead azide, generate decomposition energy by the rupture of weak linkages between nitrogen atoms. Initial detonating agents include such chemicals as lead azide, lead styphnate, mercury fulminate, and diazodinitrophenol. Non-initiating high or secondary explosives are typically ammonium nitrate or nitroglycerine.

The product mix of industrial explosives has undergone some dramatic changes in the past 20 years. Dynamite, once the main blasting explosive, began to decline in 1955 when the ANFO mixtures, consisting of 94.5% prilled ammonium nitrate and 5.5% fuel oil, were introduced. ANFO can be mixed on site in open-pit blasting and used as needed. It is also considerably less expensive than dynamite and with proper compounding it can be made with more explosive strength than trinitrotoluene.

In the 1960's, slurries consisting of an aqueous solution of ammonium nitrate, sensitized with aluminum, TNT, sulfur, or a hydrocarbon, were developed. Slurries are suitable for blasting all kinds of rock and can be used under wet conditions. Slurries now account for more than 10% of all blasting agents.

The following explosives, classified under SIC 2892, have been considered in determination of energy efficiency improvement targets in the course of the Battelle study for FEA (5).

Amatol (explosive)   Mercury azide
Azides (explosives)   Nitrocellulose powder
Blasting powder and   Nitroglycerin
  blasting caps   Nitrostarch
Carbohydrates   Nitrosugars
Cordeau detonant   Pentolite
Cordite   Permissible explosives
Detonating caps for   Picric acid
  safety fuses   Powder: pellet, smokeless
Detonators   and sporting
Dynamite   RDX
Fuse powder   Squibs, electric
Fuses, safety   Styphnic acid
Gun powder   Tetryl
High explosives   TNT
Lead azide   Well shooting torpedoes

PROCESS TECHNOLOGY INVOLVED

Explosives production consists mainly of blending operations with very minor chemical reactions. In some cases heating and cooling of explosives mixtures are also carried out. Figure 117 is a block diagram for explosive production.

FIGURE 117: EXPLOSIVES BLOCK DIAGRAM

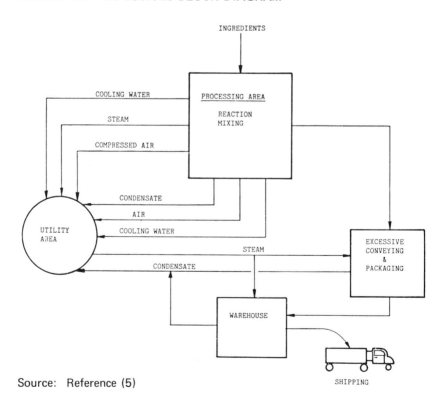

Source: Reference (5)

Since no significant processing steps are involved, hardly any process changes are expected by 1980. However, as discussed above, changes in mix of products produced by January 1, 1980 compared to 1972 is anticipated.

MAJOR ENERGY CONSERVATION OPTIONS TO 1980

A 20% savings in energy used for support services is expected by January 1, 1980. The largest saving will come from improving steam boiler operation which is the source for keeping buildings, warehouses, etc. warm, and for heating the reaction mixture. Typical other improvements include: more insulation, better maintenace of steam traps and leaks, and improving steam condensate recovery.

Individual plants are still examining improvements which may be obtained by: improving power factor, recovering heat from exhaust ventilation air, and modifying ventilation standards. Energy savings programs in explosives industry plants are not expected to be limited to any significant degree by either available funds or manpower.

GOAL YEAR (1980) ENERGY USE TARGET

The production of explosives during 1972 was 2.67 billion pounds according to the Bureau of Mines. During 1975, apparent production of explosives and blasting agents was 3.1 billion pounds. Coal mining was the only industry to show an increase in explosives consumption in 1975 compared to the previous year. Coal mining also used more than half (53.0%) of the total weight of explosives consumed in 1975. It is expected that coal production will grow at the rate of 5 to 6% per year to 1980; and explosives will also grow at 5 to 6% per year, bringing production of explosives to 3.98 billion lb/year by January 1, 1980.

The industry average specific energy consumption is estimated to have been 7,820 Btu/lb in 1972. About 80% of this energy is unrelated to production rate being used for support services (buildings, warehouses, safety systems, and the like). Only the balance of 20% varies with production rate. Since these are established processes geared to high personnel safety standards by experience, no significant change in specific process energy usage is expected by January 1, 1980.

It is expected that all increased production will come from existing plants without significant increase in the energy needed for support services. Furthermore, a decrease of about 20% in the total energy used for support services can be obtained by a combination of housekeeping, maintenance, scheduling, and related improvements. Combining the forecast increase in production with the estimated savings in energy for support services would reduce the specific energy consumption (Btu/lb) by about 37% by January 1, 1980.

The effect of government regulations such as EPA and OSHA on energy usage has been studied by the major producers. Battelle's estimate (5), which is based upon a judgement rather than definitive analysis due to limited time, is that the energy savings goal should be reduced by about 3% for this factor. The estimated energy improvement goal is, therefore, 34% in specific energy consumption by

January 1, 1980 compared to 1972. Additional information calculated on this basis is shown in Table 169.

It should be carefully noted, that the explosives industry's specific energy consumption is highly sensitive to change in production rate. If there were no growth in production volume between 1972 and January 1, 1980, the estimate is that specific energy consumption would decline by only 13% instead of the estimated goal of 34%.

TABLE 169: PRODUCTION AND ENERGY ESTIMATES FOR EXPLOSIVES, SIC 2892

	All Products SIC 2892	
	1972	1979.5
Product, 10^9 Pounds	2.67	3.98
Total Energy Required, 10^{12} BTU		
1972 Basis	20.88	31.12
1979.5 Basis	n.a.	20.51
Specific Energy Consumption BTU/lb		
1972 Basis	78.20	78.20
1979.5 Basis	n.a.	51.54
Estimated Energy Efficiency Goal, %	n.a.	34

Source: Reference (5)

PRINTING INK INDUSTRY

The ink manufacturing industry (SIC 2893) is similar to the paint industry in that it is essentially a formulation industry. Resins are made by some of the major manufacturers but resin manufacture will not be covered in this section.

Establishments covered in this industry are those primarily engaged in manufacturing printing, gravure, screen process and lithographic inks. These inks include: bronze, gold, gravure and duplicating inks, printing ink (base or finished), lithographic and screen process inks.

Printing ink production in the United States exceeds one billion pounds per year. The major components include drying oils, resins, varnish, shellac, pigments and many speciality additives. The industry comprises over 250 printing-ink producers. However, seven companies share over fifty percent of the market. These are: Inmont, Sinclair and Valentine, Sun Chemical, Cities Service, Tenneco Chemicals, Borden, and Flint Ink. Many large-volume users produce their own products as, for example, American Can and Reuben H. Donnelley.

Printing inks can be either water or oil base. Many of the raw materials are the same regardless of the fluid carrier. The inks are made with the same type equipment as in the paint industry and by the same process. The single largest volume type of ink is that used in printing of newspapers. This black ink is produced by mixing finely divided carbon black and mineral oil. Most of the colored inks are mixed on order, but many of the pigments used in them are staple quantity products such as lithol reds, chrome yellow, peacock and iron blues. A large number of more specialized inks are also used which, in aggregate, make up a considerable volume. They include vat colors and even fluorescent colors. The general trend is toward greater use of color in printing.

PROCESS TECHNOLOGY INVOLVED

Printing inks can be either oil-base or water-base. In either case, there is little difference in the production processes used. The major production difference is in the carrying agent; oil-base inks are dispersed in an oil mixture, while water-base inks are dispersed in water with a biodegradable surfactant used as the dispersing agent. All inks are generally made in batches. The major difference in the volume of ink is in the size of batches. There are three major steps in the oil-base ink manufacturing process. The flow diagram in Figure 118 illustrates these steps.

(1) Mixing and grinding of raw materials
(2) Tinting and thinning
(3) Filling operations

At most plants the mixing and grinding of raw materials for oil-base inks are accomplished in one production step. In the next stage of production, the ink is transferred to tinting and thinning tanks. Here the remaining binder and liquid, as well as various additives and tinting colors, are incorporated. The finished product is then transferred to a filling operation where it is filtered, packaged and labeled. In a large plant these operations are usually mechanized.

FIGURE 118: FLOW DIAGRAM OF MANUFACTURING PROCESS FOR OIL-BASE INKS

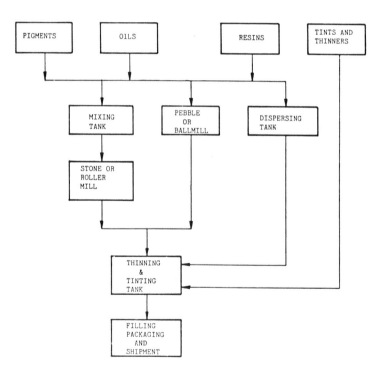

Source: Reference (5)

Water-base inks are produced by a slightly different method than oil-base inks. The pigments are usually received in proper particle size and the dispersion of the pigment, surfactant and binder into the vehicle is accomplished with a raw-toothed disperser. Once the formulation is correct, the ink is transferred to a filling operation where it is filtered, packaged and labeled in the same manner as for oil-base inks. The production process for water-base inks is shown in Figure 119. No major process changes, having significant impact on energy consumption, are expected in the manufacture of inks.

FIGURE 119: FLOW DIAGRAM OF MANUFACTURING PROCESS FOR WATER-BASE INKS

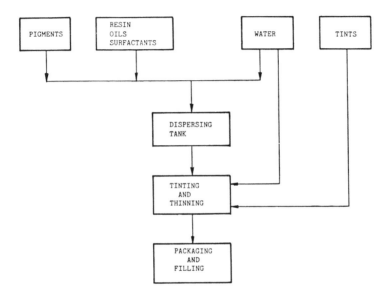

Source: Reference (5)

MAJOR ENERGY CONSERVATION OPTIONS TO 1980

Since no process changes of significance to energy consumption can be anticipated, the energy used for support services is the only significant target for conservation effort. As in other identical plants it is probable that a 20% decrease in this energy use could be made. This factor alone would result in a decrease in specific energy consumption of about 18% if there were no increase in production, or 37.5% if production increases as estimated by January 1, 1980 could all come from existing plants. This is not really possible and at least a few new plants will be built.

There are two other large offsetting factors to such an estimated goal. First, most ink plants are in metropolitan areas hence subject to the most stringent air-emission and liquid-effluent controls. A few plants have estimated a 25%

increase in total energy consumption to meet local requirements. Second, the economic incentive to reduce energy for support functions in the small establishments and the personnel and funds to study and implement it is limited.

GOAL YEAR (1980) ENERGY USE TARGET

The production of printing ink during 1972 was 1.137 billion pounds, according to industry sources. Printing ink production is expected to increase 4% per year, bringing total printing ink production, by January 1, 1980, to around 1.56 billion pounds. The specific energy consumption required in the manufacture of printing ink is estimated to have been 1,500 Btu per pound in 1972. About 90% of this energy requirement does not vary with production and is used to heat tanks, buildings and warehouses. The actual energy consumption for these support services varies significantly with geographical locations due to climate. The remaining 10% of the total energy requirements does vary with the production rate.

Overall the Battelle report to FEA (5) estimates that the maximum improvement in specific energy consumption will be 7% by January 1, 1980 as shown in Table 170.

TABLE 170: SUMMARY OF PRODUCTION AND ENERGY ESTIMATES, SIC 2893

	All Products SIC 2893	
	1972	1979.5
Production, 10^9 pounds	1.14	1.56
Total energy required, 10^{12} Btu		
1972 basis	1.71	2.34
1979.5 basis	na	2.18
Specific energy consumption, Btu/lb		
1972 basis	1,500	1,500
1979.5 basis	na	1,400
Estimated energy improvement goal, %		7

Source: Reference (5)

CARBON BLACK INDUSTRY

The carbon black industry (SIC 2895) includes establishments primarily engaged in manufacturing carbon black (channel and furnace black). Carbon black is primarily produced by the incomplete combustion of heavy aromatic oils. It is manufactured in a variety of forms which vary in particle size, surface area, surface activity, structure and electrical conductivity.

It contains 90 to 99% elemental carbon. The major use for carbon black is in the rubber industry where more than 90% of the material is consumed. It is also used as a pigment by the ink, paint and plastics industries in addition to its use in the manufacture of carbon paper.

Based on the facts that the carbon black industry

(a)   is a significant energy consumer (more than 1% of all SIC 28 in 1974— 2 times this amount of energy is consumed as feedstock which, however, is outside the scope of this study),

(b)   has a vast potential for reducing energy use compared to practice in 1972, and

(c)   produces substantially all its output from plants of a few, large, technically astute companies,

a relatively significant portion of the Battelle study for FEA (5) total effort was devoted to the carbon black industry. Furnace black accounts for 99% of all carbon black produced. It is this manufacturing process which is, therefore, examined in detail in the Battelle analysis (5).

PROCESS TECHNOLOGY INVOLVED

The oil furnace process, coming into production in November 1943, was a significant advance in the carbon black industry. It has steadily advanced in favor due to its high efficiency and versatility, taking over production from the channel and gas furnace processes. It accounted for more than 90% of carbon black production in 1972.

The oil furnace process, though similar to the gas furnace process, differs in several respects. The burner designs are different; the furnaces are generally shorter and are circular in cross section instead of rectangular. More turbulence is created in the furnaces, and because oil furnace blacks display more structure, it has been found that electrostatic precipitators are not necessarily required to collect the particles of black, as they were to collect the old original gas furnace blacks having less structure.

With these exceptions, the oil furnace and gas furnace processes are practically identical, except for, of course, the different feedstocks utilized. A block flow diagram for carbon black manufacture is shown in Figure 120. This diagram also shows energy data as will be discussed later in this section.

FIGURE 120: CARBON BLACK ENERGY CONSUMPTION DIAGRAM

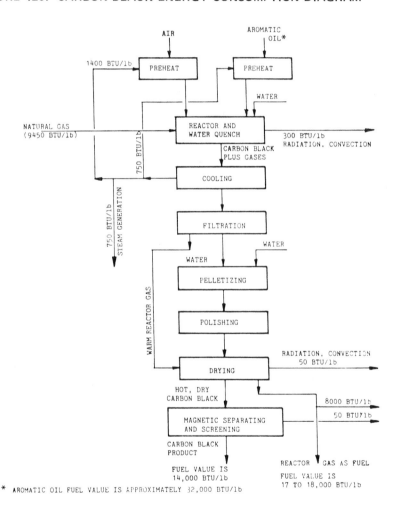

Source: References (3)(5)

Liquid hydrocarbons are used as raw material, principally aromatic refinery residual oils, in the oil furnace process. However, where fuel gases are available, such as natural gas or refinery gas, they are burned to completion to provide heat to crack the oil, thereby increasing the recovery of carbon black from the oil. The raw material oils used are rather heavy residual oils, ranging from $-2°$ to $+3°$API.

The burners are designed to atomize these liquids as they enter the hot reaction zone of the furnace, either by means of simple pressure atomization in passing through conventional swirl chamber atomizing nozzles, or with bi-fluid nozzles using compressed air as the preferred atomizing fluid.

To improve atomization by lowering the viscosity of the oil, these heavy oils are preheated to $350°$ to $500°F$. The level of preheat serves in a minor degree as a quality control of the black product. In some instances, where much lighter and less viscous oils are used, they are vaporized, instead of atomized, before they enter the reaction zone of the furnace.

The design of the furnaces and burners for effecting the conditions that produce the varied grades of black, together with the methods of operation, constitute an important part of carbon black technology. Salts of alkali metals, which are used as flame modifiers, are most effective in governing the degree of carbon black structure.

The gaseous stream from the furnace containing carbon black passes through an air-cooled flume into a spray cooler, where the temperature is reduced by water. The carbon black is separated from the combustion gases by means of a filter.

The fluffy carbon black is transported by a screw conveyor to a blower or bucket elevator, which transports it to pelleting equipment. Here the black may be pelleted dry by rotating drums or wet by several processes. In one wet process, the furnace black and water are mixed to form a pasty mass, which is then agitated in a trough by a series of pins on a rotating shaft. In a second process the black is pelleted wet in a special trough mixer, and then is given a final pelleting in a rotating cylinder. The beaded furnace black may be packaged in bags or bulk-packed in covered hopper cars.

Various yields and qualities of black are produced by controlling such factors as composition of raw materials, furnace design, degree of turbulence, amount of secondary air, temperature, and contact time. The versatility of the process has resulted in some seven different methods of effecting partial combustion or combustion-thermal decomposition of gases in furnaces of various designs.

MAJOR ENERGY CONSERVATION OPTIONS TO 1980

Almost all of the cost of manufacturing carbon black is related to energy materials. About 75% of the cost depends upon feedstock costs alone. Prior to 1973, there was little incentive for the industry to conserve in energy use, owing to the relatively low energy costs. Energy costs have, of course, increased rapidly over the last two or three years and the industry has taken a serious look at the potential for energy conservation. Based (in part) on the higher energy costs, it is judged that an energy improvement goal in specific energy consumption of 39% can be achieved by the carbon black industry by January 1, 1980, as compared to 1972. Figure 120 was a block diagram illustrating energy consumption in the manufacture of carbon black. The numbers shown are approximate, illustrating the relative magnitude of energy flows in carbon black manufacture.

The diagram illustrates that by proper utilization of reactor gas in preheating air and oil, and in steam generation, a maximum energy savings of 4,000 to 5,000 Btu/lb of product can be realized. Steam generation alone could provide a savings of around 2,500 Btu/lb. It is important to note that steam will only be generated if there is a useful outlet for its use.

Most of the carbon black plants are located at places where steam cannot be practically utilized. In this case, the heat would be removed by water quenching, thereby wasting the available energy. Figure 121 shows the quantities of energy utilized in preheating and steam generation.

FIGURE 121: EQUIPMENT DIAGRAM—REACTOR AND WASTE HEAT RECOVERY EQUIPMENT

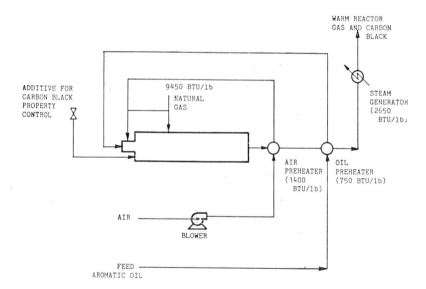

Source: References (3)(5)

The principal area where significant energy savings are feasible is in the utilization of reactor gas. The average composition of reactor gas is summarized in Table 171. Its heating value is 75 Btu/cf of dry gas which is equivalent to 17,000 to 18,000 Btu/lb of carbon black product. The heating value of 75 Btu/cf is not high enough to generate high temperature, unless it is spiked with natural gas to bring its heating value to somewhere in the area of 200 Btu/cf. It is estimated that by January 1, 1980, it is feasible for most of the heating required for drying to be met by reactor gas, thereby reducing the energy consumption from outside sources by 3,500 Btu/lb. The remaining available energy from reactor gas of 13,500 to 14,500 Btu/lb either will be discharged into the atmosphere or used in the generation of high-pressure steam where feasible. Table 172 summarizes some of the energy conservation approaches.

TABLE 171: AVERAGE DRY REACTOR GAS COMPOSITION

	Volume, %
CO	12
H_2	8
CO_2	4
N_2	75
CH_4	1
	100

Source: Reference (5)

TABLE 172: CARBON BLACK ENERGY CONSERVATION APPROACHES

Causes of Energy Losses	Approximate Magnitude of Losses, Btu/lb	Energy Conservation Approaches
Rejected heat		
Radiation and convection losses	350	Insulation
Sensible heat in generated reactor gases leaving the process	1,000	Design modification (waste heat recovery when using reactor gas as fuel)
Latent heat in generated reactor gases leaving the process	7,000	Design modification (waste heat recovery when using reactor gas as fuel)
Overall process		
Energy to heat inerts which are lost in quench water	3,000	Process modification (use higher oxygen content gas to burn natural gas)
Fuel value of reactor gas	18,000	Partly in dryer and balance in waste utilization

Source: References (3)(5)

There is another very large potential savings in energy which is not likely to be in use by January 1, 1980, but which will begin in some plants in the early 1980s. The industry will use oxygen for partial oxidation, rather than air, in the reactor. The reactor gas from the use of oxygen will have a significantly higher heating value than the approximately 200 Btu/lb required for providing higher temperatures.

In this case the balance of recycle gas most probably will be recycled back to the reactor thereby avoiding need for outside natural gas. This could make the carbon black industry self-sufficient in energy consumption. The use of natural gas could be eliminated. The energy needed could be derived entirely from electricity, oxygen and the feedstock (heavy oil).

In summary, Battelle has estimated in their study for FEA (5) that by January 1, 1980, specific energy consumption in carbon black manufacture will drop to 10,300 Btu/lb, as compared to 17,000 Btu/lb in 1972. The new plants built in the 1980s will be either self-sufficient in energy consumption or will be located near a refinery where the high-pressure steam which is generated from the carbon black plant can be consumed.

The 10,300 Btu/lb assumes that reactor gas will be used in drying, but some processors may prefer to make high-pressure steam from reactor gas. In this case, carbon black energy consumption will be 10,300 Btu/lb or less after giving proper credit to steam generation.

GOAL YEAR (1980) ENERGY USE TARGET

Estimated specific energy consumption for furnace carbon black was 17,000 Btu per pound in 1972. This excludes the energy content of the hydrocarbon feedstock required to manufacture carbon black. Some 70% of this energy was consumed in reactors. Two-thirds of the remaining energy was consumed in driers and one-third in miscellaneous uses.

The potential for energy conservation in furnace black production is rather large, owing to the common practice of water quenching for heat removal from reactants and discharging of reactor gas to the atmosphere. The recent quantum increases in energy prices have made carbon black producers more conscious of these inefficiencies and major energy improvements are considered feasible by January 1, 1980.

By January 1, 1980, it is expected that the water quenching of reaction mixture will be minimized, with the heat being recovered by air and oil preheating, or by high-pressure steam generation if steam can be utilized. This is expected to result in a net energy improvement of 3,000 Btu/lb of carbon black. Presently, this energy (not recovered) is provided by burning natural gas in the reactor.

The reactor gas normally contains carbon monoxide and hydrogen with a heating value of about 70 Btu/scf of dry gas. The total energy contained in the reactor gas varies between 17,000 and 18,000 Btu/lb of carbon black. By January 1, 1980, a portion of this gas is expected to be used in the drying of carbon black. This will result in a net savings of about 3,500 Btu/lb.

The balance of the reactor gas containing about 14,000 Btu/lb will still be discharged into the air since there is normally no place to use it. Plants located near other industry (e.g., a refinery) could possibly burn the remaining reactor gas in a waste heat boiler generating high-pressure steam for utilization in the industrial plant.

Some 5% of the energy consumed in furnace black manufacture is utilized in non-process-related areas, such as heating buildings and warehouses. The fixed energy consumption will, of course, vary, depending upon building design and climatic conditions. It is estimated that a 20% energy improvement in fixed energy demand is feasible through the application of proper housekeeping and maintenance. Additional fixed energy savings will result from increases in plant output, thereby reducing per unit energy usage. The total energy savings resulting from proper housekeeping and maintenance is expected to be about 700 Btu per pound.

The energy improvements in processing and the others just discussed can be achieved by industry by January 1, 1980. They are generally economically justifiable, and manpower and capital funds are not expected to inhibit achievement of these improvements. However, a partial offset to these improvements will result from producers satisfying government EPA and OSHA regulations. An additional 500 Btu per pound will be required between 1972 and January 1, 1980 in order to satisfy these government regulations.

Table 173 summarizes the energy improvement goal for the carbon black industry. An improvement in specific energy consumption of 39% by January 1, 1980 relative to 1972 is considered feasible.

TABLE 173: PRODUCTION AND ENERGY ESTIMATES—CARBON BLACK

	All Products SIC 2895	
	1972	1979.5
(1) Production, 10^9 pounds	2.93	3.80
(2) Total Energy Required, 10^{12} BTU		
1972 Basis	49.81	64.60
1979.5 Basis	n.a.	39.14
(3) Specific Energy Consumption, BTU/pound		
1972 Basis	17,000	17,000
1979.5 Basis	n.a.	10,300
(4) Estimated Energy Improvement Goal, percent		39

Source: Reference (5)

FATTY ACIDS AND
MISCELLANEOUS CHEMICALS INDUSTRY

Establishments included in SIC 2899 are those engaged primarily in manufacturing miscellaneous chemical preparations not elsewhere classified, such as fatty acids, essential oils, gelatin (except vegetable), sizes, bluing, laundry sours, writing and stamp pad inks; industrial compounds, such as boiler and heat insulating compounds, metal, oil and water-treating compounds, waterproofing compounds and chemical supplies for foundries. Establishments primarily engaged in manufacturing vegetable gelatin (agar-agar) are classified in SIC 2833; and dessert preparations based upon gelatin in SIC 2099.

Due to the large number of chemical products included in this industry classification and the limited time to complete the work for this study, it was necessary to develop a mechanism by which to assess specific energy consumption in this industry. The products of the industry have been divided into two separate groups. They are:

(A) Chemicals requiring chemical processing: fatty acids, gelatin, sizes, etc.

(B) Chemicals manufactured from essentially blending operations: bluing, laundry sours, writing and stamp pad inks; industrial compounds, such as boiler and heat insulating compounds, waterproofing compounds and chemical supplies for processing.

Admittedly, this is a rough way of classifying the products included in the industry, but it is based on a reasonable separation of two different methods of manufacturing (chemical processing and blending) and reduces the vast number of products included in the industry to a manageable level.

In the discussion which follows, specific mention will be made of the specific group of products, i.e., (A) or (B), which are being considered. In the case of group (A), the chemical processing of fatty acids, one of the more important miscellaneous chemicals, is specifically analyzed and energy improvements in its chemical processing are judged to be representative of other chemical products in group (A). With regard to the products included in group (B), the analysis

FIGURE 122: FATTY ACID MANUFACTURE BY FAT SPLITTING

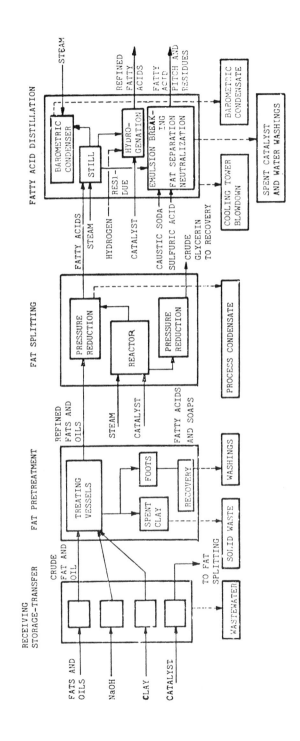

Source: Reference (5)

borrows heavily from experience gained in the examination of feasible energy improvements in the printing ink industry (SIC 2893), which involves similar blending processes.

PROCESS TECHNOLOGY INVOLVED

The flow diagram for fatty acids manufacturing is shown in Figure 122. The technology of the industry is straightforward and contains some processes analogous to petroleum refining. Included are:

(1) Caustic refining, to remove color bodies and odors.
(2) High-pressure and temperature hydrolysis, or splitting, to the basic crude acids and glycerin.
(3) Separation of unsaturated and saturated acids, similar to solvent dewaxing of lubricating oils.
(4) Hydrogenation, to reduce unsaturation.
(5) Distillation, including fractionating.
(6) Derivative manufacture, via formation of esters, amides, amines, alcohols, alkoxylates, etc.

No major changes are expected by 1980 in fatty acids manufacturing. Other miscellaneous chemicals are manufactured by blending or chemical processes. Blending operations have been discussed in previous sections of the study and fatty acid processing is taken to be representative of other chemical processes insofar as estimating an energy improvement goal.

MAJOR ENERGY CONSERVATION OPTIONS TO 1980

Fixed energy requirements can be significantly reduced (upwards of 20%) by employing proper housekeeping, maintenance and scheduling techniques. Significant savings in process energy requirements can possibly be achieved by recovery of condensate where feasible. Most old steam systems, as illustrated in Figure 123, are designed without condensate recovery.

FIGURE 123: OLD STEAM-CONDENSATE SYSTEM

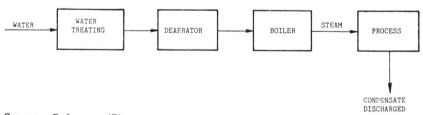

Source: Reference (5)

Condensate, depending upon steam pressure, carries a significant quantity of sensible heat which is not recovered. The fatty acids industry consumes low-pressure steam in an evaporator. If a proper steam system is designed, then significant energy savings can be realized, since almost all of the sensible heat present in condensate can be utilized (see Figure 124).

FIGURE 124: RECOMMENDED STEAM-CONDENSATE SYSTEM

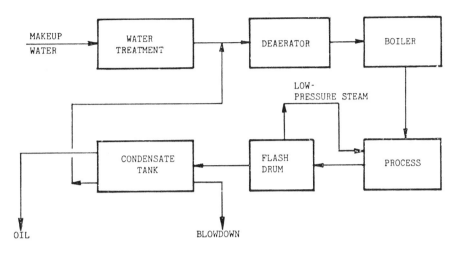

Source: Reference (5)

GOAL YEAR (1980) ENERGY USE TARGET

The sales of miscellaneous chemicals classified under SIC 2899 was valued at $2.228 billion in 1972. The growth in the value of miscellaneous chemicals, in constant 1972 dollars, is expected to be about 4% per year between 1972 and 1979.5. The specific energy consumption by the industry is estimated to have been 18,610 Btu/$ of sales in 1972.

In group A, where fatty acids and similar compounds are manufactured, about 40% of the total energy consumed is unrelated to production rates. Output of plants operating in 1972 is expected to expand over the 1972 to 1979.5 period, thereby precipitating a decline in specific energy consumption. In addition, housekeeping, maintenance and scheduling improvements could probably reduce specific fixed energy consumption by about 20% by 1979.5 relative to 1972.

The remaining 60% of the specific energy consumption of group A in 1972 varies with plant output. The variable energy is required for such things as process heating, cooling, vacuum jets, among others. No significant product mix and process changes which will affect average variable energy consumption are antici- pated by 1979.5. However, it is possible that for some plants the disposition of condensate may change. Where feasible, some plants can recycle most of the condensate, thereby using the energy therein contained. The load on water make- up and treating facilities would also be reduced.

Offsetting to some extent process and fixed energy improvements in group A will be the added energy requirements to meet government environmental and safety regulations. Sufficient information was not available to Battelle in the course of their study for FEA (5) to measure the effects of these regulations on specific energy consumption.

Considering all the factors discussed above, an energy improvement goal of 19% in energy consumption per unit of output between 1972 and 1979.5 has been estimated (5).

Group B chemicals are manufactured typically by blending operations. The fixed energy requirements vary between 70 and 90%. The total specific energy consumption varies significantly, depending upon product mix, climatic conditions and plant design. Lower fixed energy consumption per unit of output is expected as a result of expanding output of plants operating in 1972 and improved housekeeping, maintenance and scheduling. Energy use related to production rates is expected to remain at approximately 1972 levels in 1979.5. The best judgment of an energy improvement goal for the chemical products in group B is a net 10% reduction in specific energy consumption in 1979.5 relative to 1972.

The combined energy improvement goal for the products covered in groups A and B is shown in Table 174. The goal is estimated to be a 15% improvement in specific energy consumption between 1972 and 1979.5.

TABLE 174: SUMMARY OF VALUE OF PRODUCTION AND ENERGY ESTIMATES, SIC 2899

	All Products SIC 2899	
	1972	1979.5
Value of production, 10^9 $	2.228	3.029
Total energy required, 10^{12} Btu		
1972 basis	41.47	56.37
1979.5 basis	na	47.92
Specific energy consumption, Btu/$		
1972 basis	18,610	18,610
1979.5 basis	na	15,820
Estimated energy improvement goal, %		15

Source: Reference (5)

PETROLEUM

AND COAL PRODUCTS INDUSTRY

Standard Industrial Classification 29 (SIC 29)(petroleum and coal products), is dominated by the petroleum refining industry and ranks third after SIC 28 (chemical and allied products) and SIC 33 (primary metal industries) in consumption of energy. The total energy consumed within SIC 29 in 1972 was about 3 quadrillion Btu.

The industry defined as SIC 29 consists of five components: SIC 2911, petroleum refining; SIC 2951, paving mixtures and blocks; SIC 2952, asphalt felts and coatings; SIC 2992, lubricating oils and greases; and SIC 2999, products of petroleum and coal, not elsewhere classified.

In order to establish a voluntary energy efficiency improvement target for SIC 29, each of the component industries was examined for the potential for energy conservation in the relevant period. Using the base year energy consumption of each component as a weighting factor, an overall SIC 29 target was then established. The dominant component is clearly SIC 2911, which consumes 98% of all the energy used within SIC 29.

In the determination of conservation potential in the component industries, the technological feasibility of a large number of alternate operating practices and technologies was investigated. Where a measure was technologically acceptable, it was then evaluated against various economic criteria, including return on investment wherever capital investment was required. Returns on investment for specific measures were compared with the hurdle rate for the industry and thus the potential for energy efficiency contribution by the target date of January 1, 1980 was assessed in the course of a Gordian Association study for FEA (6).

PROCESS TECHNOLOGY INVOLVED

Figure 125 shows a typical integrated petroleum refining process. Refineries vary quite widely in complexity and in product mix. The process shown is representative of a refinery that is producing a high yield of gasoline from crude oil. The primary energy consumption operations include crude oil distillation, gas oil desulfurization, heavy naphtha desulfurization, naphtha desulfurization, catalytic cracking, naphtha reforming, alkylation, aromatics extraction and coking. Sources of energy for these operations are primarily natural gas, refinery produced gas, petroleum coke and fuel oil. These operations account for more than 80% of the energy consumed in the petroleum refining process shown.

FIGURE 125: PETROLEUM REFINING ENERGY CONSUMPTION DIAGRAM

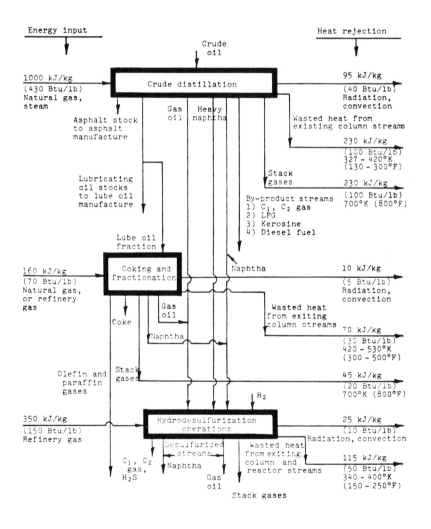

(continued)

FIGURE 125: (continued)

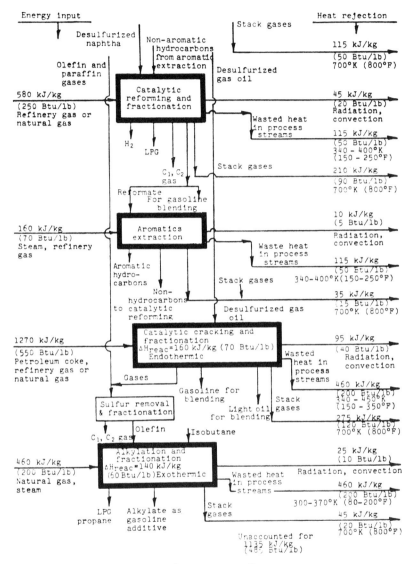

1971 U.S. production: 610×10^9 kg ($1,350 \times 10^9$ lb)
1971 energy consumption (primarily natural gas, refinery gas, petroleum coke, fuel oil): 94,000 Mw (2.8×10^{15} Btu)

Source: Reference (2)

Since there are so many different processes used in different ways in hundreds of refineries, it is impossible to define the average refinery. This study, therefore, provides data developed on the basis of process modules which can be combined in any desired sequence. The processes covered are as follows:

[1] Crude distillation   [8]   Hydrocracking
[2] Vacuum distillation   [9]   Distillate desulfurization
[3] Visbreaking   [10] Alkylation
[4] Thermal cracking   [11] Butane isomerization
[5] Coking   [12] Pentane-hexane isomerization
[6] Fluid catalytic cracking   [13] Polymerization
[7] Catalytic reforming   [14] Hydrogen production

Energy requirements for each of the above processes are summarized in Table 175 in terms of actual quantities used and also expressed as MM Btu per barrel of charge to the processing plant.

The conversion factors used were:

[1] Steam   1,333 Btu per lb steam
[2] Electricity   12,000 Btu per kwh
[3] Cooling water   3.69 Btu per gallon

Using the energy data from Table 175, a typical refinery was drawn up and material balance and energy consumption figures were calculated. The material balance is shown in Table 176 and a summary of energy requirements in Table 177. For the typical refinery as a whole, energy consumption amounts to 0.44 MM Btu per barrel of crude charged. This figure is, in fact, highly dependent on the crude oils run in the refinery and the product volume and specifications: the variations are covered in detail in reference (26), which reports the results of a study using linear programming techniques.

With regard to the cost of energy used to manufacture petroleum products, it must be noted that the bulk of the energy and the products are hydrocarbon streams which are, to a large extent, interchangeable as internal refinery fuels. The refinery will always try to burn the lowest value product as fuel; this product may be, for example, fuel gas, heavy fuel oil or propane. The cost of energy as a percentage of the product value is therefore somewhat variable, but is typically around the 5 to 10% range depending on market conditions.

The first step in the manufacture of petroleum products is the separation of the crude oil into the main conventional streams. Each stream contains many compounds and boils within a limited range. The crude oil passes through a heat exchanger train to recover heat from product streams being run down to storage. Often the crude is piped into a flash drum, where light fractions will vaporize and pass directly into the fractionating tower. The balance of the crude is ultimately raised to about 750°F in a fired heater and enters the tower. In the tower, separation of the crude oil into the required product boiling ranges takes place, controlled by temperature levels and reflux rates throughout the system.

Typical product streams are light gasoline, naphtha, kerosene, diesel or light gas oil, and a bottom product used as a fuel oil blending component.

TABLE 175: SUMMARY OF ENERGY REQUIREMENTS FOR MAJOR REFINERY PROCESSES

PROCESS	ENERGY USED PER BBL. CHARGE				EQUIVALENT ENERGY PER BBL. CHARGE				
	STEAM M LBS	FUEL MMBTU	ELEC. KWH	CW M GALS	STEAM MMBTU	FUEL MMBTU	ELEC. MMBTU	CW MMBTU	TOTAL MMBTU
Crude Distillation	0.015	0.067	0.30	0.40	0.0200	0.0670	0.0036	0.0015	0.0921
Vacuum Distillation	0.040	0.060	0.50	0.80	0.0533	0.0600	0.0060	0.0030	0.1223
Thermal Cracking	0.122	0.700	2.00	2.00	0.1627	0.7000	0.0240	0.0074	0.3941
Visbreaking	(.08)	0.260	1.80	0.29	(.1067)	0.2600	0.0216	0.0011	0.1760
Delayed Coking	0.038	0.208	1.44	1.3	0.0507	0.2080	0.0173	0.0048	0.2808
Fluid Cat. Cracking	(.1)	0.105	2.50	0.55	(.1333)	0.1050	0.0300	0.0020	0.0037
Cat. Reforming	-	0.380	3.00	0.77	-	0.3800	0.0360	0.0028	0.4138
Hydrocracking	(.006)	0.145	8.20	0.11	(.008)	0.1450	0.0984	0.9004	0.2358
Naphtha Hydrotreating	0.011	0.053	1.00	0.15	0.0147	0.0530	0.0120	0.0006	0.0803
Gas Oil Hydrotreating	0.025	0.066	1.70	0.26	0.0333	0.0660	0.0204	0.0010	0.1207
C_4 Alkylation	0.0196	1.860	6.68	6.65	0.0261	1.8601	0.0801	0.0241	0.9904
Butane Isomerization	0.065	-	1.48	1.44	0.0867	-	0.0178	0.0053	0.1098
C_5/C_6 Isomerization	-	0.180	2.40	0.75	-	0.1800	0.0288	0.0028	0.2116
Propylene Polymerization	0.311	-	2.64	0.98	0.4147	-	0.0317	0.0036	0.4500
Butylene Polymerization	0.270	-	2.27	0.84	0.3600	-	0.0272	0.0031	0.3903
Hydrogen Production	-	0.275	0.40	0.65	-	0.2750	0.0048	0.0024	0.2822

Note: ( ) signifies export of energy

Source: Reference (26)

TABLE 176: TYPICAL REFINERY MATERIAL BALANCE

PROCESSING STAGE	CRUDE OIL	LT. ENDS (FOE)	LLSR NAPHTHA	NAPHTHA	KERO	DIESEL -LGO	REDUCED CRUDE	VAC. GAS OILS	VAC. RESID.	PROPANE	PROPYLENE	ISO-BUTANE	N-BUTANE	BUTYLENES
Crude distillation	57000	(758)	(2633)	(12477)	(8727)	(6099)	(26306)							
Vacuum distillation							26306	(16052)	(10254)					
Delayed coking									10254					
Naphtha hydrotreating				(1902) 14379										
Cat. reforming											(108)	(51)	(149)	(149)
C4 alkylation										(1294)		(503) 1503	(776)	1296
Distillate hydrotreating					8727	6099								
Cat. cracking								16052						
Transfers/adjustments		758								(569)	(997)	(1008)	(297)	(1147)
PRODUCTS-CONSUMING SYSTEMS														
Fuel firing systems														
Steam raising plants														
Electricity-cooling water systems														
LPG sales										1863	1105	59	1222	
Gasoline sales			2633											
Kero-jet fuel sales														
Diesel-No. 2 oil														
Heavy fuel oil														
Coke														

ALL FIGURES:

(Out) BPCD

In BPCD

(continued)

TABLE 176: (continued)

PROCESSING STAGE	COKER GAS OIL	COKE (FOE)	FUEL GAS (FOE)	HYDROGEN (FOE)	REFORMATE	CC GASOLINE	LCO	HCO + SLURRY	LIGHT ALKYLATE	HEAVY ALKYLATE	DESULF NAPHTHA	DESULF KERO	DESULF GAS OILS	TOTALS
Crude distillation														
Vacuum distillation														
Delayed coking	(6445)	(1644)	(841)											
Naphtha hydrotreating			(935)	76							(14379)			
Cat. reforming				(690)	(10712)						14379			
C₄ alkylation			(2)						(2112)	(194)				
Distillate hydrotreating			(421)	398			4740					(8727)	(10839)	
Cat. cracking	6445		(974)	216		(11795)	(4740)	(3134)						
Transfers/adjustments										194		(194)		
PRODUCTS-CONSUMING SYSTEMS														
Fuel firing systems			3112											3112
Steam raising plants		164	61					165						390
Electricity-cooling water systems		236						236						472
LPG sales														2968
Gasoline sales					10712	11795			2112					28533
Kero-jet fuel sales												8921		8921
Diesel-No. 2 oil													10839	10839
Heavy fuel oil								2733						2733
Coke		1244												1244
ALL FIGURES:														
(Out) BPCD														
In BPCD														249T/D

Source: Reference (26)

TABLE 177: TYPICAL REFINERY ENERGY BALANCE

PROCESSING STAGE	CHARGE RATE BPCD	STEAM MLBS/D	STEAM MMBTU	STEAM %	FUEL MMBTU/D	FUEL %	ELECTRICITY KWH/D	ELECTRICITY MMBTU/D	ELECTRICITY %	COOLING WATER M GAL/D	COOLING WATER MMBTU/D	COOLING WATER %	TOTAL MMBTU/D	TOTAL %
CRUDE DISTILLATION	57000	855	1140	20.8	3819	19.3	17100	205	7.8	22800	84	23.5	5248	20.7
VACUUM DISTILLATION	26306	1052	1402	25.6	1578	8.0	13153	158	6.0	21045	78	21.9	3216	12.7
DELAYED COKING	10254	390	520	9.5	2133	10.8	14766	177	6.7	13330	49	13.7	2879	11.4
NAPHTHA HYDROTREATING	14379	158	211	3.8	762	3.8	14379	173	6.5	2157	8	2.2	1154	4.6
CAT. REFORMING	14379	-	-	-	5464	27.5	43137	518	19.6	11072	41	11.5	6023	23.8
C4 ALKYLATION	1296	26	35	0.6	2411	12.2	8651	104	3.9	8459	32	9.0	2582	10.2
DISTILLATE HYDROTREATING	19566	489	652	11.9	1291	6.5	33262	400	15.1	5087	19	5.3	2362	9.3
MISC. OFFSITES	-	1140	1522	27.8	-	-	19950	239	9.0	-	-	-	1761	7.0
		4110	5482	100.0										
		(2250)	(3000)	-										
CAT. CRACKING	22497	1860	2482	-	2362	11.9	56243	675	25.4	12373	46	12.9	83	0.3
					19820	100.0	220642	2649	100.0	96323	357	100.0	25308	100.0
EQUIVALENT BARRELS FUEL OIL PER DAY			390		3112			416			56		3974	
AS PERCENTAGE			9.8		78.3			10.5			1.4		100	

TOTAL ENERGY USE 0.444 MMBTU per bbl.

Source: Reference (26)

Figure 126 shows the distillation of crude oil. The degree of separation of the crude into components varies from one refinery to another. The scheme shown is fairly elaborate. The principal energy conservation practice is to preheat the crude oil by heat exchange with components leaving the distillation columns. Natural gas or refinery produced gas is burned to supply heat to the column feeds. Steam is also used to provide heat to the strippers and the atmospheric column.

FIGURE 126: EQUIPMENT DIAGRAM—CRUDE DISTILLATION

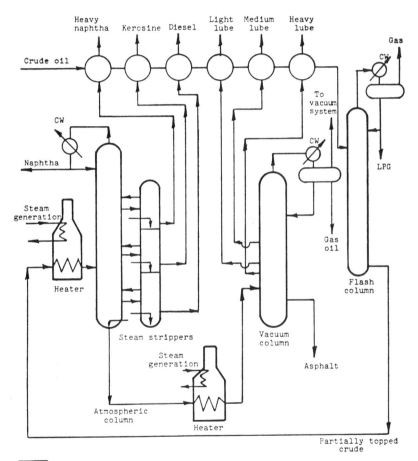

Rejected heat: radiation—95 kJ/kg (40 Btu/lb)
Heater stack gases—230 kJ/kg (100 Btu/lb) at 640°K (700°F)
Hot exiting column steams (wasted)—230 kJ/kg (100 Btu/lb) at 327° to 420°K (130° to 300°F)

Source: Reference (2)

A second stage of processing, carried out under vacuum to keep temperatures as low as possible and to avoid undue cracking of products, is used where the crude oil is to be reduced to a low percentage of bottoms. The overhead products include heavy lubricating fractions, or heavy cracking stock, in addition to the products taken overhead from a single-stage distillation unit. The residuum is a viscous pitch used for fuel oil blending; occasionally, it is a very high melting point asphalt.

Increasingly, vacuum distillation of crude oil is utilized as a means to produce a low sulfur fuel oil pool in a refinery. Processes to desulfurize vacuum gas oils are available and proven commercially. Processes to desulfurize residues are much more complex and more expensive to install and operate, as well as being somewhat limited in application due to metal contamination of catalysts, for example. The vacuum gas oils are therefore desulfurized and back blended with the vacuum bottoms to achieve a low sulfur heavy fuel oil pool. The vacuum distillation process is shown in Figure 127.

FIGURE 127: VACUUM DISTILLATION PROCESS

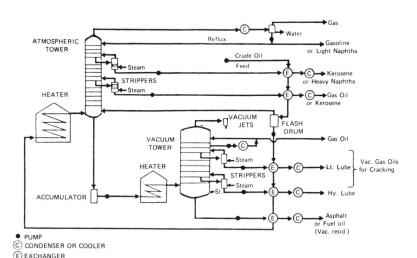

Source: Reference (26)

The visbreaking process is employed to produce a minimum of gasoline and a maximum of furnace oil (No. 2 oil), in addition to a stable heavy fuel oil component of reasonable pour point and viscosity.

The chargestock, normally vacuum residuum, is heated and thermally cracked in the visbreaker furnace. The effluent product, after quenching with light gas oil, is directed to the lower or evaporator section of the fractionator, where it is flashed. The tar accumulates in the base of the tower, the vapors being fractionated into gasoline and gas, light gas oil and a heavy gas oil in the upper part.

In some units, heavy gas oil is withdrawn from the tower and charged as recycle to a separate cracking furnace.

The tower bottoms are vacuum flashed, the distillate material being returned to the fractionator where it aids in making up the recycle charge to the heavy gas oil cracking furnace. In some refineries, a heavy gas oil stream is withdrawn for cat cracking feedstock. A flow diagram of the visbreaking operation is shown in Figure 128.

FIGURE 128: VISBREAKING PROCESS

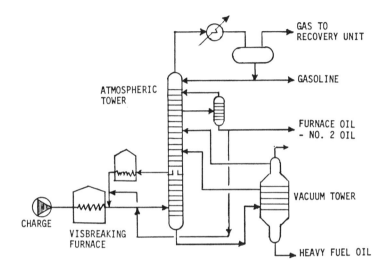

Source: Reference (26)

The thermal cracking of crude oil distillate fractions heavier than naphtha was originally practiced as a means of increasing the yields of gasoline from crude oil. Certain gas oils can be cracked thermally, i.e., without using a catalyst, to yield as much as 50% gasoline. The process, however, cannot yield more without very extensive recycling of the intermediate cracked product. The through-put of fresh feedstock is therefore low as compared with catalytic cracking where the cycle stock is discarded, or as compared with naphtha reforming which is operated on a once-through basis. In addition, thermal cracking of gas oil produces a large amount of tar whereas catalytic cracking produces a heavy gas oil distillate instead.

The operation of thermal cracking units on gas oil is usually at temperatures ranging from 900° to 1000°F and pressures ranging from 250 to 750 psig. There is a wide variety of conditions employed depending upon the quality of the feed stock, the percentage of recycle stock in the total feed and the quality of distil-late desired.

The limitations on distillate quality and yield are imposed primarily by the increasing gas and carbon yields which accompany increasing severity of operating conditions. The carbon formation occurs not only in the soaking drum or reaction chamber but it also occurs in the tubes of the furnace, thus limiting the length of runs.

The equipment for thermal cracking is similar to that for a single-stage distillation unit, often with the addition of a pair of soaking drums in which the bulk of the coke produced is accumulated. This coke is removed by mechanical or hydraulic means. The flow diagram of a thermal cracking unit is similar to the atmospheric section of crude unit or visbreaker.

The process of coking, the distillation of crude oil all the way to a coke residue instead of a liquid residue, has been practiced almost as long as petroluem has been processed. The newest continuous coking processes may be used to upgrade a wide range of low value residual stocks to naphtha, middle distillates, catalytic cracking feedstock and by-product gas and coke. Conversion of high sulfur residual reduces sulfur level in the fuel oil pool and reduces the required volume of low sulfur blend stocks.

Delayed coking is not unlike the old thermal cracking process in which the highly heated residual feedstock was passed to alternate soaking drums where coke was formed in one while being removed from the other. The process operates at temperatures around 950°F and pressures from 20 to 100 psig.

In fluid coking, the liquid feed is injected into a reactor bed of hot coke particles which are kept fluidized by gas and by stripper steam. The hot coke particles in the reactor are supplied from a burner vessel where they are kept at nearly 1200°F by partial combustion in a fluidized bed. The process operates at a pressure of 1 to 2 atmospheres. From a steam stripper at the bottom of the reactor, the coke particles, increased by the coke deposited from the feed stock, are withdrawn and airlifted into the burner, completing the cycle. The coke burned in support of the operation amounts to 4 or 5% by weight of the feed. Figure 129 is a flow diagram of a fluid coking unit.

FIGURE 129: FLUID COKING PROCESS

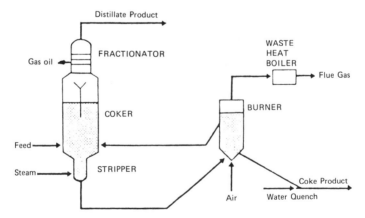

Source: Reference (26)

Figure 130 shows a drum coking operation where heavy residuals are upgraded into more valuable distillate products and coke. The residue is fed to a distillation column where light gases are flashed. The remaining material combines with recycle and is pumped to a natural gas or refinery-produced gas-fired heater where it is heated to 770°K (920°F). The liquid-vapor mixture leaving the heater passes to a coking drum. Coke builds up to a predetermined level in one drum, and then flow is diverted to the next drum. The full drum is steamed to strip out unconverted hydrocarbons, cooled by water, and then is hydraulically decoked with high pressure water jets. The coke drum overhead vapor goes to the distillation column for separation into gas, gasoline and gas oil.

FIGURE 130: EQUIPMENT DIAGRAM—DRUM COKING

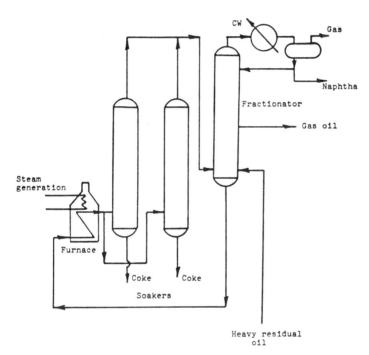

Rejected heat: radiation, convection—10 kJ/kg (5 Btu/lb)
Furnace stack gases—45 kJ/kg (20 Btu/lb) at 700°K (800°F)
Hot exiting column streams (wasted)—70 kJ/kg (30 Btu/lb) at 420° to 530°K
    (300° to 500°F)

Source: Reference (2)

The fluid catalytic cracking process converts gas oils to lower molecular weight products, such as high octane gasoline, middle distillate and olefins. Charge stock may be straight run, cracked and extracted gas oils from a wide variety of sources, ranging from light distillates to vacuum distilled gas oils and deasphalted oils.

High yields of motor gasoline with a clear octane number of 80 to 85 Motor and 92 to 99 Research are typically obtained. Isobutane and light olefins are available for alkylation, polymerization, LPG, synthetic rubber or chemical manufacture. Product distribution can be varied to meet differing market requirements and considerable desulfurization of liquid products occurs. The cracking process is extremely versatile.

The flow diagrams in Figure 131 illustrate two common designs, the Esso Model 4 "U bend" unit and the riser or transfer line cracker. The charge is mixed with the hot regenerated catalyst entering the single riser, in the simplified designs shown, where cracking occurs. Products are disengaged from the catalyst and pass to the main fractionator, where they are separated into the fractions desired. The required amount of heavy cycle oil is returned to the reactor. The unit may be designed, and older designs modified, to effectively increase the conversion to products lighter than the feed to as high as 95%.

FIGURE 131: TYPES OF FLUID CATALYTIC CRACKING UNITS

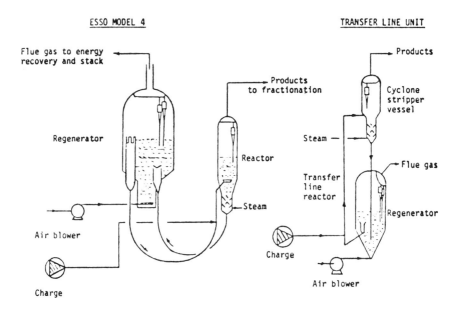

Source: Reference (26)

Figure 132 shows an overall fluid catalytic cracking operation. Gas oil is preheated in a natural gas or refinery-produced gas-fired heater. It then carries regenerated catalyst into the reactor-settler. The product comes out of the top of the reactor while spent catalyst overflows a weir and falls through a steam stripper. The steam removes entrained hydrocarbons. Then the spent catalyst goes to the catalyst regenerator where coke is burned off the catalyst. Regenerated catalyst then flows to the gas oil feed to be swept into the reactor. The product from the reactor is fed to a fractionator where gas, gasoline and light oil are obtained. The temperatures in this operation are approximately 740°K (870°F) in the reactor and 900°K (1160°F) in the catalyst regenerator. Coke combustion in the catalyst regenerator supplies heat for the reactions which occur in the reactor. The flue gas from the regenerator contains combustible carbon monoxide fuel.

FIGURE 132: EQUIPMENT DIAGRAM—FLUID CATALYTIC CRACKING

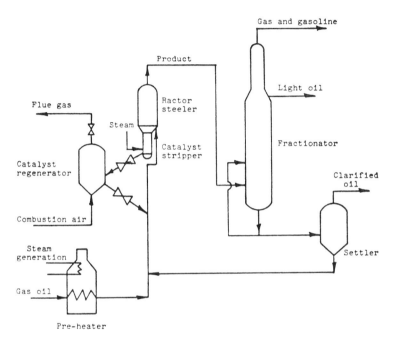

Rejected heat: radiation, convection—95 kJ/kg (40 Btu/lb)
Furnace stack gases: 115 kJ/kg (50 Btu/lb) at 700°K (800°F)
Wasted heat in process streams: 460 kJ/kg (200 Btu/lb) at 340° to 450°K (150° to 350°F)
Reactor stack gases: 160 kJ/kg (70 Btu/lb) at 700°K (800°F)

Source: Reference (2)

The catalytic reforming process is used primarily to upgrade low octane naphtha to high octane gasoline blending components containing significant quantities of aromatic hydrocarbons. Charge stocks may be straight-run, hydrocracked, hydrogenated thermal cracked, coker and FCC naphthas boiling in the range of C_6 to 400°F. The charge may be full range naphtha or selected heart cuts.

A catalytic reforming unit is composed of reaction, separation and fractionation sections. In the reaction section the charge is contacted with a tailored catalyst, containing platinum and possibly other metals, under the proper conditions for the desired reactions to occur. The principal chemical reactions involved are dehydrogenation of naphthenes to aromatics, dehydrocyclization of paraffins, hydrocracking of high molecular weight paraffins, isomerization of paraffins and of naphthenes, and desulfurization of organic sulfur compounds to form hydrogen sulfide.

In the separation section the reaction product mixture is cooled and separated into liquid and gas streams. Some of the hydrogen-rich gas is compressed and recycled to the reactors and the net gas produced is withdrawn from the system. The separator liquid is sent to suitable fractionation facilities.

The four major process variables are temperature, space velocity, pressure and hydrogen recycle rate. The reactor temperature is normally in the range of 850° to 1000°F when operating at pressures from 200 to 800 psig (125 to 350 psig with the newest bi- and multi-metallic catalysts). A hydrotreating process is the typical charge pretreatment used to remove metals and sulfur which will adversely affect reforming catalyst performance.

Figure 133 shows the catalytic reforming operation where naphthas are converted to aromatic hydrocarbons. The dehydrogenation reactions take place at high pressure [2,000 to 4,000 kN/m² (300 to 600 psi)] and at elevated temperatures [720° to 810°K (840° to 1000°F)] in a hydrogen atmosphere. Natural gas or refinery-produced gas is burned in heaters to provide heat for the endothermic reactions which occur. Hot reactor effluent is used to preheat incoming feed and to provide heat to the fractionator at the end of this operation.

FIGURE 133: EQUIPMENT DIAGRAM—CATALYTIC REFORMING

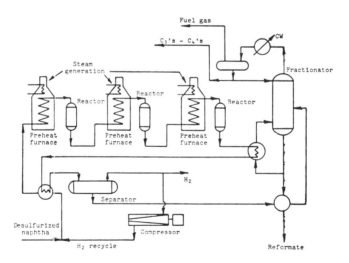

Rejected heat: radiation, convection—45 kJ/kg (20 Btu/lb)
Furnace stack gases—210 kJ/kg (90 Btu/lb) at 700°K (800°F)
Wasted heat in process streams—115 kJ/kg (50 Btu/lb) at
    340° to 400°K (150° to 250°F)

Source: Reference (2)

Figure 134 shows an aromatic extraction operation in which reformed naphtha is separated into its aromatic and nonaromatic components. A glycol-water mixture flows into an extractor and dissolves the aromatic portion of the reformed naphtha feed. The rich solvent is then taken to a stripper where the dissolved aromatics are separated from the solvent. The aromatics then go to a water wash tower where traces of dissolved glycol are removed. The aromatics then are heated in a natural gas or refinery gas-fired heater and fed to a clay tower where impurities are removed. Steam is used to provide heat to the water glycol still shown in the figure.

FIGURE 134: EQUIPMENT DIAGRAM—AROMATICS EXTRACTION

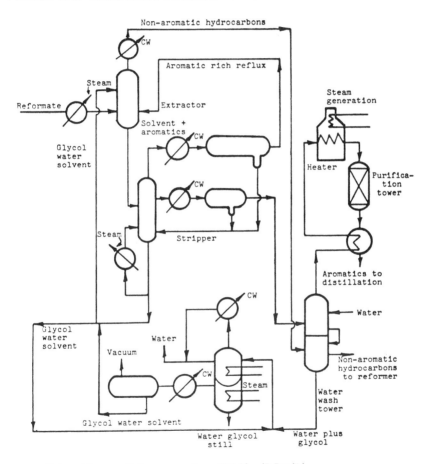

Rejected heat: radiation, convection—10 kJ/kg (5 Btu/lb)
Furnace stack gases—35 kJ/kg (15 Btu/lb) at 700°K (800°F)
Wasted heat in process streams—115 kJ/kg (50 Btu/lb) at 340° to 400°K
    (150° to 250°F)

Source: Reference (2)

The hydrocracking process is used for the conversion of a wide range of hydro-carbon feedstocks to lighter products. Typical charge stocks are naphtha, light and heavy gas oils, vacuum gas oils, cracked and coker gas oils, deasphalted re-siduum and topped crude. The refiner has wide flexibility to make different product slates that can emphasize high octane gasoline blendstocks, jet fuel, low pour point diesel, LPG or low sulfur fuel oil blendstocks.

The hydrocracking process commonly employs a fixed bed catalyst system in an environment of recycled hydrogen under elevated pressure. Due to the de-velopment of new selective and high activity catalysts, many plants employ a single-stage design as shown above. Product fractionation is tailored to individual refinery needs. Liquid yields are typically 110 to 130 volume percent C_4 and heavier material. Very low C_1 and C_2 fuel gas production and high yields of isoparaffins in the C_4, C_5 and C_6 fractions are characteristic. Figure 135 is a flow diagram of the hydrocracking process.

FIGURE 135: HYDROCRACKING PROCESS

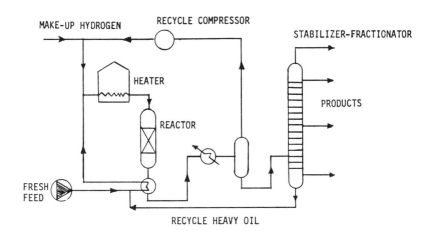

Source: Reference (26)

The catalytic desulfurization process is used to improve qualities of a wide range of petroleum stocks by removal of sulfur, nitrogen and heavy metallic contam-inants. The process hydrogenates olefinic hydrocarbons and improves the color, odor and stability of these petroleum cuts. A specific range of catalysts has been developed for aromatic hydrocarbon hydrogenation, required for instance to improve kerosene smoke point characteristics. Charge stock distillates range from light gasoline to heavy vacuum gas oil either from straight run or thermal and catalytic cracked sources. Typical products are sweet light gasoline, puri-fied catalytic reformer feedstock, desulfurized kerosene, high quality diesel oil and light fuel oil.

Figure 136 shows the catalytic hydrogenation at high pressure of gas oil or naph-tha [1,350 to 6,800 kN/m^2 (200 to 1,000 psi)] to remove sulfur. The feed is mixed with hydrogen-rich gas, heated to 585° to 730°K (600° to 850°F), and passed through a reactor containing a fixed bed of desulfurization catalyst. The feed is heated by hot reactor effluent and by the burning of natural gas or refinery-produced gas in a heater. The product is separated from gases in a high pressure separator, low pressure separator and a stripper. Heat to the stripper is supplied by hot reactor effluent.

FIGURE 136: EQUIPMENT DIAGRAM—DISTILLATE HYDRODESULFURI-
ZATION

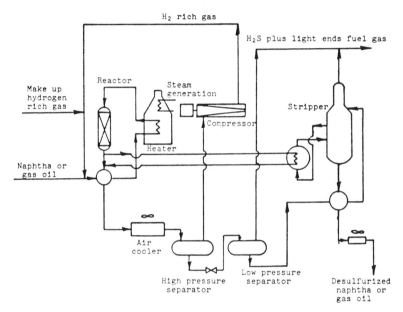

Rejected heat: radiation, convection—25 kJ/kg (10 Btu/lb)
Heater stack gases—115 kJ/kg (50 Btu/lb) at 700°K (800°F)
Wasted heat in process streams—115 kJ/kg (50 Btu/lb) at 340° to 400°K
(150° to 250°F)

Source: Reference (2)

The alkylation process combines propylene, butylenes and amylenes with isobutane in the presence of a catalyst such as strong sulfuric acid to produce high octane branched chain hydrocarbons (alkylate) for use in aviation gasoline and motor fuel.

Most of the recently installed plants are designed to process a mixture of propylene and butylenes. The total debutanized alkylate from these plants has F-1 octane numbers of 92 to 96 (unleaded). When processing straight butylenes, commercial installations can produce debutanized total alkylates with F-1 octane numbers as high as 99.0 (unleaded). Endpoints of the total alkylates from both mixed feeds and straight butylene feeds are typically between 340° and 390°F.

The feed streams along with a recycle stream of H_2SO_4 are charged to the contactor. The liquid contents of the contactor are circulated at high velocities and an extremely large amount of interfacial area is exposed between the reacting hydrocarbons and the acid catalyst. The entire volume of the liquid in the contactor is maintained at a uniform temperature. Since reaction is exothermic, the heat of reaction is removed by indirect propane refrigeration or by auto-refrigeration (allowing some vaporization in the reactor).

The hydrocarbon product is caustic and water washed and fractionated into isobutane for recycling, n-butane for discard (or isomerization) and alkylate. The alkylate goes to a rerun tower to have a small amount of heavier alkylate bottoms removed. A flow diagram of the sulfuric acid alkylation process is shown in Figure 137.

FIGURE 137: SULFURIC ACID ALKYLATION PROCESS

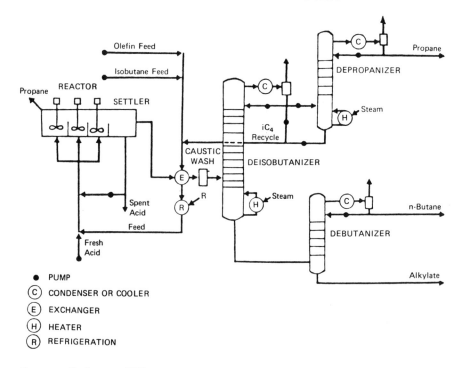

PUMP

CONDENSER OR COOLER

EXCHANGER

HEATER

REFRIGERATION

Source: Reference (26)

Figure 138 shows an alkylation operation catalyzed by hydrofluoric acid (HF) where isobutane reacts with a C_3 to C_5 olefin stream in the presence of the catalyst. The products are branch-chained C_5 to C_{10} hydrocarbons with a high octane number. The product is called the alkylate and is blended into gasoline. The reaction takes place at low temperature [285° to 320°K (50° to 110°F)] and is exothermic. The reactor products are separated by distillation. The fractionator is heated by burning natural gas or refinery-produced gas in the reboiler furnace.

FIGURE 138: EQUIPMENT DIAGRAM—HF ALKYLATION

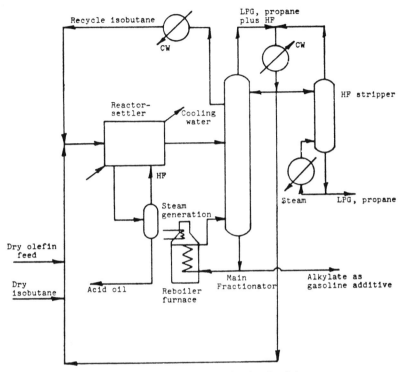

Rejected heat: radiation, convection—25 kJ/kg (10 Btu/lb)
Furnace stack gases—45 kJ/kg (20 Btu/lb) at 700°K (800°F)
Wasted heat in process streams—460 kJ/kg (200 Btu/lb) at 300° to 370°K
   (80° to 200°F)

Source: Reference (2)

As an example of a butane isomerization process, the Universal Oil Products "Butamer" process is shown, which is used for converting normal butane to iso-butane under mild processing conditions by means of a catalyst in the presence of hydrogen. The catalyst is a solid containing platinum, capable of functioning efficiently at relatively low operating temperatures for long periods of time. The once-through product approximates equilibrium concentrations of isobutane; in recycle operations, the product is high purity isobutane.

A typical unit is illustrated in the simplified flow diagram shown in Figure 139. Mixed butane feed is charged to a deisobutanizer concentrating the normal bu-tane in the bottom product. Next, the butane is mixed with hydrogen, often obtained from a reforming unit, heated and charged at moderate pressure to the reactor containing the special catalyst. The reactor effluent is cooled, hydrogen and light gases are separated from the liquid in a separator for recycling via a compressor, and the liquid is then stabilized in a conventional column. The stabilized bottoms product is deisobutanized and high-purity isobutane is taken overhead as product, together with any isobutane introduced in the fresh feed. A small chemical hydrogen consumption results from the formation of minor amounts of methane, ethane and propane.

FIGURE 139: BUTANE ISOMERIZATION PROCESS

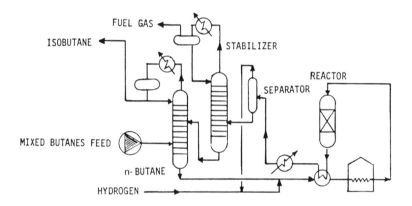

Source: Reference (26)

As an example of a pentane-hexane isomerization process, the Universal Oil Products "Penex" process is shown which improves the octane rating of pentane and/or hexane fractions from refinery naphthas and natural gasolines by isomerization over a platinum-containing catalyst in the presence of hydrogen.

A typical simplified flow diagram for Penex isomerization of a C_5 to C_6 mixture is shown which employs recycle of normal pentane (see Figure 140). If once-through processing is employed, the C_5 to C_6 splitter and deisopentanizer are not required. Hydrogen, employed to minimize formation of catalyst deposits, normally is recycled. Makeup requirements may be furnished from reforming or other sources.

Reaction conditions are mild in severity and noncorrosive. Temperatures range from 250° to 400°F; pressures from 300 to 1,000 psig. Catalyst promoter is added continuously although, because concentrations needed are only in the ppm range, recovery equipment is not installed.

Conversions approaching equilibrium are achieved in a single pass and recycling permits the achievement of virtually complete conversion.

The polymerization process is used to produce a high octane gasoline blending component from olefins, typically propylene, butylenes or mixtures of the two. Phosphoric acid is the most common catalyst used.

The simplified flow diagram (Figure 141) shows the catalyst arranged in beds within the reactor. Since the reaction is exothermic, quench is introduced between beds: the quench oil is often mixed C_3 that has passed through the reactor and is thus depleted in propylene. Another type of reactor consists of vertical tubes packed with catalyst. Cooling water jacketing of the tubes is used to control the reaction.

The reactor is held to a temperature around 400°F at a pressure sufficient to keep the hydrocarbons in the liquid phase (500 to 1,000 psig). Conversions of about 90% are achieved. Polymer gasoline F-1 clear octane number is about 98.

FIGURE 140: PENTANE-HEXANE ISOMERIZATION PROCESS

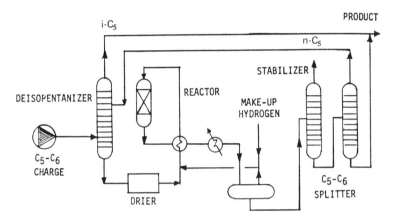

FIGURE 141: OLEFIN POLYMERIZATION PROCESS

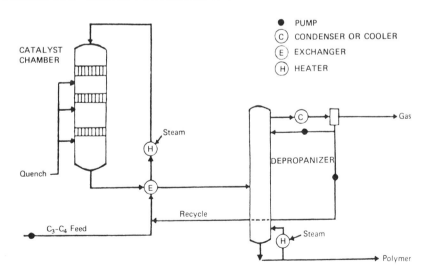

Source: Reference (26)

Steam reforming of light hydrocarbons is used for producing high purity hydrogen such as is needed for hydrogenation and hydrodesulfurization. Natural gas, refinery gas, propane, butane and naphtha are typical chargestocks. The steam-hydrocarbon reforming process includes the basic steps of desulfurization, reforming conversion, CO_2 removal and methanation, as shown in Figure 142.

Since sulfur poisons the catalysts used, the feed stock is passed through activated carbon to remove the sulfur compounds normally found in natural gas. Other desulfurization steps may be necessary for refinery gas or for heavier hydrocarbons. Superheated steam is added and the mixture flows into the reformer furnace, where reactions yielding hydrogen, carbon monoxide and carbon dioxide take place in the presence of a nickel catalyst at approximately 1400° to 1600°F. Downstream finishing of the hydrogen includes conversion of CO to CO_2 by reaction with water vapor (producing more hydrogen), removal of the CO_2 and, finally, conversion of residual CO to methane by reaction with hydrogen.

FIGURE 142: STEAM REFORMING FOR HYDROGEN PRODUCTION

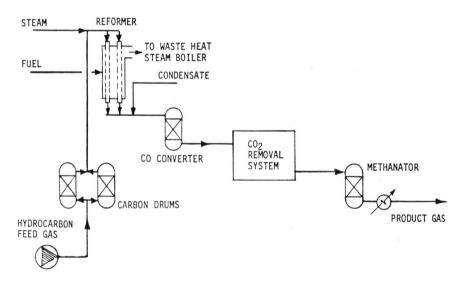

Source: Reference (26)

MAJOR ENERGY CONSERVATION OPTIONS TO 1980

Within SIC 2911, there has always been an awareness of the need for energy conservation because the conservation of fuel used internally in a refinery often represents a gain of saleable product. The other component industries are energy-intensive and must regard energy as a major cost of operation.

The American Petroleum Institute is currently encouraging and reporting progress on energy conservation through a voluntary system. Reports of progress towards an industry-set target of 15% by 1980 are made semiannually. Savings of 10.3%

over base year 1972 were reported by the API for the last half of 1975. Formal reporting programs are not developed in the other component industries, although there is ample evidence of strong efforts by individual companies to set up and manage their own formal energy conservation programs.

Table 178 shows the causes of some energy losses in the petroleum refining process. It also gives estimates of the losses and possible conservation approaches.

TABLE 178: PETROLEUM REFINING ENERGY CONSERVATION APPROACHES

Causes of Energy Losses	Approximate Magnitude of Losses	Energy Conservation Approaches
Rejected heat:		
Unrecovered heat in streams exiting energy intensive operations	1,560 kJ/kg (680 Btu/lb)	Design modification (optimize heat exchange system). Maintenance (keep heat exchange surfaces clean)
Unrecovered heat in stack gases	955 kJ/kg (415 Btu/lb)	Design modification (waste heat recovery)
Radiation, convection	305 kJ/kg (130 Btu/lb)	Insulation Maintenance
Unaccounted for	1,135 kJ/kg (485 Btu/lb)	
Overall process		
High reflux ratios in distillation columns	115 kJ/kg (50 Btu/lb)	Design modification (more plates in column). Operation modification (closer control of columns)
Unrecovered potential energy (pressure) in several operations	70 kJ/kg (30 Btu/lb)	Design modification (use hydraulic and expander turbines)
Loss of fuel value of flue gas from catalytic cracking regenerator	115 kJ/kg (50 Btu/lb)	Waste utilization (use as fuel in boiler)
Lack of integration of electrical generation with steam generation	460 KJ/kg (200 Btu/lb)	Process integration

Source: Reference (2)

There are some short-term energy conservation measures related to improvements in operating practice and to changes or additions to existing pieces of equipment to improve thermal efficiency. In general, most of the measures can be effected by the operating and maintenance departments of refiners, or in some instances by outside contractors, without disruption of production. Longer term energy conservation measures, on the other hand, require changes to systems and/or significant changes to individual pieces of equipment which may require interruption to production. In some cases, the nature of the change may make it practicable only for incorporation into new refineries.

There is a hierarchy of possible system changes, varying from changes in overall refinery configuration to changes in a steam system within a specific process. Changes to refinery configurations may be the result either of the development of new processes or to changes in the combinations and relative capacities of existing commercial processes used. The former type of change is outside of the scope of this study, whereas the latter can be studied by means of a model which simulates the refinery configuration while allowing combinations of existing commercial refinery processes.

A first attempt at such an approach was made as a corollary to the linear programming runs undertaken in this study and described by Gordian Associates (26). An interesting avenue for further research could be the addition of nonlinear optimization techniques (e.g., integer programming) to the linear program so that the optimum refinery configuration can be developed for different costs of energy. The nonlinear feature is required because the capital cost of process units is a nonlinear function of capacity.

As one narrows down the analysis from the overall refinery system to specific subsystems or specific pieces of equipment, it becomes necessary to have a sophisticated thermodynamic analysis to guide the selection of fruitful directions for energy conservation research. Though such thermodynamic analysis tends to be complex and time-consuming, it has the advantage that the conclusions drawn from a specific analysis of one process have some general applicability to all refinery processes, and also to nonrefinery processes, because the subsystems and individual pieces of equipment analyzed are common to most process operations. Similar analysis for other nonprocess industries may not have as general an applicability. In this study, an overall thermodynamic analysis of a typical refinery, and an in-depth thermodynamic analysis of the most complex process within the refinery was made [catalytic reforming, see the Gordian Association report to F.E.A. (26)] and some implications for longer term energy conservation developed from the analysis are presented.

Any process which converts feed materials into other products by effecting chemical and/or physical changes requires the addition or abstraction of a minimum quantity of energy. In order to accomplish such change, it is necessary to apply energy to the process from the surroundings (e.g., in the form of fuel fired, steam, electric energy, etc.) and to abstract energy through cooling water, export steam, etc. Theoretically, it should be possible to effect the process change by supplying from the surroundings or receiving into the surroundings only the minimum energy change required to effect the chemical and/or physical change in the process. In practice, however, quantities of energy which are considerably higher than the minimum have to be expended.

It is important to recognize that not all forms of energy are equivalent. The same quantity of energy may be described as high-grade or low-grade depending on its ability to do work. Electric energy is completely available to do work, and the heat of combustion of a fuel is essentially all available, whereas the heat content of a stream of water at ambient temperature is useless for conversion to work. An available energy analysis does account for the hierarchy of energy by accounting for all material and energy flows in terms of their available energy. Available energy is defined as that part of the total energy content of a source of energy which may be converted to work in its entirety by reducing the source of energy to the conditions of its surroundings (the sink—taken for the purpose of this study to be ambient conditions of 88°F and atmospheric pressure). A thermodynamic evaluation of the total typical refinery showed that there was a significant loss of useful work and that more detailed analysis would be fruitful. Thus, the most complex of all refinery processing modules, one containing all the major elements of the refinery (and many other process industries) was chosen for closer investigation in a Gordian Association study for F.E.A. (26). In order to perform the investigation, a computer process simulator was designed and coupled to a data bank of physical and thermodynamic properties of hydrocarbons. This general program format may be used in any hydrocarbon processing industry for similar analyses.

A summary of the available energy analysis on a typical catalytic reformer (26) shows that only 6.6% of the available energy supplied to the process (essentially all in the form of fuel fired) was usefully used in increasing the energy content of process streams and in exporting steam. Hence, 93.4% of the available energy supplied to the process is wasted. This surprising conclusion is discussed below in the even more important context of why and where the energy is lost.

The large energy loss appears to be typical of refinery processes and shows the large theoretical potential for energy conservation that exists. In order to determine fruitful areas for process improvement research, it is necessary to pinpoint what each system in the process contributes to the overall loss of available energy. A breakdown of available energy in terms of an equivalent gross heat of combustion of the average fuel fired shows that 61.7% of the available energy supplied is lost in the reactor furnace, 12.0% in the depentanizer (mainly in the reboiler), 10.9% in air and water coolers, 5.4% in other exchangers, 0.6% in the recycle gas compressor and 0.4% in mixing charge naphtha with recycle gas.

Note that a purely thermal analysis would not have shown correctly where the loss of the input high-grade energy occurs. For example, the thermal efficiency of the reactor furnace is 72.4%, whereas the efficiency of utilization of the available energy in the fuel fired is only 27.2%. The reason for this discrepancy is that the thermal efficiency does not differentiate between a Btu in the flue gas which is at 760°F and a Btu of heat liberated by combustion (which has a theoretical adiabatic flame temperature of 3600°F).

The efficiency of utilization of available energy, on the other hand, accounts for the hierarchy of the energy flows involved. Another example of the inadequacy of the purely thermal analysis is that no loss would be shown for the interchange of heat between process streams since the heat transferred from one stream is offset by the heat gained by the other stream. In fact, this interchange of heat accounts for 5.4% of the total loss of available energy introduced into the process.

The available energy analysis on the catalytic reformer showed that there is very large potential for energy savings wherever fuel is fired—in a reactor furnace, a steam boiler or a reboiler. Based on a purely thermal analysis, the emphasis for improvement of efficiency might be on reducing flue gas losses which, on an enthalpy basis, account for between 20 to 30% of losses. However, on an available energy basis, the loss in the flue gas is between 5 and 10%. For example, if heat were recovered from the reformer reactor furnace flue gas (e.g., by installing an air or water preheater in a convection section) so that it exited at

570°F instead of 670°F, only 0.493 Btu of available energy can theoretically be recovered per Btu of heat removed. This factor drops to 0.323 for recovering heat by dropping the flue gas temperature from 400° to 300°F. The real large losses of available energy in a furnace occur, not due to the loss in the flue gas, but due to not taking advantage of the potential for transferring heat at high temperatures. For example, if one considers the typical refinery boiler which generates 600 psig steam at around 780°F, efficiencies can clearly be improved by higher pressures at the expense of additional investment. If the potential for work in such higher pressure steam exceeds the work required in the refinery, then some generation of electric energy for internal use and/or export can be considered. In fact, the minimum economic size for boilers generating steam above 2,000 psia is likely to be such that they can only be economically justified if electric energy is generated.

The importance of higher steam pressures is that they result in more available energy gain by the steam for the transfer of the same quantity of heat. This is because the temperature at which water vaporizes increases with pressure while the latent heat that has to be transferred decreases with pressure. There are clearly metallurgical problems which limit the pressure and temperature of the steam which can be generated. However, most refinery furnaces and boilers are nowhere near the present technological limits. For example, supercritical electric utility boilers are being designed for operation at 3,700 psia and superheated steam pressures of 1000°F. At these conditions, the available energy increase per Btu transferred to the steam is approximately 0.489 Btu (this is the average for raising boiler feedwater at 212°F to steam at 3,700 psia and 1000°F). This factor can be compared with a factor of 0.407 which corresponds to the generation of 600 psig, 780°F steam from boiler feedwater at 212°F.

Because of the metallurgical limitations to the generation of steam at much above 3,700 psia/1000°F, some other heat transfer fluid must be utilized to take advantage of the high temperatures resulting from the combustion of fuels. Whatever fluid is used, its advantage is lost if it becomes necessary to transfer heat from it to steam in order to produce work from a steam turbine. For this reason, there is an advantage if the fluid is a gas which can be used directly to produce work in a gas turbine. Research into determining an optimum fluid (e.g., an inert gas) and work-producing cycle appears warranted.

It is possible to conceive that the heart of the future energy system for a refinery or any other large industrial complex would be a large furnace transferring heat to a series of fluids at progressively lower temperatures. The highest temperature transfer could be to a gas used to generate electric energy in a gas turbine with hot turbine exhaust gases being used to generate process steam. Transfers of heat at progressively lower temperatures could be made for additional steam generation, hot oil belts and preheating of water and air. Transfer of heat to the high temperature gas could be through use of a fluidized bed to improve overall heat transfer coefficients.

In addition to the potential for energy conservation in furnaces and boilers, there is a significant potential for energy conservation in the exchange of heat between fluids. For example, in the catalytic reforming process analyzed in the Gordian Association study for F.E.A. (26), 16.3% of the available energy input was wasted in exchangers. 67% of this loss resulted from transfer of heat to air and cooling

water. Clearly, such transfer should be minimized in favor of interchange between process streams. As far as this latter form of heat exchange is concerned,
the available energy loss can be minimized by minimizing the temperature difference between the two streams at any point where heat exchange is occurring.
This obviously requires increased heat transfer area and a more expensive exchanger.

Process designers normally design the temperature approach in exchangers based
on empirical rules of thumb. It is suggested that a simple available energy analysis be used instead to determine the optimum size and configuration of the exchanger. For different exchanger sizes and configurations, the investment cost
should be estimated and the net available energy change (i.e., the difference in
the available energy between entering and exiting streams) calculated. The available energy loss can be taken to correspond to fuel that has to be fired and can
be valued at the value of such fuel. In this way an economically optimum exchanger can be selected. Continued research into the development of cheaper
extended surface exchanges appears warranted. At present, the use of the more
sophisticated of these exchangers appears to be limited mainly to use in cryogenic service. Further investigation of direct contact heat exchange between
gases and liquids also appears warranted.

In conducting an available energy analysis, any particular system can be isolated
for study. In most refinery and chemical process plants, an analysis of the overall steam system is justified. Such an analysis can be aimed at minimizing available energy loss through selection of boiler pressure, drives to be steam-driven,
the different levels of steam pressure headers that should be used, and the
source of steam to such headers (e.g., through use of extraction from turbines).
Where efficiency can be improved by export of electric energy, this should be
explored with the local utility.

Calculations on the energy consumption of a typical refinery indicate the following usage (26): as directly fired fuel in process heaters, 78%; as electrical energy,
the equivalent of 11%; as steam, the equivalent of 10%; and for cooling water
systems, the equivalent of 1%.

The following section describes practical measures for the conservation of energy
in the petroleum refining industry. A variety of such measures are available
that can be adopted to conserve energy throughout the refinery, some of which
are normally applicable and effective in the short to medium-term future and
those which apply more appropriately to the longer term future.

The items described in this report are frequently equally applicable to existing
and to proposed plants, as well as to other industries which use similar types
of equipment. Their successful application in a majority of cases depends not
only on having the correct equipment, but on having strong management commitments to energy saving and adequate supervision of day-to-day operations,
when short-term savings of 10% should be easily achieved in most plants.

Table 179 shows a checklist for energy conservation measures with notes on
their applicability to existing or new plants and also on their effectiveness in
the short, medium or long term.

TABLE 179: SUMMARY OF ENERGY CONSERVATION MEASURES

No.	Equipment or System	Item	Remarks	Normally Effective Short, Medium or Long Term*	Normally Applicable Only to Existing or New Plants**
1	Fired heaters	Control of excess air	Control of combustion	S	E
2	Fired heaters	Fuel atomization	Control of combustion	S	E
3	Fired heaters	Flame patterns	Control of combustion	S	E
4	Fired heaters	Stack gas temperature	Control of combustion	S	E
5	Fired heaters	Install convection section	Heat recovery from stack gas	ML	EN
6	Fired heaters	Install air preheater	Heat recovery from stack gas	ML	EN
7	Fired heaters	Cleaning of equipment, items 5 and 6	Heat recovery from stack gas	S	E
8	Heat exchangers	Configuration of flows	Optimize heat recovery	ML	EN
9	Heat exchangers	Area of surface required	Optimize heat recovery	ML	EN
10	Heat exchangers	Minimize heat loss to air/water	Optimize heat recovery	SML	EN
11	Heat exchangers	Cleaning cycles	Optimize heat recovery	S	E
12	Heat exchangers	Extended surface	Improve heat transfer efficiency	SML	EN
13	Heat exchangers	Porous boiling surfaces	Improve heat transfer efficiency	SML	EN
14	Heat exchangers	Water washing	See also 11	S	E
15	Heat exchangers	Water scaling control	See also 11	S	E
16	Heat exchangers	Biocide fouling control	See also 11	S	E
17	Air coolers	Fin continuity, thermal contact	Maintain heat transfer efficiency	SML	EN
18	Air coolers	Hot air recirculation	Avoid loss of capacity	SML	EN
19	Air coolers	Inlet air distribution	Improve heat transfer efficiency	ML	N
20	Air coolers	Protection of bundles	Avoid cycling	ML	N
21	Air coolers	Temperature control, variable pitch fans	Minimize energy to drive fans while controlling process temperatures	SML	EN
22	Air coolers	Hub and tip seals	Minimize air recirculation	S	E
23	Pumps and drivers	Correct pump characteristics for required service		SML	EN
24	Pumps and drivers	Correct motor sizing for required service		SML	EN
25	Pumps	Replacement of undersized lines and control valves	Investigate equipment for internal fouling, corrosion, etc.	S	E
26	Pumps	Spillback or recirculation systems		SML	EN

(continued)

TABLE 179: (continued)

No.	Equipment or System	Item	Remarks	Normally Effective Short, Medium or Long Term*	Normally Applicable Only to Existing or New Plants**
27 28 29 30	Compressors	See above, Pumps		SML	EN
31	Fractionating towers	Optimum reflux ratio and number of trays		ML	N
32	Fractionating towers	Reduce reflux ratio	Review product purity	SML	EN
33	Fractionating towers	Feed preflash, temperature		SML	EN
34	Fractionating towers	Optimum heat exchange	Feed-bottoms, etc.	SML	EN
35	Fractionating towers	Product side draw		SML	EN
36	Fractionating towers	Operating pressure	Reflux reduction may be possible	S	E
37	Fractionating towers	Vapor recompression		ML	N
38	Fractionating towers	Tray loading	For optimum efficiency	SML	EN
39	Tanks, vessels	Mixing techniques		SML	EN
40	Tanks, vessels	Suction heater		SML	EN
41	Insulation	Increase insulation thickness		SML	EN
42	Paint	Maintain correct color in good condition		S	E
43	Instrumentation	Monitor process conditions to allow optimum operation	See other sections. Additional instruments often show good payout	SML	EN
44	Instrumentation	Correct sizing of control valves and orifice meter plates	Minimize pressure drop	SML	EN
45	Vacuum systems	Review required operating pressure	Cut down on steam rise to eductors	SML	EN
46	Vacuum systems	Eliminate leakage	Minimize unnecessary uncondensables loading	S	E
47	Vacuum systems	Efficient condenser operation		S	E
48	Vacuum systems	Review condenser type for possible improvement	Recovery of condensate may be possible	SML	EN
49	Cooling water system	Control use according to plant throughput		S	E

(continued)

TABLE 179: (continued)

No.	Equipment or System	Item	Remarks	Normally Effective Short, Medium or Long Term*	Normally Applicable Only to Existing or New Plants**
50	Cooling water system	Review system pressure drop		SML	EN
51	Steam systems	Efficient water treating	Reduce blowdown ratio	SML	EN
52	Steam systems	Condensate recovery		SML	EN
53	Steam systems	Useful work from steam let down		SML	EN
54	Steam systems	Efficient steam trap operation		S	E
55	Steam systems	Steam tracing efficiency		S	E
56	Steam systems	CO boilers on fluid catalytic cracking plants			N
57	Electrical systems	Load levelling		L	E
58	Electrical systems	Power factor improvement		S	E
59	Power recovery systems	From gases, in expansion turbines		S	N
60	Power recovery systems	From liquids, in hydraulic turbines		L	N
61	Loss control	Tank evaporation losses		L	E
62	Loss control	Many miscellaneous items, including reduced flare losses, slop oil recovery, etc.		S	E
63	Catalyst development	Improved catalysts	Lower energy requirements due to lower operating temperatures, better selectivity, less hydrogen recycle, etc.	S	E
64	Process development	Improved processes	–	ML	EN
65	Alternative process sequences	–	–	L	N

*Short term is typically less than 6 months, medium term is 6 months to 1 year and long term is over 1 year.
**The categories of existing (E) and new plants (N) are for guidance only; exceptions will be found occasionally. Where an item is identified as applicable to both new and existing plants (EN), design of the new plant should consider that item and existing plant operations should be checked for possible retrofit of equipment or use of the specified technique.

Source: Reference (26)

The descriptions of conservation measures which follow are summaries of the main characteristics of each measure as drawn from a later Gordian Associates study for F.E.A. (6).

Fired Heater Improvements

Combustion Control Instrumentation: The proper control of combustion is important for refinery energy efficiency. For any quantity of fuel to be burned in a heater, there is a theoretical quantity of air required for complete combustion. Reducing the amount of air available will lead to incomplete combustion and a rapid decrease in efficiency. To allow for small variations in fuel composition and flow rates, in air rates, and in fuel-air mixing, all of which occur even in good industrial practice, it is normal to aim for operation with a small amount of excess air, say 5 to 10% above the theoretical minimum. Adding too much excess air will lead to increasing inefficiency as the heat liberated by fuel combustion is wasted in warming up the air prior to its discharge through the stack (6).

Control of excess air may be achieved through the use of air blowers and stack dampers. The quantity of excess air may be monitored by observing the oxygen content of the stack gases, either by analyzing manually drawn samples using a portable oxygen analyzer or using fixed instruments with continuous sampling systems. Combustibles analyzers may also be used to check firing conditions.

A large number of types of stack analyzers are commercially available, and the technology is considered well proven in the industry. Analyzers can usually be justified on all sizes of heaters. Sampling systems and instruments can readily be installed on existing heaters in a short period of time, requiring no special downtime. Delivery times on instrumentation are approximately six months.

Rates of return on combustion control instrumentation are generally quite high, but vary depending on initial energy efficiency. In addition, efficiency improvement may be achieved in large measure by manpower training, reducing the potential to be gained through instrumentation. Two representative examples of instrument applications are as follows. The first example shows savings of 20.8 billion Btu per year at a cost of $18,600, the rate of return being 188.1%. The second example, for a cost of $290,000, shows savings of 1.1 trillion Btu per year and a rate of return of 39.3%.

Installation of combustion control systems is consistent with industry plans and is judged to contribute significantly to the conservation target.

Improved Burner Designs: Intimate mixing of fuel and air is essential to achieve efficient combustion and there are many proprietary burner designs available. Replacing old burners with new, more efficient, burners may provide useful fuel savings, but is highly specific to each furnace under consideration. It is therefore assumed that the burner modification will fall in the category of housekeeping for the target-setting procedure (6).

Replacement of Old Heaters with New Equipment: Prior to the early 1970s, a heater with a net efficiency of 75% was generally considered to be performing well. The latest design of process heater with a duty of over 20 million Btu/hr should achieve an efficiency of 90 to 92% (net). The life of a process heater is somewhat variable, but a figure of 30 years would seem a reasonable assumption. Thus the opportunity will occur from time to time for a refinery to perform a major furnace rebuild, or to replace the old equipment entirely. The energy impact of replacement will typically amount to an improvement in efficiency from under 70% to about 90%. Nonenergy benefits can include safer operation, reduced maintenance and reduced noise levels.

A relatively favorable heater replacement example obtained from industry contacts is that in which an investment of $1.25 million saves 314 billion Btu per year. In this case, the rate of return of 47.9% would justify replacing the heater sooner than its physical life might otherwise justify. However, such early replacement will require a change in thinking by much of the industry which is unlikely to occur prior to 1980.

This item is clearly consistent with industry plans. Replacement of a heater will take typically one to two years, including planning, engineering, equipment fabrication and installation at the plant. The item is judged a significant measure in the target setting procedure (6).

Conversion of Natural Draft to Forced Draft Operation: A primary limitation of a natural draft burner is pressure drop available across the burner, and air flow is at a relatively low velocity. Good fuel-air mixing may thus be difficult to obtain. A forced draft burner system does not have this pressure drop limitation and therefore the energy in the air stream may be used to obtain intimate mixing with the fuel. With forced draft systems, it is also possible to increase the heat release per burner (smaller flame size), and air flow control is easier to achieve. Forced draft systems are of course necessary when an air preheater is installed on a furnace.

In spite of advantages for forced draft systems, conversion from natural draft is relatively expensive and is unlikely to be considered except when burner replacement is required for the burning of alternate fuels. For this reason, this item is not considered as a separate measure for energy conservation, but is included in the application of air preheaters and conversion from gas to multifuel firing (6).

Sealing Air and Flue Gas Leaks: Air leakage into a furnace is detrimental because this air must be heated from ambient temperature to furnace stack temperature, wasting fuel and thus reducing the furnace efficiency. Serious air leakage into the heater can also reduce combustion zone temperatures and impair heat transfer. Air leaks may also give misleading readings when excess air measurements are being taken, and thus inefficient combustion can result based on a false oxygen indication in the stack gas. Possible areas of air leakage include header boxes, inspection hatches and cracks around burners and joints between duct sections. This item is considered a housekeeping measure for the target setting procedure (6).

Proper Atomization and Flame Patterns: As discussed previously, fuel must be properly atomized and intimately mixed with air to achieve efficient combustion. Often atomization is achieved in conjunction with the use of steam, or it may be achieved by spraying the fuel through a specially designed burner tip. Depending on the degrees of control exercised on combustion air, on atomization and on the positioning of the burner tip relative to air entry, flame patterns in a heater may depart from optimum conditions. Impingement of flames on tubes is a dangerous situation and must therefore be avoided. Proper flame geometry signifies efficient combustion and efficient transfer of heat by radiation to the furnace tubes. Both of these items require regular attention and are important housekeeping tasks (6).

Air Preheaters to Recover Stack Gas Waste Heat, Use of Process Streams to Pre-heat Combustion Air and Installation of Convection Sections: By far the largest part (up to about 80%) of the energy consumed in refining is in the form of fuel to process heaters or boilers. Stack gases leave the furnaces at temperatures vary-ing typically from 400° to above 1000°F in some cases. These gases represent the major area of energy loss in the combustion process, with some additional losses of the order of 3% due to radiation and convection from the outside of the firebox. The three conservation measures listed represent techniques for re-covering waste heat and applying it to preheat combustion air or to some other useful task such as process fluid heating or steam raising.

Where space allows, the use of air preheaters may often prove economic to raise heater efficiency. Based on industry contacts, it appears that retrofitting pre-heaters is unlikely to be attractive for process duties under about 60 million Btu per hour. Where installed, a preheater may be expected to improve efficiency by 10 to 15%.

The application of preheaters to new plant designs is relatively common, and the technology is basically well accepted in the industry. Lead time for a retro-fit project for engineering, fabrication and installation is typically about 2 years, with most of the installation work possible without shutting down the plant: final ducting tie-ins can usually be made at a regularly scheduled maintenance shutdown. No special manpower is needed to engineer the installation of an air preheater, and there are no environmental impacts and no special relation-ships with natural gas curtailments. Air preheaters typically cost upwards of one million dollars and yield rates of return of 20 to 40%. Air preheater instal-lations are consistent with industry plans in general, and can contribute usefully to energy conservation by 1980.

The use of hot process streams to preheat combustion air is technically feasible but it is believed that the number of opportunities for application in a refinery is strictly limited. Hot process streams are generally best used to heat other process streams using conventional heat exchange equipment. This measure is not judged relevant to target setting. Note that these comments do not apply to the use of process streams to recover heat from convection sections for sub-sequent preheating of combustion air (as employed by the patented Heat Re-search Corporation slipstream preheater system). Systems such as this are in-cluded in the same category as air preheaters for the purpose of target setting.

The installation of convection sections may be justified on heaters down to sizes in the 10 million Btu per hour range or below in specific cases. The cost of in-stallation and the use to which recovered heat can be put are highly variable items. Often a convection section can be integrated into a steam system in some way, such as by adding a preheating section for boiler feedwater or as a super-heating section for steam. There can therefore be difficulties in optimizing heat recovery systems based on convection sections tied into utility systems, and the integration of process heaters into utility systems is sometimes opposed by operating personnel. Apart from these reservations on operating flexibility, convection sections are a fully developed, proven and accepted technology, requiring about one year for installation and commissioning in a typical retrofit application. No special downtime should be required, all work being carried out during regularly scheduled maintenance shutdowns.

Two representative examples of convection sections on heaters are as follows: each costs $50,500, but one saves 7.92 billion Btu per year while the other saves 23.8 billion Btu per year. Hence, the return on investment is 30.8% in the first case and 90.4% in the second. In view of the high rates of return, their variability is not a limiting factor.

One example for a convection section on a boiler indicated a rate of return of 62%, based on a cost of $714,000 and an energy savings of 240 billion Btu per year.

Convection sections have no environmental impact, no specific relationship with natural gas curtailments, and are consistent with industry plans for conservation. They are judged an important contribution to the conservation target (6).

Replacement of Bare Tubes with Extended Surface Tubes in Convection Sections: The heat transfer coefficient in a convection section is generally quite low and therefore the recovery of a useful amount of heat will often require the installation of a large amount of heat transfer surface. Extended surface tubing may be employed, consisting of tubes with external studs or fins. While the extended surface assists heat transfer, there is also introduced the risk that particulates from the fuel (ash, catalyst particles, unburned carbon, etc.) will accumulate between the studs or fins, seriously impairing heat transfer after some period of heater operation. Regular cleaning is therefore essential.

When bare tubes have been installed in the original design, it is often possible to improve heat recovery by changing to extended surface tubes. However, it is judged that the majority of convection sections have extended surface tubes installed, and that the potential for further conservation on an industry-wide scale is therefore too small to warrant further consideration of this item (6).

Addition of Extra Tube Rows in Heaters: This measure involves the addition of heat transfer surface as tubes in the heater itself in order to increase heat absorption. The ability to add tubes in a heater is entirely dependent on the original design configuration. While offering a means of improving efficiency on some heaters, it is felt that the potential for adoption of this measure on an industry-wide basis is too small to justify inclusion in the target-setting procedure (6).

Improved Insulation: Heat loss by radiation and convection from the heater outside wall is typically in the range of 2 to 3%. Insulation may be added externally to reduce heat loss but this can lead to excessive wall temperatures. Where space is available between the internal refractory lining and the heater tubes, it is often possible to add more insulation in the form of a blanket of ceramic fiber. The technique is well known in industry, and is consistent with industry plans for conservation. Material delivery time is under six months, and the insulation may be installed during regular scheduled maintenance shutdowns. This item is judged relevant to the conservation target (6).

Soot Blowers for Furnace Tube Cleaning: The use of convection sections, and of extended surface tubes in such services, has been described previously. The extended surface tubing is particularly susceptible to fouling where liquid fuels are used, as these fuels often introduce ash forming components. A regular cleaning procedure is therefore required and soot blowers are installed on many furnaces for this purpose. The soot blower consists essentially of a tube with holes at regular intervals through which high pressure steam is passed. The steam jets blow away most of the deposits which accumulate on these tubes.

Soot blowers are standard practice in industry. For the purpose of target setting, the installation of soot blowers on furnaces being converted from natural gas firing, or being given dual firing capability, is considered as an energy conservation measure, and their continued use in existing heaters is regarded as housekeeping. Lead time for design and delivery of soot blowers is under six months for a typical retrofit application, and the soot blowers can be installed at a regular maintenance shutdown. No special resources are required. While soot blowing does generally cause the emission of particulates from a heater for short periods of time each day (1 or 2 minutes duration), there are no other environmental impacts.

Natural gas curtailments are in fact the major factor in deciding to convert a heater or apply dual firing capability. A representative industry example indicated some soot blower installations costing $698,000 and saving 482 million Btu per year, for a rate of return of 69.5%. Installation of soot blowers is consistent with industry plans and is relevant to the conservation target (6).

Cleaning of Tube Internals: Although efficient combustion may be achieved in a heater, transfer of heat to the process fluid may be severely inhibited by a buildup of coke inside the tubes. Housekeeping measures to remove the coke are well developed, although almost all methods require the plant to be shut down. Methods include: [1] steam-air decoking, in which a controlled burning of the coke deposits takes place, [2] chemical cleaning, and [3] mechanical cleaning.

Improvements in Boiler Efficiency

All of the following items correspond to similar measures discussed previously in relation to fired process heaters, and they are judged relevant to the energy conservation target: convection sections, combustion control instrumentation, replacement of old boilers with new equipment, soot blowers, and improved insulation. The following items are specific to boiler operation: boiler feed water preheating and boiler blowdown heat recovery.

Boiler Feed Water Preheating: This general item is covered in other categories, including (Convection Section) and (Boiler Blowdown Heat Recovery).

Boiler Blowdown Heat Recovery: To control the concentration of solids in a boiler water system, it is necessary to operate a regular purge stream of water. This is known as boiler blowdown, and typically requires the equivalent of about 5% of the steam make to keep water impurities under control. The temperature of the blowdown water will vary from boiler to boiler, but is typically 500°F or above. A significant amount of waste heat can therefore be represented by the blowdown, and its exchange of heat with, for example, incoming cold boiler feed water can be used to improve boiler efficiency. A representative example from industry of boiler blowdown heat recovery has a cost of $9,300 and saves 5.19 billion Btu per year. Its rate of return is a handsome 98.5%.

The use of blowdown to heat incoming boiler feed water represents a proven, accepted technology. Time for design and installation of a typical system should not exceed about 9 months, and no special downtime is required. The recovery of heat from blowdown is believed consistent with industry plans and is judged relevant to the conservation target (6).

Heat Exchanger Improvements

Recovery of Waste Heat from Product Streams, Optimum Crude Unit Preheat Systems and Reactor Feed Effluent Applications: Having transferred the heat of fuel combustion efficiently from the furnace firebox into the process fluid, it is important to continue to utilize this heat efficiently throughout the process. A typical refinery will contain large numbers of heat exchangers, used to transfer heat from one process stream to another. In a crude distillation unit, for example, the incoming crude oil is heated against various product and reflux streams before entering the fired heater to be brought to the desired fractionating column flash zone temperature. A large number of combinations of flow path are obviously possible. There are, therefore, two major considerations:

(a) the configuration of flows (the "order" of heat exchanges for the crude oil)
(b) the amount of heat exchange surface supplied within the chosen configuration.

There is usually a significant incentive to search for the optimum preheat train.

Similar considerations of stream flow configurations and heat exchanger surface area are performed to some degree for all refinery processes. The final equipment specifications are determined by the economics of each refinery situation. For example, in the early 1970s, one particular process design constructor would typically specify three to five reactor feed-effluent heat exchangers for U.S. catalytic reforming plants while specifying five to seven similar exchangers for same sized European plants, due to the greater emphasis in Europe on fuel economy. This situation is, of course, strongly dependent on fuel cost. As fuel costs rise, it is therefore most probable that existing plants could profit from the installation of additional heat exchanger surface in many cases previously considered only marginally economic. Installation of heat exchangers can often be carried out simply and quickly, affording almost immediate energy conservation benefits.

One further aspect of heat exchange to be considered in all plants is the use of cooling water, or air cooling. At some point in a process, it will become uneconomical to install an exchanger to recover heat from a relatively low temperature stream, although the stream does require cooling before transfer to storage tanks. In such a case, it is normal to cool the stream with air or water, rejecting the low level heat into the atmosphere directly or through the water which will itself subsequently lose the heat to the atmosphere in some way or other. While the heat may be at a relatively low temperature, the total quantity of Btu may be substantial. For this reason, use of water cooling (or air cooling) should be minimized and only used where unavoidable to cool streams around 200°F or below.

The amount of energy that can be saved by adding heat exchangers in a refinery, or by modifying heat exchange trains, is obviously highly specific to the plant in question. Nevertheless, this item is judged important to the energy conservation target. Installation of additional exchangers is a well accepted method of saving energy, and the lead time for a retrofit project would generally be less than one year. By appropriate prefabrication of piping and advance completion of civil work, it should be possible in most cases to complete installation during scheduled unit shutdowns. No environmental impact results from the measure, nor is there any relationship with natural gas curtailments.

Because heat exchanger applications and benefits are quite variable, several representative cases were obtained from industry contacts. The investment cost, energy savings and rates of return are shown below for each.

Representative Heat Exchanger Investments

Application	Cost, $	Energy Savings (billion Btu/yr)	Return on Investment, %
Fractionator preheater	64,600	268	663
Crude-naphtha exchanger	60,500	8.07	27.2
Waste heat recovery for feed preheat	452,000	93.2	23.2
Reactor feed effluent exchanger	122,000	66.6	96.5

Opportunities to add additional exchangers where they are now in use are quite common. Hence, heat exchangers may be applied until the incremental investment has a rate of return equal to the hurdle rate. Additional heat exchanger installations are thus consistent with industry plans and are judged relevant to target setting (6).

Improved Hot Oil Systems: In many plants, relatively small heating duties are performed over a wide area by circulating a stream of hot oil around the site and passing it through a number of heat exchangers. The oil itself is reheated by a central furnace, or by heat exchange with a hot product such as vacuum residue from a crude distillation unit. By careful consideration of the various possible heat sources and the locations where the hot oil is used, it may be possible to achieve improved heat utilization in a plant. Unfortunately, instant results are difficult to obtain: as for the optimization of steam systems (discussed subsequently), integration of heating functions across several plants requires careful consideration of factors such as reliability, relative on-stream factors of the plants concerned, and the potential loss of flexibility involved in too tight an integration of operations.

Nevertheless, one company contacted during this study indicated a multimillion dollar hot oil circulating system was being installed to integrate several heating and cooling operations at a large refinery, including boiler feed water preheating as part of the system. Large energy efficiency gains are clearly possible, subject to constraints such as availability of experienced engineers to identify and design the improvements, and time to install equipment at scheduled shutdowns of plants.

This item incorporates fully accepted techniques and is consistent with industry plans for conservation. Depending on the complexity of the system in question, 2 to 3 years may be required before finally commissioning the entire system. With a complex system, advantage would be taken to install piping and valves in each process plant at regular scheduled shutdowns, and new equipment could then be tied in at a later date. This item is judged relevant to the conservation target (6).

Lower Pressure Drop Construction: The pressure drop across a heat exchanger affects the energy needed to move process fluids through the exchanger system. Lower pressure drop therefore can lead to less horsepower being expended in compressors or pumps, but the cost of lower pressure drop equipment may well be higher than standard equipment. Installation of lower pressure drop exchangers is therefore unlikely to be observed in retrofit situations, and is more likely to be justified in completely new plants. For the purpose of energy conservation target setting, it was judged that this item should be ignored (6).

Interchange of Heat Between Different Process Units: As an example of heat integration, the heat given up by a process stream in one unit to cooling water might be usefully used in another process unit to preheat a feedstock or boiler feed water. By coupling the two units, energy efficiency can often be improved significantly, although operating flexibility may be adversely affected. In a large refinery, there will exist many opportunities for heat integration, each of which must be judged on its own merits for profitability, as well as for the less tangible factors such as reduced flexibility of operation.

This item generally requires simple equipment: the concept of heat integration is well accepted and proven in the industry. Lead time to complete a single retrofit project may, however, be 1 to 2 years, depending on complexity and on the time available at regular scheduled unit shutdowns to install piping and equipment. Heat integration is fully consistent with industry plans and is judged relevant to the target setting procedure (6).

Direct Hot Rundown-Hot Feed Systems: A specific form of heat integration between two units is the use of hot feed systems. For example, crude distillation units commonly produce materials which are to be processed further in downstream conversion units. One such material is vacuum gas oil for use as catalytic cracker feed. In many refineries, the vacuum gas oil is cooled against water and stored in a tank prior to being pumped to the catalytic cracker. In the catalytic cracker, the cold feed is heated prior to contact with the circulating catalyst. It is often possible to avoid cooling the gas oil and subsequently reheating it by by-passing the storage tank and feeding hot gas oil directly from the crude unit to the catalytic cracker. A representative industry example of a direct hot feed system had a cost of $678,000 and saved 238 billion Btu per year, the rate of return being 64.7%.

Other examples may exist in a refinery and may offer useful opportunities for energy conservation. The measure is fully accepted in the industry and is usually simple to design and install. Direct hot feed systems are consistent with industry plans and are judged relevant to the conservation target. Because of its importance, this item is considered separately from other heat integration possibilities.

Integration of Utility Generation with Process Plants: The use of process heater convection sections to heat boiler feed water or to raise steam has already been discussed. Recovery of waste heat in a process unit can often be associated with utility systems, which can usually make use of low grade as well as high grade heat. This item is considered covered in other categories.

Use of Tank Suction Heaters to Replace Steam Coils: In the case of fuel oil tanks, it is often necessary to keep the contents at a temperature high enough to maintain pumpability. Steam coils, which are frequently inefficient, may be partially or totally replaced with a combination of the appropriate insulation and a suction heater installed at the tank offtake nozzle. A suction heater is used to warm up only that part of the oil which is going to be pumped at that particular time.

This item is not believed to represent a major measure for energy saving by 1980. It is therefore judged not to be relevant to target setting as a separate item and its effects will be assumed included within general housekeeping (6).

Use of Special Surfaces for Heat Exchangers: When heat exchange between two streams is hindered by the resistance to heat flow of one of the streams, it is common to employ so-called extended surface tubing. For example, in the cooling of a hot process stream, heat transfer through the liquid film and the tube metal is relatively easy, while heat transfer through the air film is difficult. The overall efficiency of heat transfer is thus governed almost entirely by the air side and the area of tubing required to cool the stream adequately is very high. To save space and reduce the cost of equipment, some type of extended surface tubing is often used. The outside area of the tubing is increased by the use of pegs, fins, discs or other devices to substantially increase the tube area in contact with, for example, the air (6).

Porous Boiling Surfaces: Heat transfer to a boiling liquid is a common step in a refinery. Almost all fractionation equipment utilizes a reboiler to heat some of the bottom liquid and vaporize a part of it for reintroduction to the lower section of the fractionating tower. Steam generation is another example.

The mechanism of boiling is a complex subject outside the scope of this report. However, in an effort to improve boiling heat transfer, special surfaces have been developed to provide the sites needed for nucleate boiling, in which vapor forms as bubbles rather than as a film across the metal surface. Work has been continuing for many years but it is only relatively recently that the developments have been put on a fully commercial basis. The Union Carbide Corporation—Linde Division licenses "High Flux Tubing," which consists of a porous metal matrix bonded intimately to the boiling side of a standard tube. The matrix contains a large number of cavities that provide the nucleate boiling sites to improve the heat transfer rates.

A considerable number of full scale test installations of porous boiling surfaces has been studied by Union Carbide to demonstrate rates can be improved by a factor of from 3 to 8 over conventional tubing in most process heat transfer applications involving clean liquids (which do not promote fouling of the fine pores). This unfortunately restricts use of the porous surfaces in refineries, where coking, polymerization or gum formation are experienced with a variety of streams.

As a result of investigations made for the Gordian Associates study for F.E.A. (6), it became clear that extended surface tubing was generally being utilized where appropriate, and that porous boiling surfaces were currently considered inappropriate for most refinery services because of the risk of fouling. It was judged that this item was not relevant to target setting over the time period being considered.

Optimization of Exchanger Cleaning Cycles: While heat exchanger systems may be very effective when first installed, many such systems become dirty in use and heat transfer rates suffer significantly. It is therefore important to establish heat exchanger cleaning schedules, based preferably on observations of the actual efficiency deterioration of the exchangers in question. The details of the cleaning schedules will depend on an economic assessment of each situation.

A number of efficiency parameters can be used. One or more selected parameters can be determined from time to time and changes in exchanger performance followed. The most direct simple parameter is the overall heat transfer coefficient (HTC). This quantity may be calculated from the exchanger heat duty, its area of heat exchange surface and the various stream temperatures. The HTC is a useful parameter to follow, particularly if a data bank is built up for a given piece of equipment and current performance is compared to historical data.

The knowledge that a particular heat exchanger is fouled is, of course, of little value unless something can be done about it. Many companies are installing on-stream cleaning systems, which will allow an exchanger to be by-passed and taken out of service for a short period of time for cleaning. Cleaning may require opening the heat exchanger and washing down the bundle with high pressure water, or it may simply involve back flushing to remove accumulated deposits on the inlet side of the exchanger.

The concept of on-stream cleaning is well known and accepted in the industry. Installation of the necessary piping and valves is readily completed during a regular scheduled plant shutdown. From contacts made in the course of this study, it is clear that exchanger cleaning measures are consistent with industry plans. A representative example of on-stream cleaning is as follows: an investment of $44,600 shows savings of 39.2 billion Btu per year. Its highly favorable rate of return of 149.9% is not unusual for this measure. This item is judged relevant to target setting (6).

Use of Antifoulants on the Process Side and Use of Antifoulants, Biocides and Scaling Control on the Water Side of Coolers: Process equipment fouling can be a troublesome and costly problem in refining operations. Insoluble hydrocarbons, formed in process streams, deposit on heat transfer surfaces creating an insulating barrier and acting as a binding agent for entrained solids. These agglomerates limit plant capacity, reduce heat transfer and increase fuel and maintenance costs. The use of antifoulants to prevent attachment of deposits has been shown in many cases to be a successful means of improving heat transfer. However, the success of any particular antifoulant will depend on the process conditions and on the rate of antifoulant addition, and therefore a certain degree of experimentation is often needed to identify an antifoulant that will work. Action should also be taken to minimize scale formation on the water side of coolers. Most cooling water systems use water containing significant quantities of carbonates which will form scale on heat exchange equipment if: (a) the pH of the circulating water is too high (alkaline environment) or (b) the water is allowed to become excessively hot at any time.

The actual composition of the water sets the allowable temperature. For example, a particular refinery uses a well water system with a hardness of about 800 ppm calcium carbonate and does not allow cooling water to exceed 110°F at the outlet of any coolers. Acid is added periodically to the circulating system to keep the pH around 6.5 to 7. Where seawater is used, it is often necessary to take action to kill slime, algae and barnacles which would otherwise flourish in the warmed-up water. Control by injection of chlorine or proprietary biocides is widely practiced.

As with fouling of process heat exchangers, antifoulants may also be used to prevent adherence of scale deposits on tube walls. Both of these items are regarded as housekeeping for the purpose of target setting (6).

Air Cooler Improvements

Use of Induced Draft Configurations: While the advantages of induced draft configurations may be significant with respect to energy use, it is considered unlikely that conversion of forced draft coolers to induced draft will be justified for existing plants. For reference purposes, a discussion of draft systems is given in the technical appendixes to this report, but no allowance for induced draft benefits is included in the target setting procedure (6).

Automatic Fan Pitch Control: Many forced draft cooling towers and conventional fin-tube air coolers operate with constant speed, constant pitch, fans. Air flow does not vary, and a constant electrical load is drawn by the fan motors. However, this situation can lead to excessive power use because control systems are in fact available to decrease the air flow whenever maximum cooling is not needed: the power taken by the fan motors may thus be correspondingly reduced, saving in some situations about one-third of the electricity used for the no-control situation.

The degree of control needed for the system outlet temperature, whether it be cooling water in the case of a tower or a process stream in the case of a conventional air cooler, is governed by process considerations. In some cases, maximum cooling is required and therefore the air cooler or tower outlet temperature will vary according to ambient conditions. Inasmuch as ambient temperature is a function of plant location, a cooler or cooling tower could be subject to yearly ambient temperature ranges of under 30°F in tropical zones to over 100°F in northern climates. Where maximum cooling is not desired, it is therefore important to consider what type of control system is most appropriate for each situation (6).

Methods available for air cooler temperature control include the use of both manual and automatic systems, such as the following:

[1] Manually or automatically operating louvers.

[2] Switching fans on-off where several fans are used for a single duty (manual or automatic).

[3]   Two-speed motors (manual or automatic switching).

[4]   Steam coils to preheat incoming air (manual or automatic operation).

[5]   By-passing on the process stream side, using manually or automatically operating valves.

[6]   Variable speed drivers, such as steam turbines, hydraulic motors, electric motors.

[7]   Automatic variable pitch fans.

With the exception of steam coils, similar methods may generally be used for the control of mechanical draft towers. The effects on air flow of the use of two-speed motors and on-off switching are self-evident. Automatic variable pitch fans have been used in industry for many years to provide accurate outlet temperature control for air coolers under conditions of varying heat loads and ambient air temperatures, but are not believed to have been applied to cooling towers.

As applied to air coolers, automatic variable pitch fans operate with the minimum motor horsepower at all times consistent with a set process temperature since the fans only move the amount of air actually needed. The system operates automatically and is economical in terms of initial costs and operating costs. The control mechanism is similar to that on a conventional control valve and is, typically, a pneumatically activated diaphragm, which alters the blade pitch angle over a wide range while the fan is in motion. This increases, decreases or stops air flow and, for special applications, air flow can be reversed to counteract natural convection for extremely precise temperature control.

The advantages of auto-variable pitch fans for air cooler control are appreciated in industry: indeed, major fan manufacturers say that about 40% of sales are currently made up of auto-variable pitch systems. While auto-variable pitch fans have been recommended for wet cooling towers in the literature, no specific examples were revealed during the course of this study. One manufacturer of fans stated that a new model for wet tower service was under development. Thus the use of auto-variable pitch fans applied to wet cooling towers might be considered an option for energy conservation; further discussion with fan manufacturers would be required for any specific tower.

While this item can represent significant capital investment, there is much that can be done without new equipment. Indeed, it is judged that housekeeping in the control and general operation of all air coolers, including keeping proper pitch on fixed pitch fans (by adjustments at maintenance turnarounds) and preventing slippage on belt driven systems, is considerably more important than automatic pitch control in terms of energy savings. This item is therefore included in the category of general housekeeping for the purposes of target setting (6).

Optimize Balance of Water and Air Cooling: The considerations of economics which govern the choice of maximum air cooling or water cooling for a refinery or process unit affect the decision which is made for a new plant. Where an existing plant is concerned, it is unlikely that a change from water cooling to air cooling (or vice versa) could be justified on the grounds of energy conservation. For this reason, this item is excluded from consideration in the target setting procedure (6).

Improvements of Pumps, Drivers and Piping

Change Impeller to Reduce Throttling Energy Loss, Use Appropriate Motor Sizing for Actual Duty, Replace Incorrectly Sized Lines, Control Valves and Orifice Plate Meters and Replace Inefficient Pumps and Compressors with New, More Efficient Equipment: The following discussion is presented for information purposes. These items represent considerations which are most relevant for new plants, but their large scale application to existing plants will be limited by poor economic justification. Any changes which are made to existing pump systems by 1980 were considered as housekeeping in the Gordian Associates study for F.E.A. (6).

A change in design and economic philosophies by management and engineering is a necessary preliminary to increasing average pump efficiency to the level presently available. Waste arises today either because management guidelines or engineering standards have to be complied with. Often pumps and drivers are oversized to allow for future increases in flow and head, and therefore the pump cannot be selected with its maximum efficiency at the point where it will really operate. Furthermore, the extra head or flow of the oversized pump is wasted across a control valve until the time the extra pressure or flow is needed.

In the case of overdesign, economical operation of a pump is attainable by installing smaller or lower-head pump impellers. This eliminates the power loss that comes about by lowering the flow rate with a throttle valve on the pump discharge.

Clearly, when an oversized pump is installed, an oversized motor usually accompanies it. Properly sized drivers are needed to correspond to a reduced head pump operation for further energy savings.

When reviewing pump systems in general, consideration should be given to installing piping of a larger diameter to reduce unnecessary pressure drop and the corresponding need for a larger pump. It should also be noted that piping pressure drop may increase substantially over a period of time due to the formation of scale or internal corrosion products. Where the actual pressure drop in a piping system is significantly greater than the calculated pressure drop, such a situation should be suspected and appropriate action taken. This may involve internal cleaning or the entire replacement of the line. Incorrectly sized control valves and even orifice plate meters can lead to unnecessary pressure drops in a system and replacement may often be justified.

Finally, it should be noted that some old pump standards still limit the maximum inside diameter for a pump impeller to only 90% of what the pump body can accommodate, this being to defend against possible design or operational errors. Pump standards have usually been many years in development and may be updated slowly because of past pump problems in the company. Nevertheless, the improved pump efficiency and energy savings which may be earned by revising impellers is not insignificant (6).

Minimize Bypasses and Recirculating Streams: For cooling pump bearings, centrifugal pumps often use the process fluid itself. At low flow rates, however, overheating is likely to occur, leading to bearing failure in extreme cases. To prevent this, it is common for bypass or recirculation systems to be employed which guarantee a steady flow of fluid from the pump discharge back to the suction side under all flow conditions. Frequently these systems are set manually at one time and subsequently disregarded. Considerable amounts of energy can be wasted by allowing excessive recirculation around the pump, since the recirculating stream contributes nothing to the process itself. Various mechanical systems are available for the proper control of recirculation rates. For the purpose of target setting, this item is regarded as housekeeping (6).

Routine Maintenance: This includes repairing of leaks, cleaning of pump impellers to remove deposits in fouling services, proper lubrication of bearings, etc.

Improvements in Fractionating Towers

Installation of New Trays, Higher Efficiency Designs: Many examples exist of new tray or tower packing designs which can reduce the energy required to operate fractionation towers. In contacting refining companies, it was learned that few tower retrays, if any, had been performed with the object of saving energy. In some cases, new tower intervals had been installed to increase tower throughputs, and some savings in energy had been realized as a side benefit. For the purpose of target setting, this measure should not be considered a realistic option for energy conservation between now and 1980 (6).

Intermediate Reboilers on Towers: By concentrating all heat input to a fractionating tower at the bottom of the tower, the system is made simple. However, more efficient operation can often be achieved by installing an intermediate reboiler at some point higher up the tower. A lower temperature is required at that point, often allowing reboiler heat to be supplied by a lower grade heat source than used for the main, bottom reboiler. The tower internal hydraulics may well be improved, allowing an increase in capacity to be realized.

While this measure can be profitable in many situations, it is difficult to generalize on its potential. Almost all situations have their own specific characteristics, and experienced engineers are required to identify the opportunities and design the modifications. It is judged that this item represents only a small contribution to energy conservation on an industry-wide basis, and is therefore omitted from the target-setting procedure. Note, however, in these limited number of cases where an intermediate reboiler can be justified, the rate of return can be quite favorable. For example, one case resulted in savings of 130 billion Btu per year for a $177,000 investment, yielding a rate of return of 126.8% (6).

Optimum Balance Between Number of Trays and Reflux Rate: The degree of separation achieved in a fractionating tower operating for specified product purities and at a specific pressure is a function of the number of trays and the reflux ratio. Adding trays, which results in greater initial investment, can allow lower reflux ratios to be used, saving on energy input to the system.

The economic balance will undoubtedly shift towards the use of lower reflux ratios as the cost of fuel increases, but this is likely to impact only on new plants. In contacts with industry, it was learned that installation of additional trays in an existing tower to allow reflux to be reduced while maintaining the same product purity had not been found economically justified in any location. This measure, while important for new plant design, is not a practical conservation measure for most existing plants and therefore should be excluded from consideration in the target setting procedure at this time (6).

Use of Tower Side Draws: Rather than use two fractionating towers to recover two distillate products from a stream, it may often be possible to use one tower with a top product and side draw. As a conservation measure to be applied to existing plants, this measure will not prove technically feasible, nor economically justified, in any but a very few cases, and accordingly should be excluded from target setting (6).

Overhead Vapor Recompression with Heat Pump Systems: An unconventional approach to the fractionating column flow scheme is the addition of an overhead vapor recompression system (a closed heat pump). The heat pump cycle consists of adding a compressor on the overhead to compress the overhead vapor and return it to the reboiler where it is condensed. In this way, the reboiler heat duty, typically supplied by steam, is now supplied by the overhead vapor. The energy input to the system is through the motor driving the compressor. For specific light hydrocarbon fractionators, overall energy savings of 25 to 75% are quoted in the literature.

The criteria which may lead to selecting the heat pump system include the existence of small differences between the overhead and bottom temperature, and the fact that the operation is a difficult separation requiring high reflux ratios.

While heat pumps do have major advantages in a range of applications, retrofit applications are generally not justified. Discussions with refiners confirmed that the heat pump systems are not a practical measure to be considered for energy conservation in existing plants by 1980 (6).

Distribution of Feed and Reflux Return to Minimize Entrainment: Obviously, the function of a refinery fractionating tower is to separate components of a complex mixture of hydrocarbons. Entrainment of feed or reflux streams as they enter (or reenter) a tower simply makes the separation more difficult: proper mechanical design of fractionating equipment should eliminate all problems. In rare cases, modification may be necessary to some tower inlet nozzles at routine maintenance shutdowns. This item is considered housekeeping for the purpose of the F.E.A. study (6).

Control of Fractionating Towers: As indicated previously, excessive reflux and reboil on a tower are wasteful of energy. Also, the purity of a fractionation product should be monitored carefully to ensure that it meets specifications, but is not excessively pure. Towers therefore represent a major area for housekeeping to show good energy savings. Investment in analytical instrumentation may often be justified on towers, and this topic is covered subsequently under "Instrumentation."

Reduction of Tower Operating Pressure: Since low pressure improves the relative volatility of components being separated in a fractionating tower, a reduction in tower operating pressure should allow a lower reflux ratio to be adopted while achieving the same degree of separation. The lowering of pressure may be limited in practice by the capacity of the tower overhead condensing system, since the condensing temperature drops as the pressure is lowered. This item was considered housekeeping in the 1976 Gordian Associates study for F.E.A. (6).

Operation at Optimum Tray Loading and Efficiency: The trays in a fractionating tower generally operate at maximum efficiency at high loading. To achieve a given separation with a fixed number of trays, the usual situation in an existing tower, the trays should be run at maximum efficiency if reflux is to be minimized. The loading of trays may often be adjusted to the maximum efficiency point in an existing tower by such methods as: (a) changing tower throughput, (b) changing tower pressure, or (c) changing the percentage of feed vaporization. In some rare cases, retraying a tower or blocking off parts of each tray may be advantageous. As a general rule, however, this item is considered under housekeeping (6).

Use of Appropriate Feed Tray Location and Feed Condition (V/L Ratio): For efficient operation, the feed to a tower should be put in at the correct point according to its composition and should be added in the form of a vapor-liquid mixture where the proportion of vapor is about the same as the proportion of overhead product to be recovered from the feed. This is important since it is often more economical to add heat to the feed than to the tower bottom reboiler because the feed point is at a lower temperature than the tower bottom, allowing lower grade heat sources to be utilized for preheating the feed before it enters the tower. This item falls within the category of day-to-day tower control, or housekeeping (6).

Tank Mixing

The mixing of tank contents is a frequent refinery operation. A typical example is the blending of gasoline from five or six components. Some tanks have motor-powered mixers installed, while in many cases the tank contents are recirculated around a piping loop, resulting in a thoroughly mixed homogeneous blend. Often it is advantageous to install a jet mixer on the inlet line to the tank to ensure that agitation is efficient. With the use of a jet mixer, blending time can often be reduced, thus reducing the amount of time the circulating pump has to be operated. This item provides limited scope for energy conservation, requires minimum investment and is included under housekeeping for the purpose of target setting (6).

Insulation

Insulation of Storage Tanks: The use of insulation to minimize heat loss from equipment is, of course, universally practiced. The choice of a material and thickness to apply are governed by economics, and a detailed evaluation of insulation materials is not included here. However, it is clear that insulation standards are worthy of review as the price of fuel increases. New application methods, such as the use of foams which are sprayed on vessels, are available to perform relatively cheaply in situations which would otherwise incur substantial labor costs.

In the case of tankage, many refinery projects have to be stored at above-ambient temperatures to maintain pumpability (residual fuel oil, for example). Until the recent dramatic energy price rises, it was normally considered justified to insulate tanks held at a temperature of about 200°F or above. With present economics, which include increased costs for energy as well as reduced costs for effective insulation, it appears that insulation of tanks at 125°F is justified.

Many companies indicated that they were planning major tank insulation programs in the near future: many have accomplished significant conservation already by this means.

The return on investment in tank insulation depends on the tank size and temperature. Insulation should be installed up to the point that the marginal return on investment is equal to the hurdle rate. This applies both to decisions about which tanks to insulate and the selection of insulation thickness. The rate of return on the entire investment is often quite high. Two representative examples from the refining industry of tank insulation are as follows. The first costs $101,700 and saves 50.6 billion Btu per year for a rate of return of 88.6%. The second costs $188,000 and saves 73.2 million Btu per year, yielding a rate of return of 40.3%.

The insulation of tanks is clearly a well proven and fully acceptable measure to save energy. Availability of insulating materials and of contractors to perform the application appears such that a typical multitank project could be completed within about six months of initiating planning. No downtime is considered necessary. This item is judged significant in the setting of the conservation target (6).

Insulation of Process Unit Vessels, Pipes, Flanges and Other Equipment in Hot or Cold Service and Upgrading of Existing Insulation in Process Plants and Offsites: Maintaining adequate insulation on all process unit equipment makes a valuable contribution to energy conservation. In many refineries, old design standards for insulation can be shown to be below the economic optimum under today's conditions. Unfortunately, difficulties of access or equipment spacing (pipes in a piperack, for example) will often prevent existing insulation from being upgraded.

Areas where some companies are saving energy include the insulation of heat exchanger channel boxes (the end of the exchanger) and flanges, particularly in hot systems such as steam lines. Although reservations were expressed by one company about insulating flanges (and the hazard posed by hidden leakage at the flanges), another company stated that they had insulated hundreds of flanges with complete success and no problems of leakage had been reported. Because these items are generally applied to specific equipment in a refinery, piece by piece, they are considered as housekeeping for the purpose of the target setting study (6).

Instrumentation

Instrumentation and Computer Control for Complex Units: Attractive economic benefits have been shown to result from closed-loop computer control of complex units, such as fluid catalytic crackers and catalytic reformers. A computer system will typically guide operators on all product qualities, yields and energy used. Most process units can be operated to yield different sets of products from the same or different feedstocks. For each feed-product package, an optimum mode of operation can be established, and the process unit run against some physical limitation within that operating mode to maximize profits. By scanning actual operating conditions and comparing them with the optimum operation stored in the computer memory, the computer can indicate where discrepancies exist and recommend changes. Indeed, the computer system can be automated to almost any degree desired, such that changes are made without operator intervention. Direct Digital Control (DDC) can thus be used to maximize profit and minimize, within that constraint, energy consumption. The computer itself may cost $50,000 but the entire system may cost $200,000 to $500,000, depending on the jobs it is assigned. Rates of return are believed quite high, but they depend on the level of operator skill existing prior to implementation of computer controls.

Many companies contacted in the course of the Gordian Associates study for F.E.A. (6) were enthusiastic about the benefits to be gained from computer control, which were not restricted to energy savings but included capacity (and profit) increases also. This item is therefore consistent with industry plans and is judged relevant to the conservation target (6).

Analytical Instrumentation for Fractionation Tower Purity Control: When energy costs were still relatively low, fractionation columns were often operated so that their products were considerably purer than required to avoid the large economic debts associated with producing off-specification material. The difference between the purity achieved and the purity specification is directly related to profitability: it represents use of higher than necessary reflux flows. By controlling purity (i.e., reducing reflux) closer to specification, lower energy demands and a lower cost of operation will result. Many analytical instruments are available to effect this control, such as chromatographs and flash point analyzers, for example. The effect of the instrument is to achieve savings not obtainable by manual control alone.

Another benefit experienced with lower deviations of product impurities is overall column stability with a resultant improvement in tray efficiencies. Proper control of a tower will include both feedback and feed forward control. The need for feed forward control results from the fact that feedback control is based on the existence of a problem situation. In fractionation processes, disturbances in feed rate or composition have a pronounced long-term effect on the operability of the column. Thus when an error is manifested in a feedback composition controller, it is already too late to negate the effect of the disturbance. Feed forward control obviates the problem by predicting in advance the effect of a disturbance on the product composition and taking the necessary control action before the product quality can deviate.

With the combination of a feed forward and feedback control scheme, excellent fractionation control can be achieved. The forward loop will compensate for major disturbances, such as variations in feed rate and composition, while the composition feedback controller has only to contend with minor disturbances, such as variations in tray efficiency.

This item is consistent with industry plans and requires typically one year for the completion of an instrumentation system on a tower, from planning to commissioning. It is judged relevant to the target setting procedure (6).

Vacuum System Improvements

Detect and Repair Air Leaks, Routine Steam Jet Maintenance and Minimize Steam Jet Superheat: Many operations take place under vacuum: a vacuum stage is included as part of a crude distillation unit in almost all refineries. The vacuum is maintained using eductors powered by high pressure steam, which represent a significant energy input to the process. It is therefore important:

(a) to evaluate the real need for any particular degree of vacuum, and to cut back on the eductors as far as possible according to the process flexibility,

(b) to check carefully for leakage of air or other noncondensibles into the vacuum system since small leaks can have significant effects on operations,

(c) to check on the rate of wear on the eductors (maintenance should not be neglected), and

(d) to ensure that condensers are in good working order at all times, leaving only the minimum amount of uncondensed vapors to be handled by the eductors.

Many vacuum systems operate with barometric condensers, in which the steam from the eductors is condensed by direct contact with cooling water. Others use surface condensers. It may often be worth considering replacing barometric systems with surface condensers which would allow the condensate to be returned to boiler feed water service in most cases, saving treating costs and energy. All the above items are considered housekeeping for the purpose of setting the target (6).

Replace Steam Jets by Vacuum Pumps: Although vacuum pumps of all types are used in chemical process plants, the steam jet type has been widely used because of low initial cost, operating simplicity, lack of moving parts and supposed reliability. However, mechanical pumps can offer significant energy savings over steam jets, and can therefore be justified in some cases in spite of higher initial investment costs. Contacts with refiners indicated that many were indeed reevaluating the use of steam jets. No installations of mechanical pumps were reported, but suggestions were made that new plants might have mechanical pumps installed in combination with conventional steam jets. This item is not a practical consideration for existing plants at this time, however, in the judgment of Gordian Associates (6).

Boiler Feed Water Treating: The presence of impurities in water can lead to scaling and corrosion problems in boilers. The preparation of boiler feed water therefore requires careful control so that impurities which concentrate in the boiler do not exceed the limits set by the boiler design. Many processes are used to prepare feed water for boilers, including filtration, precipitation, deaeration and ion exchange. All these steps, and particularly deaeration using steam, are energy consuming and therefore should be minimized (by maximum use of recycled condensate, for example).

Once the water is in the boiler system, impurities concentrate as steam is evaporated and removed. To maintain impurities at the required level, boiler water must be blown down or discharged and replaced by fresh feed water. By careful control of blowdown, sludge deposits and scaling are prevented while waste of water and heat energy in the discharge are minimized. Good feed water treating will therefore lead directly to energy savings. This item is considered as housekeeping for target setting purposes (6).

Steam System Improvements

Optimization of Steam Balances: Since steam is required to perform mechanical work and heating duties in many locations in a plant, the design of a steam system to utilize to the maximum all the available energy at the different pressure and temperature levels can become very complicated. Plant expansions or process changes make the maintenance of such a steam system at top efficiency exceedingly difficult.

It is quite common for steam to be distributed and used within a plant at several pressure levels (say 600, 250 and 50 psig) and to have reducing stations between the levels, with condensation of excess 50 psig steam for return to boiler feed water being the ultimate pressure reduction. Any shortfall in low pressure steam availability is made up by reducing some of the next higher pressure steam, and so on. The extra demand is therefore ultimately produced from the boiler at the highest pressure and steam is cascaded down to the use point.

It is rare to find a refinery steam balance which can remain at optimum for long after the plant is built. The balance between different levels of steam (such as the 600 psig, 250 psig, 50 psig, etc.) and the use of turbine or electric drivers for rotating equipment may have been considered optimum at the time of design, but actual plant operations will not always follow design closely, and thus imperfections will exist in the system. As new plants are commissioned, the steam system will be expanded, in all probability introducing further departures from the optimum. Thus there will often exist major opportunities for energy conservation through the redesign of steam generation, distribution and consumption facilities. Four factors must be remembered, however:

[1] Redesign to achieve a new optimum is generally a highly complex task requiring knowledgeable, competent and experienced engineers.

[2] All desired alterations of a steam system may be impossible to carry out in the field because this would require a complete shutdown of all utility systems and process plants for an extended period of time. Modifications to steam systems must therefore be carefully planned and a compromise with the ideal solution must be reached.

[3] Major modifications to a steam system can often be multimillion dollar ventures, requiring the engineering and purchase of a large amount of equipment, including heat exchangers, valves, piping, instrumentation, electricity generators and turbine drivers.

[4] Each refinery has a unique situation to which some general rules may be applied, but whose optimum utility system is a function of plant process characteristics, energy costs, manpower cost and availability, historical growth and future plans.

Major steam system improvements often cost several million dollars and yield relatively low rates of return, making them sensitive to capital availability and cost. Returns are quite variable, with the better opportunities able to provide rates of return of 20 to 25%.

The optimization of steam systems is a well accepted conservation measure in industry and is consistent with industry plans. The time required to achieve close to the optimum situation may be two or three years, depending on the complexity of the utility system. Regular scheduled downtime on process units and boilers must be used to install steam system modifications section-by-section, whenever possible. No environmental impacts or relationships with natural gas curtailments is involved in this conservation measure, which is judged relevant to the target (6).

Replacement of Steam Traps: When steam is used as a heating medium, it is a normal requirement that full use be made of the latent heat of condensation. Sensible heat changes in steam or water are generally negligible compared with the heat given up when condensing steam, around 1,000 Btu/lb. For this reason, it is common practice to install steam traps in steam condensing systems: these allow only condensate to pass and prevent uncondensed steam from being wasted. Steam traps are simple devices and usually quite cheap: unfortunately, their maintenance (or replacement) is frequently neglected with the consequent loss of energy through premature release of steam.

While proper maintenance of steam traps may be considered a housekeeping measure, a major revamp of steam traps coupled with condensate return facilities can involve a significant capital investment. Energy saving from this measure will of course vary from plant to plant, depending on the general level of housekeeping in effect. This energy conservation item is judged to have the potential for saving at least 5% of the steam load in a refinery. An industry example of an investment of $62,200 for a major steam trap replacement project shows savings of 62.4 billion Btu per year, providing a rate of return of 169.6%.

This measure involves only proven and accepted technology, and does not require any special downtime for implementation. Equipment delivery times are believed under three months. The measure is consistent with industry plans and is judged relevant to the target (6).

Condensate Recovery: Condensate from steam users (such as turbines, reboilers, tank heating coils, etc.) contains energy in several forms. The sensible heat in condensate can usually be usefully recovered if the condensate is returned to the boiler feed water system. As a rough guide, for every 11°F rise in temperature of the feed water, a boiler requires about 1% less fuel to raise steam. Also, recovery of condensate reduces the quantity of new boiler feed water that is needed to raise steam: this reduces the energy and materials used to prepare boiler feed water in the refinery water treating plants. In some cases, it may be possible to recover condensate at sufficiently high pressure to install a flash drum in the system from which flash steam can be drawn for use in another part of the plant. Because each plant is a unique situation, the amount of energy to be saved from improved condensate recovery is difficult to determine without extensive investigation.

Condensate recovery can be highly profitable. A representative example from the refining industry has a cost of $9,300 and saves 11.5 billion Btu per year, for a rate of return of 207%.

Condensate recovery is of course a well accepted measure for energy conservation. Depending on specific plant circumstances, downtime for installation of equipment is generally not required outside scheduled plant shutdowns, and delivery times of equipment are unlikely to exceed six to eight months. The measure is consistent with industry plans and is judged relevant to the conservation target (6).

Better Operation of Steam Tracing Systems: A significant loss of energy often occurs from steam tracing systems, installed to protect lines from freezing in cold weather. Tracing steam should obviously not be turned on unless weather conditions dictate the heat is needed. All traced lines should be insulated and the tracing should be fitted with steam traps at intervals, from which condensate should be recovered. Use of other fluids than steam may well prove economical for tracing systems. An example is the use of Dowtherm SR-1, a commercial grade of inhibited ethylene glycol supplied in a 50% aqueous solution which freezes at –34°F. Circulating a warm solution of Dowtherm SR-1 through a tracing system eliminates the need for steam traps except at the heat exchanger which uses steam to warm the glycol solution. This item is considered as housekeeping for target setting purposes (6).

General Housekeeping Measures to Reduce Steam Use and Steam Leakage: Many housekeeping measures have been described in previous sections of this text. Other measures include minimizing the use of steam in idle equipment (used to keep turbines turning in case of the need for a rapid start up, for example), keeping tank steam coils operating at minimum levels, repairing leaking pipes, valves and flanges (steam and condensate). Results of these housekeeping activities are of course variable from plant to plant, but are usually considered significant by operations management (6).

Thermal Improvements in Catalytic Cracking Units

CO Boilers for Fluid Catalytic Crackers and High Temperature Regeneration on Fluid Catalytic Crackers: The operating temperature of typical fluid catalytic cracker regenerators runs about 1200°F. The waste gases contain, as well as nitrogen and carbon dioxide, significant quantities of carbon monoxide. A typical figure would be 9% CO by volume. As well as the energy in the gas due to its high temperature, additional energy can be recovered by burning the carbon monoxide. Either directly or after passing through a turbine to lower the gas pressure and recover energy, the hot gases pass into a boiler firebox where combustion air is added. In many installations, supplementary gas fuel is burned in the boiler to minimize the effects of swings in CO content caused by process changes: the supplementary fuel ensures ignition and flame stability.

Data obtained in the Gordian Associates study for F.E.A. (6) indicate that the energy recovered by a CO boiler typically amounts to about 90,000 Btu per barrel of FCC charge (fresh feed basis). In 1972, about 30% of FCC capacity in the United States did not have CO boiler installations. Downtime of 21 to 35 days was estimated by one company as being necessary to install a CO boiler on an existing FCC unit. With appropriate prefabrication and prepositioning of equipment, it is believed that this time could be reduced sufficiently to allow installation in many cases within the period of normal scheduled maintenance turnaround (although extra costs may be entailed).

A representative industry example of a CO boiler is as follows. For a cost of $5.09 million dollars, 1.27 trillion Btu per year are saved. Even without credits for nonenergy benefits, the rate of return is 47.7%. While higher rates of return are found on many smaller energy conservation systems, they are unusual on items of this size.

As an alternative to installing a CO boiler, a new operating technique has become available through which more complete combustion of CO to carbon dioxide occurs within the regenerator vessel. Exit gas CO concentration can be reduced to levels as low as a few parts per million, which will easily satisfy environmental regulations. The high temperature regeneration which occurs in the regenerator does require some modifications to be made to the catalytic cracker, but capital investment is said to be much lower than for a CO boiler and the regeneration route has the advantage of promoting a better yield structure for the plant. Energy recovery through this system is essentially the same as for the CO boiler system.

The hot regeneration system also has one important advantage in that there is little or no influence of a scale factor. In other words, the measure is believed applicable to essentially all units regardless of size without significant disproportionate cost penalties for small units, whereas CO boilers are not generally justifiable on small FCC units.

The technology embodied in CO boilers is now well proven and well accepted in the industry. A CO boiler is a major item of equipment and will require at least two years lead time for planning, engineering, fabrication and installation. As stated, downtime of the FCC unit may be minimized in many cases by prefabrication and prepositioning of equipment, leaving final tie-ins to be made at scheduled maintenance shutdowns.

The high temperature regeneration technique is relatively new, but it is believed that industry acceptance has been achieved. Some mechanical modifications to equipment are necessary to allow existing plants to use the technique, but it is not believed necessary for extra downtime to be used, over the normal time available at regular scheduled shutdowns.

Since environmental pollution regulations require the elimination of CO emissions to atmosphere, it is expected that essentially all FCC capacity which is not already fitted with a CO boiler or is not practicing the hot regeneration technique will employ one or other of the measures by 1980. The measures are therefore consistent with industry plans and are judged significant for the target setting procedure (6).

Electrical Load Leveling

Strictly speaking, the leveling of a plant's electrical load over 24 hours does not reduce its energy consumption. However, it is important to recognize that the electrical loads measured at the generating plant reach daily peaks, as well as seasonal peaks. At the peak times, a utility company may have to bring on line its least efficient generating equipment, frequently gas turbine powered. If the total 24-hour load were spread more evenly, it might prove possible to remain on the most efficient base-loaded coal, nuclear or hydroelectric generating plants. The rate structure for industrial consumers normally recognizes the inefficiencies of the system and is structured to discourage peak loads. A plant can therefore often balance out loads throughout the day, save energy consumption at the generating plant and save money at the same time (26).

Power Factor Improvement

The power load taken by any alternating current system is the product of voltage, current and power factor, where the power factor is a function of the phase difference between the voltage and current. This difference is itself a function of the load characteristics in terms of capacitance or inductance. When voltage and current are in phase, the most efficient situation exists with a power factor of 1. In a real situation, the plant power factor may be quite low: by installing capacitors to change the load characteristics, it is frequently possible to achieve substantial savings. The question of power factor is basically a problem for the user, not the generating plant. However, if the user operates more efficiently, savings in the power used for a given operation should be, in effect, passed back to the generating station and a real saving in fuel achieved (26).

Process Energy Recovery from Process Streams

Power Recovery Turbines on Gaseous Streams: There are many potential applications for gas expansion units in process plants. However, the economics of such energy recovery systems are by no means universally favorable. One process unit in which energy recovery has been successfully applied is the fluid catalytic cracker (FCC). Regeneration of catalyst produces large volumes of hot gases, available at pressures ranging from around 10 psig to perhaps 40 psig and temperatures up to about 1250°F. The quantity of gases available, even at such low pressures, is such that enormous energy savings are potentially available for use in driving other plant equipment such as the main air blower.

In fact, regenerator pressure is an important factor in considering application of power recovery systems to existing FCC units. Because many older units operate low pressure regenerators, the maximum power recovery may be as little as 50% of air blower horsepower requirements. In such cases, a retrofit venture is not generally justified. Thus, while this measure has proved a useful means of conserving energy in many refineries, contacts with refining companies indicated only a very few new applications are planned by 1980. One industry example of a power recovery turbine investment has a cost of $93,000 and energy savings are 1.04 billion Btu per year. The 23.1% rate of return is marginal for many companies.

Nevertheless, the measure applies proven and accepted technology. Lead time for design, equipment fabrication and installation is two to three years. Because the measure is consistent with industry plans, it is considered relevant to target setting (6).

Use of Jet Compressors to Recover Potential Energy of Gaseous Streams: This measure may be used in certain specific cases to upgrade low pressure steam using an intermediate pressure steam to power an ejector. However, because of its specific nature and the relatively small amount of energy saved, this measure has been included in the general category of housekeeping for the purpose of target setting (6).

Hydraulic Turbines on High Pressure Liquid Streams: Energy from high-pressure liquid process streams, previously lost by throttling across a control valve, may be recovered with hydraulic turbines which are essentially volute or diffuser-type pumps running backwards. These turbines may be either single or multistage depending on the pressure energy available from the process stream. Usually the turbine works in tandem with the motor driving the pump for the process, with a clutch to disengage the turbine at unit start up. A representative investment in a hydraulic turbine was quoted by one industry contact. For an investment of $298,000, savings of 4.95 million kwh per year were achieved. At a cost of 2 cents per kwh, this provides a 22.9% rate of return, which is not high enough for many companies. In addition, contacts with refiners indicated that problems of mechanical seal failure had frequently led to low on-stream factors, and that the recovery of energy using hydraulic turbines was not favored by many companies.

It is thus the judgment that hydraulic turbines will not contribute significantly to energy conservation on an industry-wide basis, and this item is therefore not relevant to the target setting procedure at this time (6).

Loss Control

In refinery processing operations, there are many possible sources of loss. Among the most common are:

[1]   Flare losses
[2]   Relief valve leaks
[3]   Filling losses from tanks
[4]   Evaporation losses from tanks
[5]   Evaporation from oil/water separators
[6]   Oil/emulsions in water from oil/water separators
[7]   Leaks in pipes, flanges, tanks, pump glands, water coolers, etc.
[8]   Oil in barometric condenser water
[9]   Tank cleaning, vessel draining for turnarounds, etc.
[10]   Loading operation spillages

Given good operating practices and well maintained equipment, it is estimated that losses will be in the range of 0.5 to 1.0% of refinery throughput, although losses with poor operation can obviously be much higher (6).

Catalyst Developments

Catalyst Developments for Reforming, Hydrocracking and Desulfurization:
Many refining processes are performed in the presence of catalysts, the development of which continues all the time in the search for better yields at lower cost. Since the role of the catalyst is normally to improve reaction rates, the energy required to complete the reaction itself is not necessarily altered. However, the conditions under which economically attractive yields are obtainable are often altered substantially in terms of temperature and pressure, leading directly to energy savings.

Because of the wide variety of new catalysts which could be tried and might prove acceptable in refinery process units, it is difficult to estimate the potential effect of new catalysts on industry energy use. Changes of catalyst in a major unit normally require considerable investigation and feasibility analysis. This item will entail relatively long lead times, and its relevance for target setting is not clearly established. For the target setting procedure, it is therefore assumed that any benefits of improved catalysts will be included in the general housekeeping category (6).

Maintenance of Higher Catalyst Activity by More Frequent Regeneration or Replacement: This item has a similar effect to the use of newer, more active catalysts in reforming and desulfurization process units. As an energy conservation measure, however, it is not judged to provide major advantages, as the decision to regenerate or replace a catalyst is made on other grounds (product quality, product yields, etc.). For this study, the measure will not be considered further (6).

Process Developments

The increased price of energy may lead to reconsideration of the choice of solvents in a number of refinery process areas, for example. Removal of acid gases (e.g., carbon dioxide, hydrogen sulfide) can be accomplished with solvents which will operate with higher solution loadings in terms of the volume of acid gas dissolved per gallon: this factor will reduce the circulation rate of solvent that is needed for a given acid gas removal rate. Also, some solvents which are regenerated by steam stripping to remove the acid gas need much less steam than others, offering useful energy economies. Similar considerations may be applied to the use of aromatics extraction solvents where used in refineries (most aromatic extractions being a chemical plant activity). This item is regarded as housekeeping in connection with the target setting procedure (6).

Most of the current processes for removing ammonia and hydrogen sulfide from foul refinery wastewaters require the use of steam as a stripping source. A process which will reduce the need for a steam stripping source is thus highly desirable because of energy shortages and resultant high steam costs. Petroleum refineries usually produce a quantity of residual gas which is available at pressures of between 100 and 220 psig. Ordinarily the energy contained in this residual gas is not utilized because the gas is released to a header operating at 40 to 80 psig. This gas subsequently is burned in various furnaces.

A process developed by G.I. Worrall, D.A. Strege and G.D. Myers (60) is one in which refinery wastewaters containing dissolved hydrogen sulfide and ammonia are stripped of hydrogen sulfide by a countercurrent stream of refinery absorber gas which removes substantially all of the hydrogen sulfide and none of the ammonia. The dissolved ammonia is subsequently removed by steam stripping, leaving a wastewater sufficiently clean to be nonpolluting.

Alternative Processing Sequences: The subject of alternative processing sequences is addressed in the 1974 Gordian Associates Report to F.E.A. (26), which uses a linear program refinery model to explore the effect on energy consumption of different processes, different product specifications and different product patterns.

Housekeeping Measures, N.E.C.: A variety of housekeeping measures has been covered already in previous categories as noted in each description of specific conservation measures. Some of the more important items are:

[1]   Combustion control, especially excess air
[2]   Proper operation of steam systems

[3] Fractionation tower control, especially reduction of reflux
[4] Repair of steam leaks, steam traps and condensate leaks
[5] Repair of insulation on lines, tanks and vessels

The importance of housekeeping cannot be overemphasized. Attainment of energy conservation in manufacturing operations is impossible without the motivation and commitment of all employees. A continuing emphasis is needed to maintain the level of gains already reported achieved by the end of 1975.

GOAL YEAR (1980) ENERGY USE TARGET

On the basis that operations remain essentially the same as in base year 1972, a gross energy efficiency improvement target has been set for each component industry and weighted (according to base year energy consumption) to provide an overall SIC 29 target. The individual component targets were found to be as follows:

	Gross Target, %
SIC 2911 Petroleum Refining	19.4
SIC 2951 Paving Mixtures and Blocks	47.0
SIC 2952 Asphalt Felts and Coatings	25.0
SIC 2992 Lubricating Oils and Greases	25.0
SIC 2999 Miscellaneous	25.5

It is particularly interesting to examine the aggregation of contributions from individual conservation measures for SIC 2911 as shown in Table 180 and as described in the paragraphs which follow.

It is clear that the target is made up from a large number of items which, for the most part, do not provide individually any major contribution to the overall target. Significant energy conservation is thus achievable only by thorough attention to many small items.

TABLE 180: CONTRIBUTION OF SPECIFIC CONSERVATION MEASURES, SIC 2911

Item Identification	Percent
1. Air preheater installation (process heaters)	1.0
2. Convection sections on process heaters	0.5
3. Sootblowers on process heaters	0.3
4. Combustion control instrumentation for process heaters	0.4
5. Improved insulation for process heaters	0.1
6. Replacement of old heaters with new equipment	0.7
7. Convection section for boilers	0.1
8. Combustion control instrumentation for boilers	–*
9. Boiler blowdown heat recovery	0.1
10. Replacement of old boilers with new equipment	0.1
11. Sootblowers for boilers	–*
12. Improved insulation for boilers	–*
13. Optimization of steam balances	0.4

(continued)

TABLE 180: (continued)

Item Identification	Percent
14. Condensate recovery (and replacement of steam traps)	0.5
15. Additional heat exchangers, hot oil systems Interchange of heat between different process units	2.3
16. Optimization of heat exchanger cleaning cycles	0.5
17. Insulation of storage tanks	1.0
18. Power recovery turbines (FCC systems)	_*
19. Instrumentation and computer control of complex units (including fractionating tower controls)	0.3
20. Direct hot rundown/hot feed systems	1.1
21. CO boilers for FCC units and High temperature regeneration on FCC units	1.3
22. Refinery loss control	1.0
23. Housekeeping	7.1
24. Other small items not included as separate contributions	0.2
25. Capacity expansions	0.4
Total	19.4

*Less than 0.05%, see "Other small items."

Source: Reference (6)

Air Preheater Installation on Process Heaters: Typical heater efficiency improvement is taken as 10%. Based on a heater efficiency/size distribution for a major integrated oil company, the potential for preheater installation was estimated. It was assumed that 60% of the industry will accomplish air preheater retrofitting to the potential extent estimated by January 1, 1980, representing a target contribution of 1.0%.

Convection Sections on Process Heaters: Based on the heater efficiency/size distribution mentioned previously, the potential for convection section installation was estimated. Energy savings correspond to an efficiency improvement of 10% and therefore, assuming 50% of the potential retrofitting projects are actually accomplished by January 1, 1980, the target contribution of this item is 0.5%.

Sootblowers on Process Heaters: The contribution of this item to conservation is based on the need for sootblower installation on process heaters currently firing natural gas and being converted to heavy liquid fuel firing. The use of natural gas outside Louisiana and Texas is currently about 18% of Btu fired: assuming total conversion of the industry outside these two states (60% of industry capacity), and an efficiency improvement of 3% of installation of sootblowers, this item is estimated to contribute 0.3% to the target.

Combustion Control Instrumentation for Process Heaters: Based on the extent of monitoring by continuous analyzers and of closed loop control existing in 1972 and the projected situation for 1980, both estimated by the instrumentation specialists of a major design contractor to the industry, the contribution to the target of this item was calculated at 0.4%. This corresponds to a total efficiency improvement of 1% for fully closed loop control with instruments, over and above diligent manual operation.

Improved Insulation for Process Heaters: Based on a maximum potential of 1% savings of the heater fired duty, the contribution of this item is judged to be 80% of that potential by January 1, 1980, which amounts to a contribution of 0.1% towards the target.

Replacement of Old Heaters with New Equipment: On the basis of 10% of heater capacity being candidates for replacement, but only 70% of this actually being replaced directly (as opposed to being retired altogether), and an efficiency improvement of 15% in each situation, the contribution of this item is estimated to be 0.7%.

Convection Sections for Boilers: Through a similar procedure as adopted for item 2, as shown in Table 180, the target contribution of this item was estimated to be 0.1%.

Combustion Control Instrumentation for Boilers: A similar procedure to item 4 was used, but the potential for installation on boilers was judged somewhat less than for fixed heaters due to a generally higher standard of instrumentation installed on the original plant. A target contribution of less than 0.05% was estimated.

Boiler Blowdown Heat Recovery: On the basis that blowdown amounts to 5% of steam make, and that energy recovery corresponds to a blowdown temperature drop of 250°F, the percentage of total refinery energy use recoverable from boiler blowdown is about 0.14%. About 10% of this potential was being recovered in 1972 and a further 70% will be recovered by the target date of January 1, 1980, making the target contribution of this item 0.1%.

Replacement of Old Boilers with New Equipment: A similar procedure to item 6 was adopted, giving a target contribution of 0.1%.

Sootblowers for Boilers and Improved Insulation of Boilers: Based on the procedures used for items 3 and 5, the target contributions of these items are both less than 0.05%.

Optimization of Steam Balances: This is a long lead time item, often requiring major engineering effort to identify and design changes to an existing steam system. For those projects that proceed, an estimated energy saving of 40% of the steam-raising energy is achieved. However, since it is felt that only 10% of industry capacity will complete a steam system reorganization with such large savings between 1972 and 1980, the target contribution for this item becomes 0.4% of total energy input.

Condensate Recovery and Replacement of Steam Traps: Together, these two items are estimated to provide savings of 5% of steam system energy use, representing a target contribution of 0.5%.

Additional Heat Exchangers, Improved Hot Oil Systems and Interchange of Heat Between Different Process Units: Taken together, these items represent a major area of better energy management in the industry. Based on savings to date by the industry, the target contribution of this item is estimated as 2.3%.

Optimization of Exchanger Cleaning Cycles: Contacts with industry indicated a number of onstream cleaning systems had been installed since 1972 and more are planned. In a typical refinery, there are several key areas for onstream exchanger cleaning, and it is estimated that somewhat less than 50% of potential systems will be installed between 1972 and 1980. The target contribution from this item amounts to 0.5%.

Insulation of Storage Tanks: While many residual fuel oil tanks are well insulated, industry contacts indicated a big potential for energy saving by insulating some residual fuel oil tanks and a large number of distillate component and product tanks not now insulated. Including the insulation of some crude oil tanks, the contribution of this item to the conservation target is estimated to be 1.0%.

Power Recovery Turbines for FCC Systems: From industry contacts, Gordian Associates (6) estimated that a total of six power recovery systems will be installed between 1972 and 1980. Although each item is a significant energy-saver for the refinery involved, the conservation target contribution of this item, on an industry-wide basis, will be less than 0.05%.

Instrumentation and Computer Control of Complex Units, Fractionating Tower Control and Analytical Instrumentation for Fractionating Tower Purity Control: Taken together, these items represent useful energy savings measures to most companies. About 50% of energy use in a typical refinery is expended on complex units (such as catalytic reformers, FCC units, hydrocrackers, alkylation units, etc.) and 1% of this may be saved by comprehensive automatic monitoring of key variables, or by their closed loop control. Some 50% of the industry will install comprehensive instrumentation and computer control between 1972 and 1980, giving a target contribution for this item of 0.3%.

Direct Hot Rundown/Hot Feed Systems: Based on the well-known example of hot vacuum gas oil being transferred from a crude distillation unit to a catalytic cracker directly (by-passing cooling, storage and reheating steps), this item represents a target contribution for the industry of about 1.1%.

CO Boiler for FCC Units and High Temperature Regeneration on FCC Units: Based on industry contacts, the savings from each of these items appear comparable, in energy terms, at 90 MBtu per barrel FCC charge. The high temperature regeneration has the added advantage, however, of an improved FCC yield structure. On the basis that 25% of FCC capacity accomplishes one or the other of these measures by January 1, 1980, the target contribution for this item is 1.3%.

Refinery Loss Control: The more common sources of processing losses have been described previously in this section. An improvement of 0.1% of throughput is attainable as an industry average by January 1, 1980, which corresponds to a 1.0% improvement in energy efficiency (6).

Housekeeping: Approximately 5% of the 10.3% energy savings reported through the API voluntary reporting system are attributable to housekeeping. The potential contribution for the period 1972 to January 1, 1980 amounts to 7.1%.

Other Small Items Not Included as Separate Contributions: Together with the contributions of items which were calculated to be less than 0.05%, a contribution of 0.2% from this item is estimated.

Capacity Expansions: Because of the commissioning of energy-efficient new crude distillation units and new downstream process units between 1972 and January 1, 1980, the energy efficiency of the industry as a whole may be expected to improve. The impact of new capacity expansions is shown to amount to a contribution of 0.4% towards the target for January 1, 1980.

Bearing in mind that operations on the target date of January 1, 1980 may be affected by many extraneous factors, a series of target offsets and sensitivities has been developed. Offsets include allowance for increased energy use for environmental pollution control, and sensitivities include changes in the target which could result from, for example, quality changes in traditional industry products. Taking energy offsets into account, the following net targets were developed:

	Net Targets, %
SIC 2911 Petroleum Refining	11.6
SIC 2951 Paving Mixtures and Blocks	42.0
SIC 2952 Asphalt Felts and Coatings	7.5
SIC 2992 Lubricating Oils and Greases	20.0
SIC 2999 Miscellaneous	20.5

The weighted net target for SIC 29, rounded to the nearest integer, is 12%.

From contacts with industry and from a review of conservation efforts and results to date, it is believed that the voluntary energy efficiency improvement targets derived in the Gordian Associates study for F.E.A. (6) represent a challenge to industry, but the targets are by no means unattainable by diligent efforts in housekeeping and by appropriate investment in conservation measures. The required conservation expenditures are considered to meet the criteria of industry for prudent capital investment and do not require channeling of funds into conservation to the detriment of overall company profitability.

Maintenance of achievements to date, especially in the field of housekeeping, will undoubtedly require vigilance by supervisory staff. Further progress will undoubtedly require the commitment of top management to ensure that adequate funds are budgeted for training, for competent engineering personnel and for the necessary capital investments. It is fair to claim that the easy conservation has been achieved: further progress needs more investment of capital and more investment of the technical talent to uncover opportunities for energy efficiency improvement and to exploit them. Other references on specific energy conservation measures in petroleum processing are given in references (27) to (35) inclusive.

SOME PROJECTIONS BEYOND 1980 TO 1990

Projecting per unit and total energy requirements for the refining industry is

TABLE 181: PROJECTED ENERGY REQUIREMENTS

Year	EXISTING PLANTS Capacity (MMBD)	Efficiency Improvements (% cumulative)	Gross Energy Requirements (MBTU/B)	NEW PLANTS Net Additions (MMBD)	Gross Energy Requirements (MBTU/B)	ENERGY REQUIREMENTS Weighted Average (MBTU/B)	Total (10^{12} BTU)
1958	8.2	--	660	--	--	662	1976
1962	9.2	--	674	--	--	678	2266
1967	10.6	--	641	--	--	625	2481
1971	12.4	--	616	--	--	612	2790
1975	14.5	7	573	.81	516	570	3185
1977	14.5	10	555	1.56	516	551	3230
1980	14.5	15	531	2.79	516	529	3336
1985	14.5	15	531	4.89	516	527	3731
1990	14.5	15	531	7.19	516	526	4165

Source: Reference (10)

extremely complex. There are a number of important variables, such as U.S. petroleum import policy, the composition of crude oils and the future octane quality of gasoline which are difficult to forecast accurately. Table 181 provides one projection of future refining energy requirements. It is based on a number of assumptions:

[1]   Short-term noncapital intensive housekeeping efforts will reduce energy requirements by 5% per unit of output;

[2]   Longer-term capital improvements, e.g., air preheaters, optimization of heat exchangers on existing plants, will ultimately reduce energy consumption in existing facilities by an additional 10%;

[3]   New refineries coming on line between 1975 and 1990 will have an average complexity rating of 9. These plants will be operated at a relatively high level of efficiency, consuming 560 MBtu per barrel of output;

[4]   The energy impact of the changing composition of crude oil processed in U.S. refineries and the mandatory reduction of lead additives in gasoline are reflected by the increase in the average complexity of U.S. refinery capacity;

[5]   The U.S. will adopt an import policy which discriminates against imported refined product, but will continue to import residual fuel oil from Caribbean refineries.

The net result of these assumptions is that energy consumption per unit of output decreases by approximately 97 MBtu over the period 1971 to 1980 and then levels off as the effect of the industry's efforts to improve efficiency is offset by the more complex new refineries coming on line after 1971. In the absence of efforts to improve the efficiency of energy utilization in existing facilities and if new refineries were not being built to meet high energy efficiency levels of operation, energy consumption in 1990 would have been around 5.4 quadrillion Btu. The projection contained in Table 181 shows that energy requirements should be approximately 1 quadrillion Btu less.

Therefore, the efficiency gains in the case of the petroleum industry will result in a savings of roughly the equivalent of 440,000 barrels of oil per day. These savings are fairly evenly distributed between improvements made to existing facilities and the difference in energy requirements reflected by the medium versus high efficiency levels projected for new refineries.

REFERENCES

(1) Reding, J.T. and Shepherd, B.P. (Dow Chemical Co., Texas Div.), *Energy Consumption: Fuel Utilization and Conservation in Industry,* Report EPA-650/2-75-032d, Washington, D.C., U.S. Environmental Protection Agency (September 1975).

(2) Reding, J.T. and Shepherd, B.P. (Dow Chemical Co., Texas Div.), *Energy Consumption: The Primary Metals and Petroleum Industries,* Report EPA-650/2-75-032b, Washington, D.C., U.S. Environmental Protection Agency (April 1975).

(3) Reding, J.T. and Shepherd, B.P. (Dow Chemical Co., Texas Div.), *Energy Consumption: The Chemical Industry,* Report EPA-650/2-75-032a, Washington, D.C., U.S. Environmental Protection Agency (April 1975).

(4) Lownie, H.W., Jr., Mayruth, D.J., McLeer, T.J., Kura, J.G., Varga, J., Jr. and Griffith, W.I. (Battelle Columbus Laboratories), *Draft Target Report on Development and Establishment of Energy Efficiency Improvement Targets for Primary Metal Industries,* Washington, D.C., Federal Energy Administration (August 13, 1976).

(5) Battelle Columbus Laboratories, *Draft Target and Support Document on Developing a Maximum Energy Efficiency Improvement Target for SIC 28: Chemical & Allied Products,* Washington, D.C., Federal Energy Administration (July 1, 1976).

(6) Gordian Associates, Inc., *An Energy Conservation Target for Industry SIC 29,* Washington, D.C., Federal Energy Administration (June 25, 1976).

(7) Gordian Associates, Inc., *The Data Base: The Potential for Energy Conservation in Nine Selected Industries, Vol. 6: Steel,* Conservation Paper No. 14, Washington, D.C., Federal Energy Administration (June 1974).

(8) Hall, E.H., Hanna, W.T., Reed, L.D., Varga, J., Jr., Williams, D.N., Wilkes, K.E., Johnson, B.E., Mueller, W.J., Bradbury, E.J. and Frederick, W.J. (Battelle Columbus Laboratories), *Final Report on Evaluation of the Theoretical Potential for Energy Conservation in Seven Basic Industries,* Washington, D.C., Federal Energy Administration (July 11, 1975).

(9) Resource Planning Associates, *Energy Requirements for Environmental Control in the Iron and Steel Industry,* Washington, D.C., Office of Environmental Affairs, U.S. Department of Commerce (January 1976).

(10) Federal Energy Administration, *Project Independence Blueprint—Final Task Force Report—Energy Conservation in the Manufacturing Sector 1954-1990,* Volume 3, Washington, D.C., Federal Energy Administration (November 1974).

(11) Gordian Associates, Inc., *The Potential for Energy Conservation in Nine Selected Industries,* Washington, D.C., Federal Energy Administration (June 1974).

(12) Battelle Columbus Laboratories, *Energy Use Patterns in Metallurgical and Nonmetallic Mineral Processing (Phase 4—Energy Data and Flowsheets, High Priority Commodities),* Report BuMines OFR 80-75, Washington, D.C., U.S. Bureau of Mines (June 27, 1975).

(13) Nerkervis, R.J. and Hallowell, J.B. (Battelle Columbus Laboratories), *Metals Mining and Milling Process Profiles with Environmental Aspects,* Report EPA-600/2-76-167, Washington, D.C., U.S. Environmental Protection Agency (June 1976).

(14) Myers, J.B., Gelb, B.A., Nakamura, L., Preston, N., Parker, P.A., Wehle, M., Levmore, S., Kolatch, B., Rabitsch, E.K., Elliott-Jones, M.F., Chiba, H., Apostolides, A.D. and Garvey, N. (Energy Policy Project of the Ford Foundation), *Energy Consumption in Manufacturing,* Cambridge, Mass., Ballinger Publishing Co. (1974).

(15) Department of Energy, *Energy Saving: The Fuel Industries and Some Large Firms,* London, Her Majesty's Stationery Office (1975).

(16) National Economic Development Office, *Energy Conservation in the United Kingdom: Achievements, Aims & Options,* London, Her Majesty's Stationery Office (1974).

(17) Institution of Mechanical Engineers, *Energy Recovery in Process Plants,* London, Mechanical Engineering Publications, Ltd. (1976).

(18) Gordian Associates, Inc., *The Data Base: The Potential for Energy Conservation in Nine Selected Industries, Vol. 5: Aluminum,* Conservation Paper No. 13, Washington, D.C., Federal Energy Administration (June 1974).

(19) Martin, D.J., (York Research Corp. for Flynn & Emrich Co.), *Energy Conservation Techniques for the Iron Foundry Cupola,* Report EPA-600/2-76-071, Washington, D.C., U.S. Environmental Protection Agency (March 1976).

(20) Gordian Associates, Inc., *The Data Base: The Potential for Energy Conservation in Nine Selected Industries, Vol. 5: Copper,* Conservation Paper No. 12, Washington, D.C., Federal Energy Administration (June 1974).

(21) Rosenkranz, Rodney D., *Energy Consumption in Domestic Primary Copper Production,* Information Circular 8698, Washington, D.C., U.S. Bureau of Mines (1976).

(22) Gordian Associates, Inc., *The Data Base: The Potential for Energy Conservation in Nine Selected Industries, Vol. 1: Selected Plastics,* Conservation Paper No. 9, Washington, D.C., Federal Energy Administration (June 1974).

(23) Snell, Foster D., Inc., *Industrial Energy Study of the Plastics and Rubber Industries, SIC's 282 and 30,* Washington, D.C., Federal Energy Administration (May 1974).

(24) Gordian Associates, Inc., *The Data Base: The Potential for Energy Conservation in Nine Selected Industries, Vol. 9: Styrene Butadiene Rubber,* Conservation Paper No. 17, Washington, D.C., Federal Energy Administration (June 1974).

(25) Versar, Inc., *Industrial Energy Study of the Drug Manufacturing Industries,* Washington, D.C., Federal Energy Administration (September 30, 1974).

(26) Gordian Associates, Inc., *The Data Base: The Potential for Energy Conservation in Nine Selected Industries, Vol. 2: Petroleum Refining,* Conservation Paper No. 10, Washington, D.C., Federal Energy Administration (June 1974).

(27) Hughes, T.R., Jacobson, R.L., Gibson, K.R., Schornak, L.G. and McCabe, J.R. (Chevron Research Co.), "To Save Energy When Reforming", *Hydrocarbon Processing* 55, No. 5, 75-80 (May 1976).

(28) Ryskamp, C.J., Wade, H.L. and Britton, R.B. (The Foxboro Co.), "Improve Crude Unit Operation", *Hydrocarbon Processing* 55, No. 5, 81-86 (May 1976).

(29) Prather, B.V. and Young, E.P. (Williams Brothers Waste Control, Inc.), "Energy for Wastewater Treatment", *Hydrocarbon Processing* 55, No. 5, 88-91 (May 1976).

(30) Thomson, S.J. and Crow, R.H. (Fluor Engineers & Constructors, Inc.), "Energy Cost of NO_x Control", *Hydrocarbon Processing* 55, No. 5, 95-97 (May 1976).

(31) Taylor, R.I. (Exxon Co.), "Refiners Can Save Energy Too", *Hydrocarbon Processing* 55, No. 7, 91-95 (July 1976).

(32) Gyger, R.F. and Doerflein, E.L. (Union Carbide Corp.), "Save Energy–Use O_2 on Wastes", *Hydrocarbon Processing* 55, No. 7, 96-100 (July 1976).

(33) Fleming, J., Duckham, H. and Styslinger, J. (Pullman Kellogg Co.), "Recover Energy With Exchangers", *Hydrocarbon Processing* 55, No. 7, 101-104 (July 1976).

(34) Duckham, H. and Fleming, J. (Pullman Kellogg Co.), "Better Plant Design Saves Energy", *Hydrocarbon Processing* 55, No. 7, 78-84 (July 1976).

(35) Wells, G.L., Hodgkinson, M.G., Al-Kadhi, H. and Wardle, I. (University of Sheffield), "Energy Considerations During Flowsheeting", *Chemistry & Industry,* pp. 943-97 (November 6, 1976).

(36) Kemmetmueller, R., U.S. Patent 3,888,742; June 10, 1975; assigned to American Waagner-Bird Co.

(37) Sato, C., Yamada, Y. and Takenaka, Y., U.S. Patent 3,869,537; March 4, 1975; assigned to Showa Denko KK.

(38) Sato, C., Yamada, Y. and Takenaka, Y., U.S. Patent 3,966,537; June 29, 1976; assigned to Showa Denko KK.

(39) Bravard, J.C., Flora, H.B. and Portal, C., *Energy Expenditures Associated with the Production and Recycle of Metals,* Oak Ridge National Laboratory Report No. ORNL-NSF-EP-24 (November 1972).

(40) Chapman, P.F., "The Energy Costs of Producing Copper and Aluminum from Primary Sources", *Metals and Materials,* February, 1974, p. 107; "Energy Conservation and Recycling of Copper and Aluminum", ibid, June, 1974, p. 311.

(41) Kellogg, H., "Energy Consumption for Production of Aluminum", paper prepared at Henry Crumb School of Mines, Columbia University, Nov. 1972.

(42) Smith, F.R.A., "Aluminum Reduction and Refining", *Materials and Methods,* March, 1974, p. 182.

(43) Kellogg, H.H., "Energy Efficiency in the Age of Scarcity", *Journal of Metals,* p. 25, June, 1974; also, with Tien, J., "Energy Considerations in Metal Production Selection and Utilization", paper presented at AIME Meeting in Chicago, Illinois, October 2, 1973.

(44) "Energy Conservation in the Copper Industry", Foster D. Snell, Inc., discussion paper on Contract No. C-04-50090-00 with FEA, March 27, 1975.

(45) Kellogg, H.H., and Henderson, J.M., "Energy Use in Sulfide Smelting of Copper", paper presented at 105th Annual AIME Meeting, Las Vegas, Nevada, February 23, 1976.

(46) Kellogg, H.H., "Melting Cathode Copper—A Case Study in Process Energy Efficiency", paper submitted to AIME for publication, 1975 and cited in Reference (4) above.

(47) *Energy Use Patterns in Metallurgical and Nonmetallic Mineral Processing (Phase 5—Energy Data and Flowsheets, Intermediate-Priority Commodities),* Interim Report to United States Bureau of Mines, Contract No. SO144093, Battelle's Columbus Laboratories, Columbus, Ohio (September 16, 1975).

(48) Doerr, R.A., "Six Ways to Keep Score in Energy Savings", *Oil & Gas Journal* 74, No. 20, (May 17, 1976).

(49) Prengle, W.H., Jr., *Potential for Energy Conservation in Industrial Operations in Texas,* Report No. NSF-RANN-74-231, Springfield, Virginia, Nat. Tech. Information Service (November 1974).

(50) Stanford Research Institute, *The Plastics Industry in the Year 2000,* New York, Society of the Plastics Industry (April 1973).

(51) Environmental Protection Agency, *Development Document for Effluent Levitation Guidelines and New Performance Standard for the Soap and Detergent Manufacturing Point Source Category,* Report EPA-440/1-74-0182, Washington, D.C. (April 1974).

(52) Anderson, E.V., *Chem. & Eng. News* 54, No. 19, 35 (1976).

(53) Stanford Research Institute, *Chemical Economics Handbook,* Menlo Park, California.

(54) Dwyer, F.G., Lewis, P. and Schneider, F., *Chemical Engineering* 83, No. 1, 90 (1976).

(55) Hughes, R.E., U.S. Patent 3,691,020; September 12, 1972; assigned to The Badger Co.

(56) Pettman, M.J. and Humphreys, G.C. (Davy Powergas Ltd.), "Improved Design to Save Energy" *Hydrocarbon Processing* 54, No. 1, 77-81 (January 1975).

(57) Mehta, D.D. (Chemical Construction Corp.), "Use Methanol Converter Reaction Heat", *Hydrocarbon Processing* 56, No. 5, 165-168 (May 1976).

(58) Brownstein, A.M. (Chem Systems Inc.), "Energy Crisis Impacts on Ethylene Glycol Trends", *Hydrocarbon Processing* 53, No. 6, 129-132 (June 1974).

(59) Kaupas, P.F., Bress, D.F., U.S. Patent 3,985,523; October 12, 1976; assigned to Foster Wheeler Energy Corp.

(60) Worrall, G.I., Strege, D.A. and Myers, G.D., U.S. Patent 3,984,316; Oct. 5, 1976; assigned to Ashland Oil, Inc.

HOW TO SAVE ENERGY AND CUT COSTS IN EXISTING INDUSTRIAL AND COMMERCIAL BUILDINGS 1976

An Energy Conservation Manual

by Fred S. Dubin, Harold L. Mindell and Selwyn Bloome

Energy Technology Review No. 10

This manual offers guidelines for an organized approach toward conserving energy through more efficient utilization and the concomitant reduction of losses and waste.

The current tight supply of fuels and energy is unprecedented in the U.S.A. and other countries, and this situation is expected to continue for many years. Never before has there been as pressing a need for the efficient use of fuels and energy in all forms.

Most of the energy savings will result from planned systematic identification of, and action on, conservation opportunities.

Part I of this manual is directed primarily to owners, occupants, and operators of buildings. It identifies a wide range of opportunities and options to save energy and operating costs through proper operation and maintenance. It also includes minor modifications to the building and mechanical and electrical systems which can be carried out promptly with little, if any, investment costs.

Part II is intended for engineers, architects, and skilled building operators who are responsible for analyzing, devising, and implementing comprehensive energy conservation programs. Such programs involve additional and more complex measures than those in **Part I**. The investment is usually recovered through demonstrably lower operating expenses and much greater energy savings.

A partial and much condensed table of contents follows here:

Much of the technology required to achieve energy savings is already available. Current research is providing refinements and evaluating new techniques that can help to curb the waste inherent in yesteryear's designs. The principal need is to get the available technology, described here, into widespread use.

ISBN 0-8155-0638-4

725 pages

ENERGY FROM BIOCONVERSION
OF WASTE MATERIALS 1977

by Dorothy J. De Renzo

Energy Technology Review No. 11
Pollution Technology Review No. 33

One of the chief gaseous products of the anaerobic decomposition of organic matter is methane, CH_4. This is how natural gas was formed in prehistoric times along with other fossil fuels.

By applying this principle today in environmentally acceptable fashion it is possible to bioconvert municipal solid sewage, animal manure, agricultural and other organic wastes into substitute natural gas (95% CH_4). In its simplest essentials the process consists of loading the material into a digester (a closed tank with a gas outlet). Given favorable thermal and chemical conditions, the appropriate biological processes will then take their course.

The bioconversion of waste materials to methane provides at least partial solutions not only to the energy problem, but also to the solid waste disposal problem. The harvesting of heretofore undesirable vegetations, such as algae, water hyacinths, and kelp as "energy crops" offers unconventional opportunities for supplementary utilization of natural resources.

This book describes practical methods for the bioconversion of waste matter. It is based on reports of academic and industrial research teams working under government contracts. A partial and condensed table of contents follows here. Chapter headings and important subtitles are given.

Note: Each chapter is followed by bibliographic reference lists in order to provide the reader with easy access to further information on these timely topics.

ISBN 0-8155-0656-2

223 pages

SOLAR ENERGY
FOR HEATING AND COOLING
OF BUILDINGS 1975

by Arthur R. Patton

Energy Technology Review No. 7

Solar energy can be used for indirect heating purposes in many ways. The information in this book has been limited to so-called low temperature solar thermal processes. Designs requiring photocells or other thermoelectric generators and lenses or reflecting mirrors plus tracking equipment have been excluded.

Low temperatures are the easiest to obtain, and the necessary collectors are fairly simple in construction. A black surface is used to absorb the sun's rays, this surface is usually covered with glass and the collector is insulated on the back and sides against heat loss. Water or some other heat transfer fluid is passed through the collector and can reach temperatures from 60°C (140°F) to about 95°C (203°F). The thermal energy is then stored in a heat storage system (perhaps based on the latent heat of fusion of selected salts). Coupled to the heat storage system are heating loops to furnish heat by convection and to operate an air conditioning system. In most temperate zones an auxiliary heater, operated with conventional fuels, must also be connected.

Large scale applications designed for schools and similar building are beginning to appear or are in the planning stage. This book describes in detail several large scale feasibility studies with designs suitable for institutions and industrial plants.

Descriptions are based on studies conducted by industrial or engineering firms or university research teams under the auspices of various government agencies. A partial and condensed table of contents follows here.

ISBN 0-8155-0579-5

328 pages

HOW TO REMOVE POLLUTANTS AND TOXIC MATERIALS FROM AIR AND WATER A PRACTICAL GUIDE 1977

by Marshall Sittig

Pollution Technology Review No. 32

This book is designed to provide a one volume ready reference for the handling of noxious materials emerging into the air and water as a result of industrial processes or from running machinery of any kind.

The descriptions are based almost entirely on U.S. patents dealing with practical environmental control methods and systems. The book surveys some 500 patents in the 1973 to 1976 period with exhaustive coverage up to November 1st, 1976. Since environmental patents are given priority handling by the U.S. Patent Office, some applications for these patents were filed only in late 1975 and even early 1976. This book therefore contains substantial technical information on very late developments in their field.

This book is addressed to industrialists who must keep abreast of the latest pollution removal techniques, to legislators and public health officials intent upon sensible rules and regulations, to Interested conservationists and to those eager students who can foresee permanent and brilliant careers in the fields of pollution abatement and engineering.

Subject entries are In alphabetical sequence. Because of the encyclopedic nature of the book, only those entries which are at the beginning of the alphabet can be shown here without bias.

Introduction

ACETONE CYANOHYDRIN
Removal from Water

ACID MINE WATERS

ACROLEIN PROCESS EFFLUENTS
Removal from Water

ACRYLIC RESIN EMISSIONS
Removal from Water

ACRYLONITRILE EFFLUENTS
Removal from Air
Removal from Water

ADIPIC ACID EFFLUENTS
Removal from Water

ALDEHYDES
Removal from Air or Water
Removal from Water only

ALKALI
Removal from Air

ALKALI CYANIDES
Removal from Air

ALKYL IODIDES
Removal from Air

ALKYLATION PROCESS EFFLUENTS
Removal from Air

ALUMINUM
Removal from Water

ALUMINUM CELL EFFLUENTS
Removal from Air

ALUMINUM CHLORIDE EFFLUENTS
Removal from Air

ALUMINUM ETCHING LIQUORS
Removal from Water

ALUMINUM REFINING EFFLUENTS
Removal from Air

AMINES
see **Foundry Casting Effluents**

AMMONIA
Removal from Water

AMMONIA-SODA EFFLUENTS
Removal from Water

AMMONIA SYNTHESIS EFFLUENTS
Removal from Water

AMMONIUM PHOSPHATE EFFLUENTS
Removal from Air

AMMONIUM SULFATE
Removal from Air

AMMONIUN SULFIDE
Removal from Water

AMMONIUM SULFITE
Removal from Water

ANTIFOULING PAINTS
Disposal of Residues

ASBESTOS
Removal from Air
Removal from Water

ASPHALT VAPORS
Removal from Air

AUTOMOTIVE EXHAUST
Removal from Air

BATTERY CHARGING EFFLUENTS
Removal from Air

ETC.

The book contains a total of 293 subject entries. The subject name refers to the polluting substance and the text underneath each entry tells how to combat pollution by said substance.

ISBN 0-8155-0654-6

621 pages